Birkhäuser

Pathways in Mathematics

Each "Pathways in Mathematics" book offers a roadmap to a currently well developing mathematical research field and is a first-hand information and inspiration for further study, aimed both at students and researchers. It is written in an educational style, i.e., in a way that is accessible for advanced undergraduate and graduate students. It also serves as an introduction to and survey of the field for researchers who want to be quickly informed about the state of the art. The point of departure is typically a bachelor/masters level background, from which the reader is expeditiously guided to the frontiers. This is achieved by focusing on ideas and concepts underlying the development of the subject while keeping technicalities to a minimum. Each volume contains an extensive annotated bibliography as well as a discussion of open problems and future research directions as recommendations for starting new projects. Titles from this series are indexed by Scopus.

The Virtual Series on Symplectic Geometry

Yannis Bähni

Lectures on Twisted Rabinowitz-Floer Homology

Birkhäuser

Yannis Bähni
MINT Lernzentrum
ETH Zürich
Zürich, Switzerland

ISSN 2367-3451 ISSN 2367-346X (electronic)
Pathways in Mathematics
ISBN 978-3-032-10672-8 ISBN 978-3-032-10673-5 (eBook)
https://doi.org/10.1007/978-3-032-10673-5

Mathematics Subject Classification: 57R58, 37J46, 37D15

This book is published under the imprint Birkhäuser, www.birkhauser-science.com by the registered company Springer Nature Switzerland AG
The registered company address is: Gewerbestrasse 11, 6330 Cham, Switzerland

To Dean, Colin, Neil, and Jil. In memory of Will J. Merry. A brilliant teacher and a guiding light.

Preface

The aim of this book is to give a concise and effective introduction to a generalisation of Rabinowitz–Floer homology called twisted Rabinowitz–Floer homology as introduced in [27]. Rabinowitz–Floer homology is the semi-infinite dimensional Morse–Bott homology in the sense of Floer associated with the Rabinowitz action functional constructed by Kai Cieliebak and Urs Frauenfelder in 2009. The Rabinowitz action functional was used in [157] to prove an existence result about periodic orbits on star-shaped hypersurfaces in $\mathbb{R}^{2n}$. Despite the fact that Jürgen Moser declared this functional useless for proving such existence results [144, (1.5)], it has since been very useful and led Alan Weinstein to prompt his famous conjecture about the existence of periodic Reeb orbits on contact manifolds [194]. This conjecture is still open in general but has been proven in some interesting cases. For example, in [27, Theorem 1.2] I proved the Weinstein conjecture for even lens spaces. At the same time, it was proven for all lens spaces in [10]. My proof used a generalisation of Rabinowitz–Floer homology for spaces admitting discrete symmetries. Floer homology is an active area of research with many different flavours and thus is hard to access because it combines many different areas.

This book is intended for interested graduate students and doctoral students entering the field. To my knowledge, there is currently no book treating Rabinowitz–Floer homology except the excellent foundational papers [51] and [12] as well as the book in preparation on periodic orbits [191]. The present book offers a pathway for aspiring researchers from classical mechanics formulated in the language of symplectic geometry to current research in Rabinowitz–Floer homology and neighbouring areas. Of course, the exposition is based on my taste, but I tried to include many relevant open questions. As this topic is complex, I think the reader should consult additional resources during reading if necessary. As a bare minimum, the reader should be familiar with the basic formalism of symplectic geometry including Cartan calculus of differential forms in [129, Chapter 22] or [200, Chapter V] as well as the basics of nonlinear functional analysis, especially Fredholm theory as in [136, Appendix A] and the theory of Sobolev spaces [136, Appendix B]. A concise treatment of some linear symplectic geometry towards the modern era of symplectic geometry starting with Gromov's landmark paper [96] from 1985 is treated in the Appendix A. But at this stage of the academic path, readers should be able to consult the appropriate literature on their own and not get stuck on details when not needed. This is an important skill in (mathematical) research!

I included many exercises and problems which are an essential part of the exposition. Some ideas and techniques are intentionally hidden to gain a deeper understanding of the subject. Therefore, I have decided to include detailed solutions to almost every exercise and some of the problems in the back of this book. However, the reader is encouraged to think about them first. Maybe you come up with a more elegant solution or have a different idea. In contrast to mathematics taught at university, research requires creativity and acceptance of failure [182]. So, methods from teaching and learning science, as in [93] and [169], are applied to optimise the reader's learning process. I implemented these techniques in some of the exercises and problems. For linear algebra, this was done partially in [32]. The main idea is that the reader gets actively involved in the material and, for example, tries to solve a problem before the needed theory behind it is introduced, and thus creates a theory on his own. At school or university level, we often learn a mathematical topic as a completed theory, albeit there has been a lot of cumulative effort by many to arrive at such results, often including mistakes and failure. In the spirit of this educational viewpoint and to prepare the reader for mathematical research, we also transparently provide learning goals and encourage the reader to think further! After reading this book, you should be able to answer the following questions. You can always go back during your journey:

1. First part about classical mechanics:
   (a) What are the differences and similarities of the Hamiltonian and the Lagrangian formulation of classical mechanics? Is there a connection between the two? Can you name three examples of such systems?
   (b) What are the strengths and limitations of Hamiltonian and Lagrangian systems?
   (c) Can you state Hamilton's and the Euler–Lagrange equations? Do you know how to derive them from a variational principle?
   (d) Explain what a symmetry group of a Hamiltonian or Lagrangian system is. What results do you know about the symmetries of Hamiltonian systems? What is the importance of such symmetries?
   (e) Consider a compact regular energy hypersurface in a Hamiltonian system. What can we say about the flow of the Hamiltonian vector field restricted to this hypersurface? Are there any interesting phenomena to look at?
   (f) Do you know what a stable Hamiltonian structure is and can you motivate its definition by a dynamical explanation? Can you name two examples?
   (g) Can you explain the term "blue sky catastrophe"?
   (h) Can you compare and contrast the Rabinowitz and the symplectic action functional? What do the corresponding critical sets look like?
   (i) In the first part about classical mechanics, what was easy for you to understand and what concept did you not fully grasp? What topic did you find particularly interesting and want to know more about it?

   (j) Are there practical applications of the abstract theory of Hamiltonian and Lagrangian systems?
2. Second part about Hamiltonian Floer homology:
   (a) Can you state the definition of a Morse–Bott function and give two examples? Why are Morse–Bott functions useful?
   (b) Given a Morse–Bott function, how does one compute the Morse–Bott homology of a compact smooth manifold via the cascade approach? Can you give a recipe?
   (c) What are the technical difficulties in defining the Morse–Bott homology with coefficients in $\mathbb{Z}_2$? Can you also consider Morse–Bott homology with coefficients in $\mathbb{Z}$?
   (d) Why is the language of polyfolds useful in order to formally define Morse–Bott homology? What is an important property of negative gradient flow lines of a Morse–Bott function?
   (e) What additional obstacle occurs in the semi-infinite dimensional case for the Morse–Bott homology of the symplectic action functional?
   (f) Why is the Floer homology of a symplectically aspherical compact symplectic manifold interesting?
3. Third part about twisted Rabinowitz–Floer homology:
   (a) Can you give three reasons why the study of twisted loop spaces is useful?
   (b) Can you explain the most geometrical setup for defining twisted Rabinowitz–Floer homology?
   (c) Can you state and prove two important applications of (twisted) Rabinowitz–Floer homology showing the existence of periodic Reeb orbits?
   (d) What are the Riemannian and contact geometric viewpoints of systolic geometry? What is a symplectic capacity?
   (e) What can you say about the symmetric ratio of a convex symmetric hypersurface?

Zürich, Switzerland Yannis Bähni
August 2025

Acknowledgements First, I thank my former advisors, Urs Frauenfelder and Will J. Merry. They have introduced me to the fascinating world of Rabinowitz–Floer homology and its usefulness in celestial mechanics. Unfortunately, the untimely death of Will J. Merry in 2022 has shattered the whole community working in symplectic geometry and is a tragic loss. It is the second tragic death in Floer theory after its founder Andreas Floer took his life in 1991 at a young age; see the book [108]. I am grateful to Ralph Schumacher and Elsbeth Stern for giving me the opportunity to complete this book at the institute of behavioural sciences and the MINT Lernzentrum at the Swiss Federal Institute of Technology in Zurich (ETHZ) and enrich it with valuable didactical ideas coming from empirical teaching and learning research. They definitely shaped the way I think about teaching mathematics in general. In the early stages of this project, Dorothy Mazlum and the two anonymous referees commenting on an early draft were of invaluable help. A lot of feedback on the mathematics presented in this book was given by Ana Cannas da Silva, Kai Cieliebak, Seongchan Kim, Otto van Koert, Agustin Moreno, Dietmar Salamon, Felix Schlenk, Joa Weber, Thomas O. Weinmann, and Lei Zhao. Finally, I thank Michelle Hartman and Patrick Oberholzer for all of the beautiful graphics. Without them, this book would have been a much more dry experience. A special thanks goes to Joé Brendel; without him, the teapot for illustrating the mechanics behind Morse–Bott homology would have not been invented!

Contents

Introduction

1

According to [14, p. 169], periodic orbits are the fundamental building blocks of Hamiltonian systems. The existence of closed orbits is important in the study of celestial mechanics. Indeed, as in [77], consider the restricted three-body problem, that is, a massless satellite moving in the gravitational field of two planets. In order to search for extraterrestrial life, one needs to know if the satellite can orbit around one of them. Finding periodic orbits is no easy task, as we illustrate in this chapter!

1.1 The Search for Periodic Orbits

Imagine that you are planning an exploration mission to Jupiter's moon Europa as depicted in Fig. 1.1. This is not fiction because in the year 2024 NASA sent the satellite named "Europa Clipper" on a journey to explore possible life. Jupiter's moon Europa is around 40,000 times lighter than the planet Jupiter and therefore has a much weaker gravitational pull. This makes it difficult to plan a suitable route, because if you change the satellite's orbit just a little, it will orbit Jupiter and no longer Europa! This is a so-called three-body problem that recently gained public attention through a television adaptation of the book [131].

Exercise 1.1 Consider Fig. 1.1. Draw a physically meaningful route to Jupiter's moon Europa for your satellite and explain your decision!

The three-body problem is described deterministically by the Newtonian equations of classical physics [141, Section 3]. Assume that the position of Jupiter is $J := (\mu, 0, 0)$ and that of Europa is $E := (\mu - 1, 0, 0)$ for the mass ratio $\mu \in [0, 1]$. The gravitational force of the two celestial bodies determines the trajectory of the massless satellite via the total energy function $H\colon \mathbb{R}^3 \setminus \{J, E\} \times \mathbb{R}^3 \to \mathbb{R}$ given by

Y. Bähni, *Lectures on Twisted Rabinowitz-Floer Homology*, Pathways in Mathematics, https://doi.org/10.1007/978-3-032-10673-5_1

Fig. 1.1 The Jupiter-Europa-Ganymede-Satellite system

$$H(q, p) := \frac{1}{2}\|p\|^2 - \frac{\mu}{\|q - E\|} - \frac{1-\mu}{\|q - J\|} + p_1 q_2 - p_2 q_1. \tag{1.1}$$

If an initial position $q_0 \in \mathbb{R}^3 \setminus \{J, E\}$ and an initial momentum $p_0 \in \mathbb{R}^3$ are given, the position and momentum of the satellite in the future can be calculated with great accuracy as integral curves of the Hamiltonian vector field

$$X_H := \sum_{j=1}^{3} \left(\frac{\partial H}{\partial p_j}\frac{\partial}{\partial q_j} - \frac{\partial H}{\partial q_j}\frac{\partial}{\partial p_j} \right) \in \mathfrak{X}(\mathbb{R}^3 \setminus \{J, E\} \times \mathbb{R}^3).$$

This is the spatial circular restricted three-body problem. If we consider the subset of the phase space $\mathbb{R}^3 \setminus \{J, E\} \times \mathbb{R}^3$ where $q_3 = p_3 = 0$, the problem is said to be planar. These equations can usually only be solved using a computer. An important property of the Hamiltonian function (1.1) is that if the satellite has a specific constant energy, then it can only stay in distinguished regions. If $\mu < \frac{1}{2}$, there are precisely five critical points $L_1, \ldots, L_5$ of H, called Lagrange points, such that

$$H(L_1) < H(L_2) < H(L_3) < H(L_4) = H(L_5).$$

As the Lagrange points are stationary, one can place a satellite there for observations. For example, this is the case for the James–Webb telescope. If the energy value is below the first critical value $H(L_1)$, the satellite can stay in the vicinity of Europa or Jupiter, or far away from both. If the energy is slightly above $H(L_1)$, then the satellite may transfer between Jupiter and Europa. It is also possible that the satellite collides with either planet. This is the case if q either converges to E or J and $\|p\| \to +\infty$ for the momentum. To calculate the path of a collision, many calculation steps are required at very short time intervals as the numerical error also explodes otherwise. Furthermore, many mathematical results assume some form of compactness, and therefore cannot be applied to objects involving collisions. With the help of a so-called regularisation as in [175, Chapter 3] or [77, Chapter 4], the three-body problem can be modelled in such a way that in the event of a collision, the satellite simply bounces off like a rubber ball. Even if such a reformulation is not physically meaningful, it makes it possible to calculate the orbits efficiently. What does the regularised environment of Jupiter's moon in phase space look like? According to [141, Theorem K], the regularised hypersurface in the planar problem below the first critical value $H(L_1)$ is precisely the unit cotangent bundle

$$S^*\mathbb{S}^2 := \{(q, p) \in T^*\mathbb{R}^3 : \|q\| = \|p\| = 1, \langle q, p\rangle = 0\}.$$

The aim of the next exercise is to get to know some equivalent geometrical descriptions of the unit cotangent bundle $S^*\mathbb{S}^2$ of the unit sphere $\mathbb{S}^2 \subseteq \mathbb{R}^3$.

Exercise 1.2 Consider the special orthogonal group

$$\mathrm{SO}(3) := \{A \in \mathbb{R}^{3\times 3} : AA^T = I_3 \text{ and } \det A = 1\}.$$

1. Show that there is a diffeomorphism $S^*\mathbb{S}^2 \cong \mathrm{SO}(3)$.
2. Show that there is a diffeomorphism $\mathrm{SO}(3) \cong \mathbb{RP}^3$, where the real projective space $\mathbb{RP}^3 := \mathbb{S}^3/\mathbb{Z}_2$ is the smooth quotient manifold obtained from the sphere $\mathbb{S}^3$ by identifying antipodal points z and $-z$.

By examining the long exact sequence of homotopy groups of a fibration [104, Theorem 4.41], there is a short exact sequence

$$0 \longrightarrow \pi_1(\mathbb{S}^3, z) \longrightarrow \pi_1(\mathbb{S}^3/\mathbb{Z}_2, \pi(z)) \longrightarrow \pi_0(\mathbb{Z}_2) \longrightarrow 0,$$

implying $\pi_1(\mathbb{RP}^3) \cong \mathbb{Z}_2$ by [128, Corollary 12.9]. Thus we may even ask for the existence of non-contractible periodic orbits in the real projective space $\mathbb{RP}^3$. An example of a noncontractible loop in $\mathrm{SO}(3)$ can be illustrated by a gymnastic exercise as in [121, Satz 6.7]. However, this loop is not a periodic orbit in general as it does not necessarily satisfy Newton's equations for the Hamiltonian function (1.1). The question of the existence of periodic orbits on manifolds like the sphere $\mathbb{S}^3$ was already solved in 1978 by Paul Rabinowitz from the University of Wisconsin in Madinson. In 1979, Alan Weinstein from the University of California in Berkeley formulated a general conjecture about the existence of periodic orbits on compact odd-dimensional spaces admitting a distinguished volume form like the sphere $\mathbb{S}^3$ or the real projective space $\mathbb{RP}^3$. This *Weinstein conjecture* has so far only been proven in special cases and is an important open question in symplectic geometry. Symplectic geometry is the formal language of classical mechanics and a relatively young field of active research. Most of the existence proofs are not constructive, since these results are obtained by making use of the abstract machinery of Floer theory. Thus, we do not get explicit orbits as with the Birkhoff shooting method [77, Section 8.3.2] or the Ganymede-Europa orbit constructed in [122]. Recently, there has also been a lot of interest in tackling questions from space mission design via machine learning and artificial intelligence. See the survey article [163] for an overview of the relevant literature, as well as [188], [186], and [86] for a recent approach through generative AI techniques. However, we will not discuss any numerical implementations in this book, as the aforementioned papers on machine learning and the book [142] are a much better reference. In this book, we give a pathway from the very analytical basics to the frontiers of current research. To start our journey exploring the abstract machinery behind existence results of periodic orbits and why it is actually needed, we try to prove the Weinstein conjecture for certain deformed spheres $\mathbb{S}^3$ ourselves! On the way, we will learn why searching for periodic orbits is a highly non-trivial problem, and thus existence results are indeed not obvious in general.

1.2 Star-Shaped Hypersurfaces

Let $n \geq 1$ be a natural number and denote by

$$\mathbb{S}^{2n-1} := \left\{ (z_1, \ldots, z_n) \in \mathbb{C}^n : \sum_{j=1}^{n} |z_j|^2 = 1 \right\}$$

the odd-dimensional unit sphere in $\mathbb{C}^n$. We can also consider $\mathbb{S}^{2n-1} \subseteq \mathbb{R}^{2n}$ via the coordinates $z_j = x_j + iy_j$.

Definition 1.3 (Star-Shaped Hypersurface) Let $\rho \in C^\infty(\mathbb{S}^{2n-1})$ be a positive function. A ***star-shaped hypersurface*** $\Sigma = \Sigma_\rho \subseteq \mathbb{R}^{2n}$ is defined to be the subset

$$\Sigma_\rho := \{\rho(z)z : z \in \mathbb{S}^{2n-1}\}.$$

Let $\Sigma = \Sigma_\rho \subseteq \mathbb{R}^{2n}$ be a star-shaped hypersurface. If we can write $\Sigma = H_\Sigma^{-1}(0)$ for some smooth function $H_\Sigma \in C^\infty(\mathbb{R}^{2n})$ such that $\mathrm{Crit}(H_\Sigma) \cap \Sigma = \varnothing$, then Σ is a well-behaved smooth submanifold of $\mathbb{R}^{2n}$ by [129, Corollary 5.14] of dimension $\dim \Sigma = 2n - 1$ with tangent space $T\Sigma = \ker(dH)|_\Sigma$. In order to do that, let

$h \in C^\infty(\mathbb{R})$ be a sufficiently small mollification of the piecewise-linear function

$$h(r) := \begin{cases} r & r \in [-\frac{1}{2}, \frac{1}{2}], \\ -\frac{1}{2} & r \in (-\infty, -\frac{1}{2}], \\ \frac{1}{2} & r \in [\frac{1}{2}, +\infty). \end{cases}$$

This just means smoothing the two corners of h! Then the smooth function

$$H_\Sigma \colon \mathbb{C}^n \to \mathbb{R}, \qquad H_\Sigma(z) := \begin{cases} h\left(\|z\|^2 \rho(\frac{z}{\|z\|})^{-2} - 1\right), & z \neq 0, \\ -\frac{1}{2}, & z = 0, \end{cases} \tag{1.2}$$

satisfies $\Sigma = H_\Sigma^{-1}(0)$. That $\mathrm{Crit}(H_\Sigma) \cap \Sigma = \varnothing$ can be seen as follows. By [77, Lemma 12.2.2] we have that

$$X_{H_\Sigma}|_\Sigma \in \ker d\lambda_0|_\Sigma \qquad \text{and} \qquad \lambda_0(X_{H_\Sigma})|_\Sigma = 1, \tag{1.3}$$

for the differential form

$$\lambda_0 := \frac{1}{2}\sum_{j=1}^n (x_j dy_j - y_j dx_j) = \frac{i}{4}\sum_{j=1}^n (z_j d\bar{z}_j - \bar{z}_j dz_j) \in \Omega^1(\mathbb{R}^{2n}) \tag{1.4}$$

and the vector field

$$X_{H_\Sigma} := \sum_{j=1}^n \left(\frac{\partial H_\Sigma}{\partial y_j}\frac{\partial}{\partial x_j} - \frac{\partial H_\Sigma}{\partial x_j}\frac{\partial}{\partial y_j}\right) \in \mathfrak{X}(\mathbb{R}^{2n}).$$

The definition of the function (1.2) looks rather unintuitive. We will get to know explanations for this choice during the course of the book (Fig. 1.2).

Exercise 1.4 Show that the vector field X_{H_Σ} satisfies $d\lambda_0(X_{H_\Sigma}, \cdot) = dH_\Sigma$ and conclude that the restriction $X_{H_\Sigma}|_\Sigma$ is tangent to Σ defining a vector field $R_\Sigma \in \mathfrak{X}(\Sigma)$. Can you explain why the vector field R_Σ does not have any stationary points?

Exercise 1.5 Maybe you are more familiar with the following definition of a star-shaped hypersurface. A ***star-shaped domain*** is an open subset $W \subseteq \mathbb{C}^n$ with a smooth boundary ∂W, such that for every point $z \in \mathbb{C}^n$ the line segment σz for all $\sigma \in [0, 1]$ is contained in W. A star-shaped hypersurface Σ is then the boundary ∂W of a star-shaped domain W. Show that this definition is equivalent to Definition 1.3.

Let $\Sigma \subseteq \mathbb{R}^{2n}$ be a star-shaped hypersurface and denote by $R_\Sigma \in \mathfrak{X}(\Sigma)$ the vector field of Exercise 1.4. This vector field vanishes nowhere and thus there are no resting points for the satellite in Fig. 1.3. The next thing to ask for is the existence

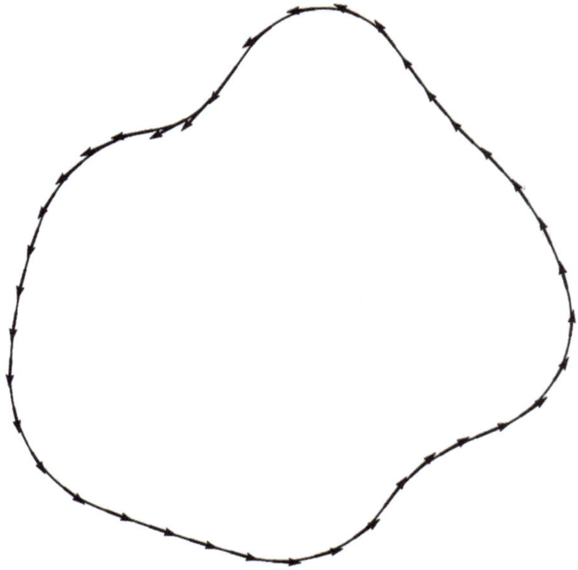

Fig. 1.2 A star-shaped hypersurface in $\mathbb{R}^2$ and its distinguished vector field

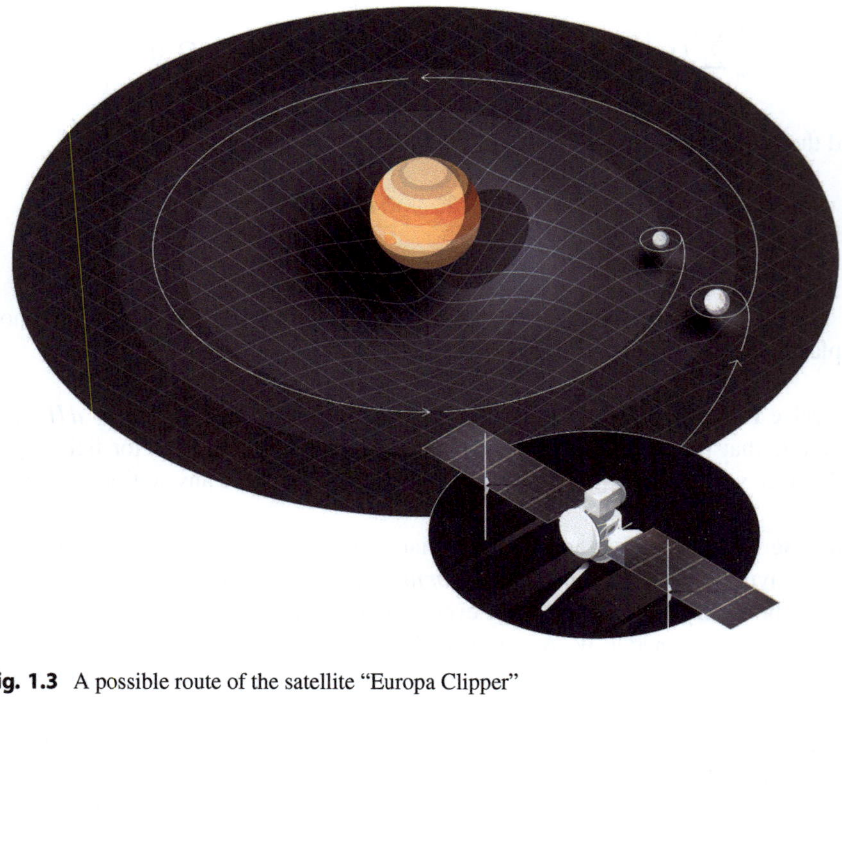

Fig. 1.3 A possible route of the satellite "Europa Clipper"

of periodic orbits. Even if all star-shaped hypersurfaces are diffeomorphic to the sphere $\mathbb{S}^{2n-1}$, the flow of the vector field R_Σ can be quite different!

Exercise 1.6 Determine the closed orbits of R_Σ for $\Sigma = \mathbb{S}^{2n-1}$.

Exercise 1.7 For real numbers $a_1, \ldots, a_n > 0$, consider the ellipsoids

$$E(a_1, \ldots, a_n) := \left\{ (z_1, \ldots, z_n) \in \mathbb{C}^n : \sum_{j=1}^{n} \frac{\pi |z_j|^2}{a_j} = 1 \right\}.$$

Determine the closed orbits of R_Σ for $\Sigma = E(a_1, \ldots, a_n)$ depending on a_j.

Part I
Classical Mechanics

In the mathematical treatment of classical mechanics one is interested in the qualitative behaviour of motions rather than the quantitative behaviour. This is because in most cases the equations of motion can only be solved numerically. Thus we may ask, for example, if closed orbits exist or orbits of minimal period. For an introduction to Hamiltonian dynamics, see [9] as well as [107] for a view towards symplectic invariants. Moreover, the books [22] and [63] provide a much more in-depth introduction to the mathematical treatment of classical mechanics with many concrete examples. As a short introduction, we recommend [176, Chapter 1], which also uses the formalism of classical mechanics to introduce quantum mechanics. Our main goal will be to introduce the necessary variational principle we need later to study periodic orbits in lens spaces. For an introduction including all necessary prerequisites, we recommend [120].

2 The Hamiltonian Formalism

The Hamiltonian description of classical mechanics admits a natural formulation in the language of symplectic geometry [199]. The other modern formulation of the laws of classical mechanics is provided by the Lagrangian formalism, which can be viewed as a dual viewpoint to the Hamiltonian one. For a concise historical overview, we recommend [102, Chapter VI]. In order to state the Hamiltonian formalism, we need to recall some basic notions from Cartan's calculus of differential forms on a smooth manifold M. For a short review of the most important concepts coming from the theory of smooth manifolds, we recommend the appendices in [129] and [130]. A mixed tensor field of type (l, k) is a $C^\infty(M)$-multilinear map

$$\underbrace{\Omega^1(M) \times \cdots \times \Omega^1(M)}_{l} \times \underbrace{\mathfrak{X}(M) \times \cdots \times \mathfrak{X}(M)}_{k} \to C^\infty(M),$$

where $\Omega^1(M)$ denotes the real vector space of all smooth differential 1-forms, and the notation $\mathfrak{X}(M)$ stands for the Lie algebra of all smooth vector fields on M with Lie bracket $[X, Y] := XY - YX$ for all $X, Y \in \mathfrak{X}(M)$. We denote the vector space of all tensor fields of type (l, k) by $\mathcal{T}^{l,k}(M)$. For any vector field $X \in \mathfrak{X}(M)$ and tensor field $A \in \mathcal{T}^{l,k}(M)$ we define the ***Lie derivative*** of A with respect to X to be the tensor field $L_X A \in \mathcal{T}^{l,k}(M)$ given by

$$L_X A := \left.\frac{d}{dt}\right|_{t=0} \phi_t^* A,$$

where ϕ denotes the smooth local flow of X and $\phi_t^* A$ denotes the pull-back of A by the local diffeomorphism ϕ_t. The Lie derivative generalises the directional derivative of a function to arbitrary tensor fields. As can be verified, the Lie derivative satisfies

$$L_X f = Xf \qquad \text{and} \qquad L_X Y = [X, Y] \tag{2.1}$$

Y. Bähni, *Lectures on Twisted Rabinowitz-Floer Homology*, Pathways in Mathematics, https://doi.org/10.1007/978-3-032-10673-5_2

for all $f \in C^\infty(M)$. In other words, one can define the Lie derivative as the unique tensor derivation satisfying the two conditions (2.1). This follows from the repeated application of [129, Proposition 12.32]. Similarly, the exterior differential generalises the differential of a function to arbitrary differential forms in $\Omega^k(M) := \mathcal{T}^{0,k}(M)$. Let $\omega \in \Omega^k(M)$ be a differential k-form. Then we can locally write $\omega = \sum_I \omega_I dx_I$, where the sum runs over all increasing multiindices I. We define the ***exterior differential of*** ω to be the differential $k+1$-form

$$d\omega := \sum_I d\omega_I \wedge dx_I \in \Omega^{k+1}(M).$$

Of course, one must check that this definition is well defined, that is, the differential form $d\omega$ does not depend on the choice of coordinate charts. More elegantly, the exterior differential is the unique graded derivation of degree one such that

$$df(X) = Xf = L_X f \qquad \text{and} \qquad d \circ d = 0$$

for all $f \in C^\infty(M)$ and $X \in \mathfrak{X}(M)$. This is the content of [129, Theorem 14.24]. A somehow "magical" result that combines the exterior differential and the Lie derivative is due to Cartan.

Lemma 2.1 (Cartan's Magic Formula, [129, Theorem 14.35]) *Let M be a smooth manifold. Then*

$$L_X = i_X \circ d + d \circ i_X$$

holds for all $X \in \mathfrak{X}(M)$, where

$$i_X : \Omega^{*+1}(M) \to \Omega^*(M), \qquad i_X\omega := \omega(X, \cdot, \dots, \cdot)$$

denotes interior multiplication.

Another useful formula coming from the calculus of differential forms is "sailing along the flow of a vector field", a terminology introduced in [21].

Lemma 2.2 (Fisherman's Formula, [129, Proposition 22.14]) *Let M be a smooth manifold. Then for all $\omega \in \Omega^k(M)$ and time-dependent vector fields X it holds*

$$\frac{d}{dt}\phi_t^*\omega = \phi_t^* L_{X_t}\omega.$$

In the following two multiple choice questions we recall some basic properties of the exterior differential d, the Lie derivative L and the interior multiplication i. The exterior differential gives rise to the de Rham cohomology groups [129, Chapter 17].

For every integer $k \geq 0$, we define the ***de Rham cohomology group of degree*** k, to be the real quotient vector space

$$\mathrm{H}^k_{\mathrm{dR}}(M) := \frac{\ker\left(d\colon \Omega^k(M) \to \Omega^{k+1}\right)}{\operatorname{im}\left(d\colon \Omega^{k-1}(M) \to \Omega^k\right)} = \frac{\{\text{closed } k\text{-forms on } M\}}{\{\text{exact } k\text{-forms on } M\}}.$$

This is indeed well defined, as every exact k-form is closed by the identity $d \circ d = 0$. By the de Rham Theorem [129, Theorem 18.14], we have an isomorphism between the k-th de Rham homology group $\mathrm{H}^k_{\mathrm{dR}}(M)$ and the singular cohomology $\mathrm{H}^k(M; \mathbb{R})$ of M with coefficients in $\mathbb{R}$ for all $k \geq 0$.

Exercise 2.3 Let M be a smooth manifold. Which statements are correct?

- ☐ $L_X(\omega \wedge \eta) = L_X\omega \wedge \eta + \omega \wedge L_X\eta$ for all $\omega \in \Omega^k(M)$, $\eta \in \Omega^l(M)$ and $X \in \mathfrak{X}(M)$.
- ☐ $L_X(\omega \wedge \eta) = L_X\omega \wedge \eta + (-1)^{kl}\omega \wedge L_X\eta$ for all $\omega \in \Omega^k(M)$, $\eta \in \Omega^l(M)$ and all vector fields $X \in \mathfrak{X}(M)$.
- ☐ $d(\omega \wedge \eta) = d\omega \wedge \eta + \omega \wedge d\eta$ for all $\omega \in \Omega^k(M)$ and $\eta \in \Omega^l(M)$.
- ☐ $d(\omega \wedge \eta) = d\omega \wedge \eta + (-1)^{kl}\omega \wedge d\eta$ for all $\omega \in \Omega^k(M)$ and $\eta \in \Omega^l(M)$.

Exercise 2.4 Let M be a smooth manifold. Which statements are correct?

- ☐ $i_{[X,Y]} = L_Y i_X - i_X L_Y$ for all $X, Y \in \mathfrak{X}(M)$.
- ☐ $i_{[X,Y]} = L_X i_Y - i_Y L_X$ for all $X, Y \in \mathfrak{X}(M)$.
- ☐ $i_X(\omega \wedge \eta) = i_X\omega \wedge \eta + \omega \wedge i_X\eta$ for all $\omega \in \Omega^k(M)$, $\eta \in \Omega^l(M)$ and $X \in \mathfrak{X}(M)$.
- ☐ $i_X(\omega \wedge \eta) = i_X\omega \wedge \eta + (-1)^{kl}\omega \wedge i_X\eta$ for all $\omega \in \Omega^k(M)$, $\eta \in \Omega^l(M)$ and all vector fields $X \in \mathfrak{X}(M)$.

A ***symplectic manifold*** is a tuple (M, ω), where M is a nonempty smooth finite-dimensional manifold and $\omega \in \Omega^2(M)$ is closed and nondegenerate, that is, $d\omega = 0$ and the map

$$\Psi\colon \mathfrak{X}(M) \to \Omega^1(M), \qquad \Psi(X)(Y) := \omega(X, Y) \qquad \forall Y \in \mathfrak{X}(M),$$

is an isomorphism, called a ***symplectic form***. As ω is nondegenerate, every tangent space (T_xM, ω_x) is a symplectic vector space for all $x \in M$. Hence $\dim M = 2n$ is even for some $n \geq 1$ according to the Lemma A.3 and the fact that the manifold dimension coincides with the dimension of each tangent space [129, Proposition 3.10]. Another consequence of Lemma A.3 is that the top exterior power

$$\omega^n = \underbrace{\omega \wedge \cdots \wedge \omega}_{n \text{ times}} \in \Omega^{2n}(M)$$

is a volume form and thus nowhere vanishing on M. Consequently, every symplectic manifold is orientable. The following result is one of the most fundamental results in symplectic geometry and can be thought of a nonlinear analogue of Lemma A.3.

Lemma 2.5 (Darboux Theorem, [137, Theorem 3.2.2]) *Let $\omega \in \Omega^2(M)$ be a nondegenerate differential form on a smooth manifold M^{2n}. Then $d\omega = 0$ if and only if for every point $x \in M$ there exists a chart $\big(U, (q_j, p_j)\big)$ about x with*

$$\omega|_U = \sum_{j=1}^{n} dp_j \wedge dq_j.$$

*If ω is a symplectic form, such coordinates are called **Darboux coordinates**.*

In particular, the symplectic form ω is locally diffeomorphic to the standard symplectic form $\omega_0 := d\lambda_0$ on $\mathbb{R}^{2n}$ with $\lambda_0 \in \Omega^1(\mathbb{R}^{2n})$ defined as in (1.4). Another consequence of the Darboux Theorem 2.5 is an answer to the local equivalence question: For any geometric structure on a smooth manifold, are all examples of the structure locally equivalent to each other under an appropriate notion of local equivalence? The most important technique for proving that two geometric structures are not locally equivalent is to find local invariants, which are quantities that must be preserved by local equivalences. For example, the round 2-sphere and the Euclidean plane are not locally isometric by [130, Problem 5-5]. This is due to the fact that the round 2-sphere is not flat because it has a local nonvanishing invariant called curvature [130, Theorem 7.10]. By the Darboux theorem, any two symplectic manifolds of the same dimension look locally like $(\mathbb{R}^{2n}, \omega_0)$ and thus in symplectic geometry one is interested in finding global invariants!

Exercise 2.6 Let (M^{2n}, ω) be a symplectic manifold. Which statements are correct?

- ☐ For the second de Rham cohomology group we have $\mathrm{H}^2_{\mathrm{dR}}(M) \neq 0$.
- ☐ If M is compact, then $[\omega] \neq 0$ in $\mathrm{H}^2_{\mathrm{dR}}(M)$, which means that ω is not exact.
- ☐ Every sphere $\mathbb{S}^{2n}$ admits a symplectic form for $n \geq 1$.
- ☐ The cotangent bundle $M = T^*Q$ of any smooth manifold Q (orientable or not!) admits a symplectic form and is therefore orientable.

As ω is closed by definition, we can always consider its de Rham cohomology class $[\omega] \in \mathrm{H}^2_{\mathrm{dR}}(M)$ for any symplectic manifold (M, ω). The case where ω is exact gets its own terminology.

Definition 2.7 (Liouville Form and Liouville Vector Field) An ***exact symplectic manifold*** is a tuple (M, λ), such that $(M, d\lambda)$ is a symplectic manifold. The differential form λ is called ***Liouville form***. The vector field $X = X_\lambda \in \mathfrak{X}(M)$

implicitly defined by the condition $i_X d\lambda = \lambda$ is called the ***Liouville vector field*** associated with the exact symplectic manifold (M, λ).

Example 2.8 (Star-Shaped Hypersurface) The tuple $(\mathbb{R}^{2n}, \lambda_0)$ for $\lambda_0 \in \Omega^1(\mathbb{R}^{2n})$ as in (1.4) is an exact symplectic manifold with Liouville vector field

$$X_{\lambda_0} = \frac{1}{2}\sum_{j=1}^{n}\left(x_j \frac{\partial}{\partial x^j} + y_j \frac{\partial}{\partial y_j}\right) = \frac{1}{2}\sum_{j=1}^{n}\left(z_j \frac{\partial}{\partial z_j} + \bar{z}_j \frac{\partial}{\partial \bar{z}_j}\right) \in \mathfrak{X}(\mathbb{R}^{2n}).$$

Exercise 2.9 Let (M, λ) be a compact exact symplectic manifold. Show that

$$\phi_t^* \lambda = e^t \lambda \qquad \forall t \in \mathbb{R}$$

holds, where ϕ denotes the smooth global flow of the Liouville vector field X_λ.

2.1 Classical Observables

We start with the main object that we will study throughout the book.

Definition 2.10 (Hamiltonian System) A ***Hamiltonian system*** is a symplectic manifold (M, ω), called ***phase space***, together with a smooth function $H \in C^\infty(M)$, called a ***Hamiltonian function***. We write (M, ω, H) for a Hamiltonian system.

Example 2.11 (Magnetic Hamiltonian System) Let (Q, m) be a pseudo-Riemannian manifold and denote by $\pi_{T^*Q} \colon T^*Q \to Q$ its cotangent bundle. For a smooth potential function $V \in C^\infty(Q)$ define a Hamiltonian function $H \in C^\infty(T^*Q)$ by

$$H(q, p) := \frac{1}{2}\|p\|_{m^*}^2 + V(q), \tag{2.2}$$

where (q, p) denote the standard coordinates on the cotangent bundle T^*Q. For a closed differential form $\sigma \in \Omega^2(Q)$, the form $\omega_\sigma := d\lambda + \pi_{T^*Q}^*\sigma$ is a symplectic form, where $\lambda \in \Omega^1(T^*Q)$ is the ***canonical Liouville form***

$$\lambda_{(q,p)}(v) := p\left(D\pi_{T^*Q}(v)\right) \qquad \forall v \in T_{(q,p)}T^*Q. \tag{2.3}$$

Locally, we have that $\lambda = pdq$. The symplectic manifold (T^*Q, ω_σ) is called a ***magnetic cotangent bundle***, and the corresponding Hamiltonian system (T^*Q, ω_σ, H) is called a ***magnetic Hamiltonian system***. If $\sigma = 0$, the system is called a ***mechanical Hamiltonian system***.

The dynamics of a Hamiltonian system is governed by a distinguished vector field. The integral curves of this vector field describe the motion of a physical system.

Definition 2.12 (Hamiltonian Vector Field) Let (M, ω, H) be a Hamiltonian system. The ***Hamiltonian vector field*** is the vector field $X_H \in \mathfrak{X}(M)$ given implicitly by the requirement

$$i_{X_H}\omega = -dH.$$

Exercise 2.13 Explain, why for every Hamiltonian system there does exist a unique Hamiltonian vector field.

Example 2.14 (Magnetic Hamiltonian Systems) Let $(T^*Q^n, \omega_\sigma, H)$ be a magnetic Hamiltonian system. Then the Hamiltonian vector field X_H is given by

$$X_H(q, p) = \sum_{i=1}^{n} \frac{\partial H}{\partial p_i}\frac{\partial}{\partial q_i} + \sum_{i=1}^{n}\left(\sum_{j=1}^{n} \sigma_{ij}(q)\frac{\partial H}{\partial p_j} - \frac{\partial H}{\partial q_i}\right)\frac{\partial}{\partial p_i}, \tag{2.4}$$

where

$$\sigma = \frac{1}{2}\sum_{i,j=1}^{n} \sigma_{ij}(q)dq_i \wedge dq_j, \qquad \sigma_{ji} = -\sigma_{ij}.$$

Exercise 2.15 (Hamiltonian Equations) Let (M^{2n}, ω, H) be a Hamiltonian system. Write down the local formula for the Hamiltonian vector field X_H in the Darboux coordinates (q_j, p_j). What equations do integral curves of the Hamiltonian vector field solve? What is the connection to the equations of motion described by Newton's second law if we consider a mechanical Hamiltonian system? Prove directly using Darboux coordinates that the flow of a Hamiltonian vector field is volume preserving!

Exercise 2.16 (Symplectomorphism) Let (M, ω) be a symplectic manifold. A ***symplectomorphism*** is a diffeomorphism $\varphi \in \mathrm{Diff}(M)$ such that $\varphi^*\omega = \omega$. We denote the group of all symplectomorphisms on the symplectic manifold (M, ω) by $\mathrm{Symp}(M, \omega)$. Let $H \in C^\infty(M)$ be a Hamiltonian function. Show that the flow ϕ of the Hamiltonian vector field X_H is a local symplectomorphism. Prove that the local flow preserves volume!

An immediate consequence of Exercise 2.16 is that the flow of any Hamiltonian vector field preserves volume in the phase space (M, ω). More generally, this is true for any symplectomorphism $\varphi \in \mathrm{Symp}(M, \omega)$, for example a physical transformation as in Problem 2.64. Moreover, Hamiltonian vector fields behave well under symplectomorphisms.

Lemma 2.17 (Jacobi, [9, Theorem 3.3.19]) *Let* (M, ω, H) *be a Hamiltonian system and let* $\varphi \in \mathrm{Symp}(M, \omega)$ *be a symplectomorphism. Then*

$$\varphi^* X_H = X_{\varphi^* H}.$$

Proof We compute

$$i_{X_{\varphi^* H}}\omega = -d\varphi^* H = -\varphi^* dH = \varphi^*(i_{X_H}\omega) = i_{\varphi^* X_H}(\varphi^*\omega) = i_{\varphi^* X_H}\omega.$$

Thus, we conclude by the uniqueness of the Hamiltonian vector field. □

Lemma 2.18 *Let* (M, ω, H) *be a Hamiltonian system and let* $\varphi \in \mathrm{Symp}(M, \omega)$ *be a symplectomorphism. Then*

$$\phi_t^{\varphi^* H} = \varphi^{-1} \circ \phi_t^H \circ \varphi,$$

whenever either side is defined, where ϕ^H *denotes the smooth flow of the Hamiltonian vector field* X_H.

Proof Using Lemma 2.17 we compute

$$\begin{aligned}
\frac{d}{dt}\varphi^{-1} \circ \phi_t^H \circ \varphi &= D\varphi^{-1} \circ X_H \circ \phi_t^H \circ \varphi \\
&= D\varphi^{-1} \circ X_H \circ \varphi \circ \varphi^{-1} \circ \phi_t^H \circ \varphi \\
&= \varphi^* X_H \circ \varphi^{-1} \circ \phi_t^H \circ \varphi \\
&= X_{\varphi^* H} \circ \varphi^{-1} \circ \phi_t^H \circ \varphi,
\end{aligned}$$

and the result follows by the uniqueness of integral curves. □

Hamiltonian mechanics is all about studying smooth functions on phase spaces!

Definition 2.19 (Algebra of Classical Observables, [176, p. 46]) Let (M, ω) be a symplectic manifold. The real commutative algebra $C^\infty(M)$ of all smooth functions in M is called ***algebra of classical observables***.

Remark 2.20 In quantum mechanics, the algebra of observables is not commutative. This gives rise to very interesting phenomena. For more details, see [176, p. 66].

Definition 2.21 (Poisson Bracket, [129, p. 578]) Let (M, ω) be a symplectic manifold. Define the ***Poisson bracket on the algebra of classical observables*** to be the map

$$\{\cdot,\cdot\}: C^\infty(M) \times C^\infty(M) \to C^\infty(M), \qquad \{f, g\} := \omega(X_f, X_g).$$

Definition 2.22 (Poisson Algebra, [64, Definition 18.7]) A ***Poisson algebra*** is defined to be a real commutative algebra $\mathfrak{p}$ together with a Lie bracket $\{\cdot,\cdot\}$ on $\mathfrak{p}$ satisfying the ***Leibniz rule***

$$\{f, gh\} = h\{f, g\} + g\{f, h\} \qquad \forall f, g, h \in \mathfrak{p}.$$

Exercise 2.23 Let (M, ω) be a symplectic manifold.

1. Show the identity

$$X_{\{f,g\}} = [X_f, X_g] \qquad \forall f, g \in C^\infty(M). \tag{2.5}$$

2. Show that $(C^\infty(M), \{\cdot,\cdot\})$ is a Poisson algebra.

Lemma 2.24 *Let (M, ω) be a symplectic manifold. Then*

$$(C^\infty(M), \{\cdot,\cdot\}) \to (\mathfrak{X}(M), [\cdot,\cdot]), \qquad f \mapsto X_f$$

is a Lie algebra homomorphism.

Lemma 2.25 *Let (M, ω) be a symplectic manifold and $\varphi \in \mathrm{Symp}(M, \omega)$. Then*

$$(C^\infty(M), \{\cdot,\cdot\}) \to (C^\infty(M), \{\cdot,\cdot\}), \qquad f \mapsto \varphi^* f$$

is a Lie algebra homomorphism.

Exercise 2.26 Prove Lemmas 2.24 and 2.25.

Given an observable, we want to describe its evolution under the flow of the Hamiltonian vector field.

Lemma 2.27 (Evolution Equation, [9, Corollary 3.3.15]) *Let (M, ω, H) be a Hamiltonian system. Then*

$$\frac{d}{dt} f \circ \phi_t^{X_H} = \{H, f\} \circ \phi_t^{X_H} \qquad \forall f \in C^\infty(M),$$

whenever either side is defined.

Proof We compute

$$\frac{d}{dt} f \circ \phi_t^H = df \left(\frac{d}{dt} \phi_t^H \right) = df(X_H \circ \phi_t^H) = \omega(X_H, X_f) \circ \phi_t^H = \{H, f\} \circ \phi_t^H$$

for all $f \in C^\infty(M)$. □

An important consequence of the evolution equation (Lemma 2.27) is that energy is conserved in a Hamiltonian system or, equivalently, that there is no friction.

Corollary 2.28 (Conservation of Energy, [77, Theorem 2.2.2]) *Let (M, ω, H) be a Hamiltonian system. Then*

$$H\left(\phi_t^H(x)\right) = H(x) \qquad \forall x \in M,$$

whenever the left side is defined.

Proof Using Lemma 2.27 we compute

$$\frac{d}{dt} H \circ \phi_t^H = \{H, H\} \circ \phi_t^H = 0$$

by antisymmetry of the Poisson bracket. □

Exercise 2.29 Let $H \in C^\infty([0, 1] \times M)$ be a time-dependent Hamiltonian function. Show that the energy H is not conserved in general. What could be the physical interpretation of such a time-dependent Hamiltonian function H?

The conservation of energy (Corollary 2.28) has many important applications in pure and applied mathematics.

Example 2.30 (The Mathematical Pendulum) As in [63, Introduction], consider the mechanical Hamiltonian system $(T^*\mathbb{R}, dp \wedge dq, H)$, where

$$H\colon \mathbb{R} \times \mathbb{R} \to \mathbb{R}, \qquad H(q, p) := \frac{1}{2}p^2 - \cos(q)$$

models the motion of a frictionless pendulum. The corresponding integral curves of the Hamiltonian vector field X_H are given by

$$\dot{q} = \frac{\partial H}{\partial p} = p \qquad \text{and} \qquad \dot{p} = -\frac{\partial H}{\partial q} = -\sin(q).$$

The Hamiltonian function H and thus also the Hamiltonian vector field X_H naturally descend to the cylinder $\mathbb{S}^1 \times \mathbb{R}$. By conservation of energy (Corollary 2.28),

Fig. 2.1 Some level sets $H^{-1}(c)$, $c \geq 0$, of the mathematical pendulum together with an exact solution corresponding to the initial value $(-\pi/3, 0)$ depicted in red

we conclude that the integral curves of X_H are contained in the level sets $H^{-1}(c)$, $c \geq 0$. Some of these level sets as well as a numerical solution are depicted in Fig. 2.1. The integral curves can only be calculated numerically, as the corresponding Hamiltonian equations are not solvable by analytical functions [102, Chapter 1].

The simplest method for solving Hamiltonian equations is provided by the ***explicit Euler method*** due to Euler (1768)

$$\begin{cases} q_{n+1} = q_n + h\dfrac{\partial H}{\partial p}(q_n, p_n), \\ p_{n+1} = p_n - h\dfrac{\partial H}{\partial q}(q_n, p_n), \end{cases}$$

for a fixed initial value (q_0, p_0) and step size h. Another discretisation method for obtaining an approximate solution is given by the ***implicit Euler method***

$$\begin{cases} q_{n+1} = q_n + h\dfrac{\partial H}{\partial p}(q_{n+1}, p_{n+1}), \\ p_{n+1} = p_n - h\dfrac{\partial H}{\partial q}(q_{n+1}, p_{n+1}). \end{cases}$$

In contrast to the explicit method, this method requires the numerical solution of a non-linear system of equations. As one sees from Fig. 2.2a, the solutions obtained via the explicit method spiral outwards, whereas the numerical approximations

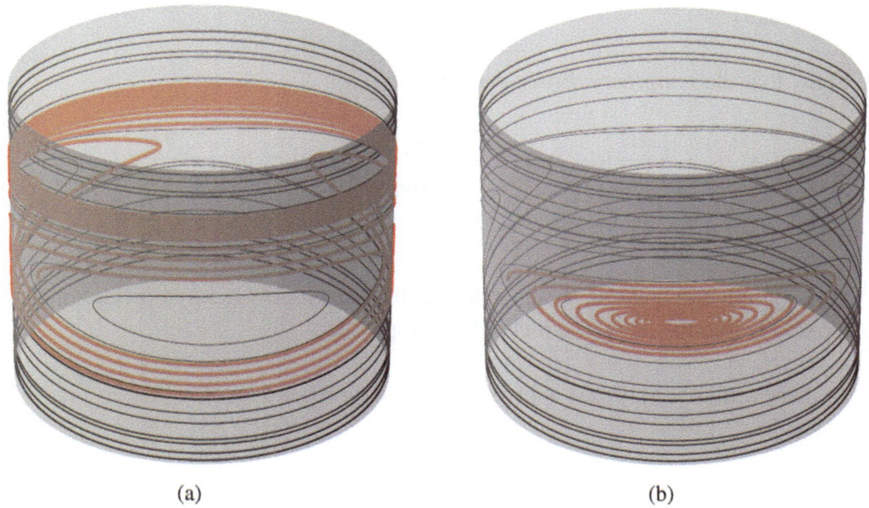

Fig. 2.2 (a) Numerical solution of the mathematical pendulum obtained via the explicit Euler method for the initial value $(-\pi/3, 0)$. (b) Numerical solution of the mathematical pendulum obtained via the implicit Euler method for the initial value $(-\pi/3, 0)$

obtained via the implicit method spiral inwards as depicted in Fig. 2.2b. Hence both methods are physically incorrect as they do not conserve energy! The physically correct solution in Fig. 2.1 is obtained via the ***leap-frog method*** [102, Theorem 3.4]

$$\begin{cases} p_{n+1/2} = p_n - \dfrac{h}{2}\dfrac{\partial H}{\partial q}(q_n, p_{n+1}), \\ q_{n+1} = q_n + \dfrac{h}{2}\left(\dfrac{\partial H}{\partial p}(q_n, p_{n+1/2}) + \dfrac{\partial H}{\partial p}(q_{n+1}, p_{n+1/2}\right), \\ p_{n+1} = p_{n+1/2} - \dfrac{h}{2}\dfrac{\partial H}{\partial q}(q_{n+1}, p_{n+1/2}). \end{cases}$$

In contrast to the explicit and the implicit Euler method, the leap-frog integrator is a symplectic integrator and does conserve the energy of a Hamiltonian system. For general numerical methods on manifolds see [102, Section IV.4–IV.9]. Due to the simplicity and exactness of the leap-frog method, it is used for practical implementations of Hamiltonian Monte Carlo methods in Bayesian statistics [39, Section 5.1], see also Example 2.34 below.

Exercise 2.31

1. Show that the ***symplectic Euler methods***

$$\begin{cases} p_{n+1} = p_n - h\dfrac{\partial H}{\partial q}(q_n, p_{n+1}), \\ q_{n+1} = q_n + h\dfrac{\partial H}{\partial p}(q_n, p_{n+1}), \end{cases} \quad \text{and} \quad \begin{cases} q_{n+1} = q_n + h\dfrac{\partial H}{\partial p}(q_{n+1}, p_n), \\ p_{n+1} = p_n - h\dfrac{\partial H}{\partial q}(q_{n+1}, p_n), \end{cases}$$

are a symplectic method, that is, that the discretised flow is a symplectic map.
2. Conclude that the leap-frog method is symplectic.

Is there a more analytical approach for describing the level curves in Fig. 2.1? One checks

$$\mathrm{Crit}(H) = \{(k\pi, 0) : k \in \mathbb{Z}\},$$

and that the Hessian $\mathrm{Hess}\, H(q, p)$ is invertible for every $(q, p) \in \mathrm{Crit}(H)$. In particular, $\ker \mathrm{Hess}\, H(q, p)$ is the trivial vector space and thus H is a ***Morse function***. The ***Morse index*** of H at a critical point (q, p) is defined to be the number of negative eigenvalues of the Hessian matrix $\mathrm{Hess}\, H(q, p)$ counted with multiplicity. If k is even, then the Morse index of $(k\pi, 0)$ is zero and thus the critical points $(k\pi, 0)$ are relative minima. If k is odd, then the Morse index of $(k\pi, 0)$ is one, and thus the critical points $(k\pi, 0)$ are saddle points. By the Morse lemma [29, Lemma 3.11], we can locally write via a change of variables

$$H(q, p) = H(2k\pi, 0) + \frac{1}{2}\mathrm{Hess}\, H_{(2k\pi,0)}\left((q, p), (q, p)\right) = \frac{1}{2}\left(q^2 + p^2\right).$$

Exercise 2.32 How does the Hamiltonian function H look like near the critical points $((2k + 1)\pi, 0)$ for $k \in \mathbb{Z}$?

Exercise 2.33 Compute the equation of motion for the mathematical pendulum using Hamilton's equations and Newton's second law. Compare these two methods. Which method is easier to handle?

Example 2.34 (Bayesian Statistics) Bayesian statistics is an approach to data analysis based on Bayes' theorem [187], where available knowledge about parameters in a statistical model is updated with the information of observed data. The background knowledge is expressed as a prior distribution and combined with observational data in the form of a likelihood function to determine the posterior distribution. Sampling such a posterior distribution is a highly non-trivial task and makes use of Markov Chain Monte–Carlo methods for stochastically exploring a unimodal probability distribution. A computationally robust implementation is, for example, provided by the Hamiltonian Monte–Carlo method [40] developed in the

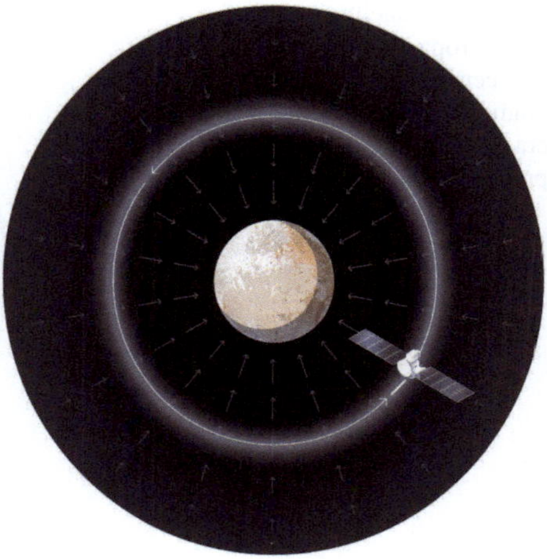

Fig. 2.3 An orbit around Jupiter's moon Europa lying in the region of high probability

late 1980s based on conservation of energy (Corollary 2.28) allowing for large jumps in exploring the probability distribution [39, Section 3]. The Hamiltonian Monte–Carlo method makes use of symplectic integrators [102, Chapter VI] such as the leap-frog method already introduced in Remark 2.30 for solving high-dimensional systems of non-linear, coupled ordinary differential equations, because these integrators remain accurate even over long integration times. An accessible implementation of the Hamiltonian Monte–Carlo method is provided by the R-package brms [45] or [46]. Moreover, there do exist generalisations of the Hamiltonian Monte–Carlo method suitable for multimodal target distributions [38] provided by contact geometry and recent quantum Hamiltonian Monte–Carlo methods [133]. We explain now how the Hamiltonian Monte–Carlo method works by an analogy with a satellite in an orbit around Jupiter's moon Europa following [39, Section 3]. Performing a random walk in a subset of high probability mass of a unimodal target distribution in a multidimensional setting is very inefficient without any further restrictions. However, by assuming that the target probability density function is smooth, we can apply the Hamiltonian framework. The gradient of the smooth target probability density function defines a vector field in parameter space sensitive to the structure of the target distribution, like the gravitational field of Europa, as depicted in Fig. 2.3.

Exploring regions of high probability then corresponds to finding a set of periodic orbits around Europa! This analogy exhibits already catastrophic outcomes in both the physical and the probabilistic system. If the tangential momentum is too small, the satellite will eventually crash into Europa due to the gravitational attraction as depicted in Fig. 2.4. In contrast, if the tangential momentum of the satellite is too large, it will disappear into space as illustrated in Fig. 2.5. As elaborated on in Exercise 2.30, only very few Hamiltonian systems in practice are analytically solvable, and one needs to stick to numerical approximations. Even for symplectic integrators like the leap-frog method, these approximations are prone to error in the long run. So the two scenarios of a satellite crashing into Europa or disappearing

Fig. 2.4 A satellite crashing into Europa corresponds to the scenario where any gradient-driven trajectory crashes into the mode of the probability distribution

Fig. 2.5 A satellite disappearing into space

into space correspond in the numerical setting to the scenario when the exact trajectories lie in regions of high curvature. Then even symplectic integrators can be divergent due to computational errors, and approximated trajectories diverge to infinite energies. So to successfully explore a probabilistic system via a random walk, one has to carefully correct errors in the integration process made by a symplectic integrator. This results in Fig. 2.6 for a corrected symplectic integrator oscillating around the exact trajectory over long times. In the physical system, this corresponds to the satellite accelerating in different directions along its orbit. This is also reminiscent of the output of a generative artificial intelligence model generating periodic orbits in the restricted three-body problem [86, Figure 3].

By energy conservation (Corollary 2.28), the flow of a Hamiltonian vector field X_H does not change the total energy H and hence conserves every preimage $H^{-1}(c)$, $c \in \mathbb{R}$. It follows that we can study the flow of the Hamiltonian vector

Fig. 2.6 A Markov-chain path of a satellite around Europa

field restricted to such a preimage. This is particularly useful, as we may have a compact hypersurface in a noncompact Hamiltonian system!

Definition 2.35 (Regular Energy Hypersurface, [193]) A ***regular energy hypersurface in a Hamiltonian system*** (M, ω, H) is defined to be an embedded hypersurface $\Sigma = H^{-1}(0)$ such that $\text{Crit}(H) \cap \Sigma = \varnothing$.

Remark 2.36 The set of critical values $H(\text{Crit}(H)) \subseteq \mathbb{R}$ has Lebesgue measure zero according to Sard's Theorem [129, Theorem 6.10].

Compact regular energy hypersurfaces satisfy a strong recurrence property for the flow of the Hamiltonian vector field as in Problem 2.65. This is because every symplectomorphism conserves the volume in phase space by definition. However, the flow of a time-dependent Hamiltonian vector field does not necessarily conserve energy by Exercise 2.29 and thus models dissipative systems. Moreover, the Hamiltonian formalism is not suitable for modeling systems with friction in general, as the next exercise shows.

Exercise 2.37 The equation of motion that models the motion of a damped pendulum with linear friction is given by

$$\ddot{q} = -\mu\dot{q} - \sin q, \qquad \mu > 0.$$

Show that this equation cannot be derived from a time-dependent Hamiltonian function.

How should one interpret a time-dependent Hamiltonian function instead? Given a time-dependent Hamiltonian function $H \in C^\infty(M \times [0, 1])$ such that $\Sigma_0 := H_0^{-1}(0)$ and $\Sigma_1 := H_1^{-1}(0)$ are regular energy hypersurfaces, then H can be interpreted as a homotopy from Σ_0 to Σ_1. Thus, a time-dependent Hamiltonian function does "deform" regular energy hypersurfaces. Another geometrical interpretation is as follows. If (M, ω) is a compact symplectic manifold, then the flow of certain time-dependent Hamiltonian functions is contained in the Lie algebra to the group of symplectomorphisms $\mathrm{Symp}(M, \omega)$. The symplectomorphisms form an infinite-dimensional Lie group, but we will not delve deeper into the technicalities.

Definition 2.38 (Symplectic Vector Field) Let (M, ω) be a symplectic manifold. A vector field $X \in \mathfrak{X}(M)$ is called ***symplectic***, if $di_X\omega = 0$. We denote the space of all symplectic vector fields on (M, ω) by $\mathfrak{X}(M, \omega)$.

Example 2.39 (Hamiltonian Vector Fields) Let (M, ω) be a symplectic manifold and let $H \in C^\infty(M \times [0, 1])$ be a time-dependent Hamiltonian function. Then the time-dependent vector field X_{H_t} belongs to $\mathfrak{X}(M, \omega)$ for all $t \in [0, 1]$.

Proposition 2.40 *Let (M, ω) be a compact symplectic manifold. Then the Lie algebra of the Lie group $\mathrm{Symp}(M, \omega)$ is $\mathfrak{X}(M, \omega)$.*

Proof Let $(\varphi_t)_{t\in(-\varepsilon,\varepsilon)}$ be a smooth path in $\mathrm{Diff}(M)$ for some $\varepsilon > 0$ starting at the identity, that is, we have $\varphi_0 = \mathrm{id}_M$. Then φ_t satisfies

$$\frac{d}{dt}\varphi_t = X_t \circ \varphi_t, \qquad X_t := \frac{d}{dt}\varphi_t \circ \varphi_t^{-1} \in \mathfrak{X}(M).$$

For the time-dependent vector field X_t we compute

$$\frac{d}{dt}\varphi_t^*\omega = \varphi_t^*(L_{X_t}\omega) = \varphi_t^*(di_{X_t}\omega + i_{X_t}d\omega) = \varphi_t^*(di_{X_t}\omega). \tag{2.6}$$

By assumption, our path belongs to $\mathrm{Symp}(M, \omega)$ and so $\varphi_t^*\omega = \omega$ for all $t \in (-\varepsilon, \varepsilon)$. Consequently, the vector field X_t is symplectic for every t and in particular

$$X_0 = \left.\frac{d}{dt}\right|_{t=0} \varphi_t \in \mathfrak{X}(M, \omega) = T_{\mathrm{id}_M} \mathrm{Symp}(M, \omega),$$

implying $\mathfrak{X}(M, \omega) \subseteq T_{\mathrm{id}_M} \mathrm{Symp}(M, \omega)$. The reverse inclusion also follows from the computation in (2.6), as the flow of any symplectic vector field in $\mathfrak{X}(M, \omega)$ consists of symplectomorphisms. Lastly, $\mathfrak{X}(M, \omega)$ is a Lie algebra under the usual Lie bracket for vector fields. Indeed, for all $X, Y \in \mathfrak{X}(M, \omega)$ we compute

$$di_{[X,Y]}\omega = d(L_X i_Y\omega - i_Y L_X\omega) = d(di_X i_Y\omega + i_X di_Y\omega) = 0$$

by Exercise 2.4. □

An important subgroup of the group of symplectomorphisms $\mathrm{Symp}(M, \omega)$ consists of all symplectomorphisms arising as a fixed-time flow of a time-dependent Hamiltonian vector field.

Definition 2.41 (Hamiltonian Symplectomorphism, [137, Definition 3.1.13]) A symplectomorphism $\varphi \in \mathrm{Symp}(M, \omega)$ is called ***Hamiltonian***, if there exists a time-dependent Hamiltonian function $H \in C^\infty(M \times [0, 1])$ such that the flow ϕ of the time-dependent vector field X_{H_t} satisfies $\varphi = \phi_1$. We denote the space of all Hamiltonian symplectomorphisms by $\mathrm{Ham}(M, \omega)$. In addition, we define the group of all compactly supported Hamiltonian symplectomorphisms by

$$\mathrm{Ham}_c(M, \omega) := \{\varphi_H \in \mathrm{Ham}(M, \omega) : H \in C_c^\infty(M \times [0, 1])\}.$$

Remark 2.42 With some work, one can show that the Lie algebra of the group of all Hamiltonian symplectomorphisms $\mathrm{Ham}(M, \omega)$ is the algebra of all Hamiltonian vector fields, see [137, Proposition 10.2.12]

An interesting question stated in [155, Chapter 2] is what the minimal amount of energy needed to generate a given Hamiltonian symplectomorphism is. This question leads to an interesting geometry on the infinite-dimensional Lie group $\mathrm{Ham}_c(M, \omega)$, called ***Hofer's geometry***. A related question is what is the minimal energy to displace a compact regular energy hypersurface from itself by a compactly supported Hamiltonian symplectomorphism? First, we need to define a "length" of a compactly supported time-dependent Hamiltonian function.

Definition 2.43 (Hofer Norm, [137, p. 466]) Let (M, ω) be a symplectic manifold. Define the ***Hofer norm*** of $F \in C_c^\infty(M \times [0, 1])$ by

$$\|F\| := \|F\|_+ + \|F\|_-,$$

where

$$\|F\|_+ := \int_0^1 \max_{x \in M} F_t(x) dt \qquad \text{and} \qquad \|F\|_- := -\int_0^1 \min_{x \in M} F_t(x) dt.$$

Guided by our question, which is an optimisation problem, we can define the notion of displacement energy of a subset of a symplectic manifold.

Definition 2.44 (Displacement Energy, [137, p. 469]) Let (M, ω) be a symplectic manifold. For a compact subset $K \subseteq M$ define the ***displacement energy*** of K by

$$e(K) := \inf_{\substack{F \in C_c^\infty(M \times [0,1]) \\ \varphi_F(K) \cap K = \varnothing}} \|F\| \in [0, +\infty],$$

where $\varphi_F := \phi_1^{X_F}$ denotes the time-1-map of the smooth flow of the time-dependent Hamiltonian vector field X_{F_t}. If $A \subseteq M$ is an arbitrary subset, we define

$$e(A) := \sup_{K \subseteq A \text{ compact}} e(K).$$

A subset $A \subseteq M$ is called ***displaceable***, if $e(A) < +\infty$.

Exercise 2.45 Let (M, ω) be a symplectic manifold. Show that if $A \subseteq B$ are two subsets, then $e(A) \leq e(B)$. Moreover, show that $e(A) = e(\varphi(A))$ for every symplectomorphism $\varphi \in \mathrm{Symp}(M, \omega)$.

In general, it is notoriously difficult to explicitly compute the displacement energy. Consider the closed ball of radius r in $\mathbb{R}^{2n}$ defined by

$$B^{2n}(r) := \{(x, y) \in \mathbb{R}^{2n} : \|x\|^2 + \|y\|^2 \leq r^2\}.$$

We will see that $e(B^{2n}(r)) = \pi r^2$ holds for all $r > 0$, independent of the dimension! So, the ball $B^{2n}(r)$ is displaceable in $\mathbb{R}^{2n}$. Likewise, any compact subset of $\mathbb{R}^{2n}$ should be displaceable by considering a translation. However, this only provides an upper bound for the displacement energy. Turning to the next intuitive setting of the cotangent bundle T^*Q of a compact configuration space Q, the question of displaceability becomes even more complicated. In fact, by [139] we have $e(Q) = +\infty$ in $(T^*Q, dp \wedge dq)$ for the zero section Q in T^*Q, and thus Q is not displaceable. However, if we look at magnetic cotangent bundles (T^*Q, ω_σ) for a nowhere-vanishing closed form $\sigma \in \Omega^2(Q)$, we see by [154] and [127] that the zero section Q is displaceable if the Euler characteristic of Q vanishes.

2.2 Noether's Theorem

In this section, we look at physical systems with symmetries. For example, the Hamiltonian function of the Kepler problem from Problem 2.66 is invariant under rotational symmetries. These symmetries sometimes allow us to actually solve the corresponding Hamiltonian equations analytically. For a much more detailed study of this approach, see [63]. Motivated by Lemma 2.27, we give the following definition of a conserved quantity in a Hamiltonian system.

Definition 2.46 (Integral of Motion) An ***integral of motion*** for a Hamiltonian system (M, ω, H) is defined as a smooth function $I \in C^\infty(M)$ such that $\{H, I\} = 0$.

Remark 2.47 It is easy to check that the integrals of motion form a (nonempty) Lie subalgebra of the Poisson algebra of classical observables $(C^\infty(M), \{\cdot, \cdot\})$.

A natural object representing the symmetries of a Hamiltonian system are Lie groups. Let $\psi\colon G \times M \to M$ be a smooth left action of a Lie group G on a smooth manifold M and denote by $\mathfrak{g} := T_eG \cong \mathfrak{X}_L(G)$ the corresponding Lie algebra. Every vector $\xi \in \mathfrak{g}$ determines a smooth global flow on M by $(t, x) \mapsto \psi_{\exp(-t\xi)}(x)$, where

$$\exp\colon \mathfrak{g} \to G, \qquad \exp(\xi) := \gamma_\xi(1)$$

denotes the exponential map and γ_ξ is the integral curve of the left-invariant vector field X_ξ starting at e with $\dot{\gamma}_\xi(0) = \xi$. Note that if there exists a biinvariant Riemannian metric on G, then this exponential map coincides with the exponential map of the associated Levi–Civita connection at the identity [130, Problem 5–8. (c)]. Define a vector field $\widehat{\xi} \in \mathfrak{X}(M)$ to be the infinitesimal generator of this flow, that is,

$$\widehat{\xi}_x := \frac{d}{dt}\bigg|_{t=0} \psi_{\exp(-t\xi)}(x) \qquad \forall x \in M.$$

Then

$$(\mathfrak{g}, [\cdot,\cdot]) \to (\mathfrak{X}(M), [\cdot,\cdot]), \qquad \xi \mapsto \widehat{\xi}$$

is a Lie algebra homomorphism by [129, Theorem 20.15].

Definition 2.48 (Weakly Hamiltonian Action) Let (M, ω) be a symplectic manifold. A smooth action $\psi\colon G \times M \to M$ of a Lie group G is said to be ***weakly Hamiltonian***, if the map $\psi_g \in \mathrm{Diff}(M)$ is a symplectomorphism for all $g \in G$ and there exists a linear map

$$\mu\colon \mathfrak{g} \to C^\infty(M),$$

called a ***momentum map***, such that the diagram

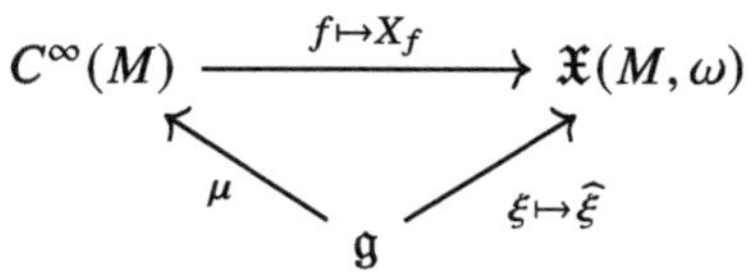

commutes.

Remark 2.49 The moment mapping we defined is sometimes also referred to as a comoment map in the literature. Lie groups, Lie algebras and (co)moment maps play an important role in formulating and analysing concrete Hamiltonian systems. We will not dive deeper into this topic since there are already excellent textbooks such as [63] on that subject.

Using the notion of a weakly Hamiltonian action, we can specify the concept of a symmetry group of a Hamiltonian system.

Definition 2.50 (Symmetry Group) A Lie group G is said to be a ***symmetry group of a Hamiltonian system*** (M, ω, H), if there exists a weakly Hamiltonian action ψ of G on (M, ω), such that $H \circ \psi_g = H$ for all $g \in G$.

We next discuss the symmetry group of the Kepler problem in detail.

Example 2.51 (Homogeneous Singular Potentials) Let $m \geq 2$ be an integer. Following [120, Exercise 11.25], consider the mechanical Hamiltonian function

$$H\colon T^*(\mathbb{R}^n \setminus \{0\}) \to \mathbb{R}, \qquad H(q, p) := \frac{1}{2}\|p\|^2 - \frac{1}{\|q\|^{2-2/m}}. \tag{2.7}$$

If $m = 2$, then this Hamiltonian function is the Hamiltonian describing the ***Kepler problem***. Define an $\mathrm{SO}(n)$-action on $\mathbb{R}^n \setminus \{0\}$ by

$$\psi\colon \mathrm{SO}(n) \times \mathbb{R}^n \setminus \{0\} \to \mathbb{R}^n \setminus \{0\}, \qquad \psi_A(q) := Aq.$$

By Problem 2.64, we get an induced $\mathrm{SO}(n)$-action on $T^*(\mathbb{R}^n \setminus \{0\})$ and one can check that the momentum map is given by

$$\mu\colon \mathfrak{so}(n) \to C^\infty((\mathbb{R}^n \setminus \{0\}) \times \mathbb{R}^n), \qquad \mu(B)(q, p) = q^T Bp,$$

where

$$\mathfrak{so}(n) = \left\{A \in \mathbb{R}^{n\times n} : A^T = -A\right\}.$$

Remark 2.52 (Regularising Homogeneous Singular Potentials) The mechanical Hamiltonian function (2.7) has a singularity for $q \to 0$. This corresponds to a collision of the two masses. Collisions can be regularised, meaning that there exists a symplectic embedding $\Sigma \hookrightarrow \overline{\Sigma}$ of the noncompact regular energy hypersurface Σ containing a collision into a compact regular energy hypersurface $\overline{\Sigma}$ of a regularised Hamiltonian function. For $n = 2$, the generalised ***Levi–Civita regularisation***

$$\Phi\colon T^*(\mathbb{C} \setminus \{0\}) \to T^*(\mathbb{C} \setminus \{0\}), \qquad (z, w) \mapsto \left(z^m, \frac{w}{m\bar{z}^{m-1}}\right) \tag{2.8}$$

is defined to be the cotangent lift (Problem 2.64) of the map $z \mapsto z^m$. Following [59, Section 4.1], let $c < 0$ be fixed and consider the level set $\Sigma_c = H^{-1}(0)$ for the mechanical Hamiltonian function (2.7). We change the time parametrisation on Σ_c by rescaling

$$K(z, w) := \|z\|^{2m-2}(H(\Phi(z, w)) - c) = \frac{1}{2m^2}\|w\|^2 - c\|z^{m-1}\|^2 - 1.$$

By this transformation of the Hamiltonian function H, the collision locus $\{q = 0\}$ in the closure of Σ_c is transformed to $\{z = 0\}$ in the regular energy hypersurface $K^{-1}(0)$, which is no longer singular and, therefore, collisions are regularised.

How does one find integrals of motion corresponding to a symmetry group? The answer is given by a famous result by Emmy Noether from 1918 reproduced in [151] or in the book [171].

Lemma 2.53 (Noether's Theorem, Hamiltonian Version) *Let G be a symmetry group of a Hamiltonian system (M, ω, H). Then $\mu(\xi)$ is an integral of motion for all elements $\xi \in \mathfrak{g}$ in the Lie algebra.*

Proof For $\xi \in \mathfrak{g}$ we compute

$$\{\mu(\xi), H\} = X_{\mu(\xi)}H = \widehat{\xi}H = \left.\frac{d}{dt}\right|_{t=0} H \circ \psi_{\exp(-t\xi)} = \left.\frac{d}{dt}\right|_{t=0} H = 0.$$

This proves the statement. □

Exercise 2.54 Write down the three integrals of motion in the Kepler problem given by Noether's theorem (Lemma 2.53) for the case $n = 3$. What is the physical interpretation of these integrals of motion?

For a much more detailed analysis of the Kepler problem as well its connection to the restricted three-body problem see [77] and the excellent survey article [142]. Compare also Problem 2.66. The momentum map of a weakly Hamiltonian action can have additional properties.

Definition 2.55 A weakly Hamiltonian action of a Lie group G on a symplectic manifold (M, ω) is called

- ***Hamiltonian***, if the momentum map $\mu\colon \mathfrak{g} \to C^\infty(M)$ is G-equivariant with respect to the adjoint action of G on its associated Lie algebra $\mathfrak{g}$ and the induced action of G on $C^\infty(M)$, that is

$$\mu(\mathrm{Ad}_{g^{-1}}(\xi)) = \mu(\xi) \circ \psi_g \qquad \forall g \in G, \xi \in \mathfrak{g},$$

  where

$$\mathrm{Ad}_{g^{-1}}(\xi) := \left.\frac{d}{dt}\right|_{t=0} g^{-1}\exp(t\xi)g.$$

- ***Poisson***, if the associated momentum map

$$\mu\colon (\mathfrak{g}, [\cdot,\cdot]) \to (C^\infty(M), \{\cdot,\cdot\})$$

  is a Lie algebra homomorphism.

Remark 2.56 For a connected Lie group, a weakly Hamiltonian action is Hamiltonian if and only if it is Poisson by [77, p. 33–34].

In applications, the Hamiltonian system is often exact. In this case, an explicit formula for the momentum map can be given. In particular, this applies to the momentum map of homogeneous singular potentials in Example 2.51.

Lemma 2.57 (Momentum Lemma, [137, Exercise 5.2.2]) *Let $\psi\colon G \times M \to M$ be a smooth Lie group action on an exact symplectic manifold (M, λ) with $\psi_g^*\lambda = \lambda$ for all $g \in G$. Then the action is Hamiltonian and Poisson with momentum map*

$$\mu(\xi) = i_{\widehat{\xi}}\lambda, \qquad \forall \xi \in \mathfrak{g}.$$

Proof We show the result in five steps.
Step 1: ψ is an action by symplectomorphisms. For all $g \in G$ we compute

$$\psi_g^* d\lambda = d\psi_g^*\lambda = d\lambda$$

Thus $\psi_g \in \mathrm{Diff}(M)$ is a symplectomorphism.
Step 2: ψ is a weakly Hamiltonian action. For all $\xi \in \mathfrak{g}$ we see

$$\begin{aligned} i_{\widehat{\xi}} d\lambda &= L_{\widehat{\xi}}\lambda - d i_{\widehat{\xi}}\lambda \\ &= \frac{d}{dt}\bigg|_{t=0} \psi^*_{\exp(-t\xi)}\lambda - d\mu(\xi) \\ &= \frac{d}{dt}\bigg|_{t=0} \lambda - d\mu(\xi) \\ &= -d\mu(\xi). \end{aligned}$$

Step 3: $\psi_g^*\widehat{\xi} = \widehat{\mathrm{Ad}_{g^{-1}}(\xi)}$ for all $g \in G$ and $\xi \in \mathfrak{g}$. We have a commutative diagram

$$\begin{array}{ccc} \mathfrak{g} & \xrightarrow{\mathrm{Ad}_{g^{-1}}} & \mathfrak{g} \\ {\scriptstyle \exp}\downarrow & & \downarrow{\scriptstyle \exp} \\ G & \xrightarrow[C_{g^{-1}}]{} & G, \end{array}$$

where $C_{g^{-1}}(h) := g^{-1}hg$ denotes the conjugation action on G. We compute

$$\begin{aligned} \psi_g^*\widehat{\xi} &= D\psi_{g^{-1}} \circ \widehat{\xi} \circ \psi_g \\ &= D\psi_{g^{-1}} \circ \frac{d}{dt}\bigg|_{t=0} \psi_{\exp(-t\xi)} \circ \psi_g \end{aligned}$$

$$
\begin{aligned}
&= \frac{d}{dt}\bigg|_{t=0} \psi_{g^{-1}} \circ \psi_{\exp(-t\xi)} \circ \psi_g \\
&= \frac{d}{dt}\bigg|_{t=0} \psi_{g^{-1}\exp(-t\xi)g} \\
&= \frac{d}{dt}\bigg|_{t=0} \psi_{\exp(-t\,\mathrm{Ad}_{g^{-1}}(\xi))} \\
&= \widehat{\mathrm{Ad}_{g^{-1}}(\xi)}.
\end{aligned}
$$

Step 4: ψ is a Hamiltonian action. Using Step 2, for all $g \in G$ and $\xi \in \mathfrak{g}$ we compute

$$
\begin{aligned}
\mu(\mathrm{Ad}_{g^{-1}}(\xi)) &= i_{\widehat{\mathrm{Ad}_{g^{-1}}(\xi)}}\lambda \\
&= i_{\psi_g^*\widehat{\xi}}\lambda \\
&= \lambda(D\psi_{g^{-1}} \circ \widehat{\xi} \circ \psi_g) \\
&= \psi_{-g}^*\lambda(\widehat{\xi} \circ \psi_g) \\
&= i_{\widehat{\xi}}\lambda \circ \psi_g \\
&= \mu(\xi) \circ \psi_g.
\end{aligned}
$$

Step 5: ψ is a Poisson action. For $\xi, \eta \in \mathfrak{g}$ we compute

$$
\begin{aligned}
\mu[\xi, \eta] &= i_{\widehat{[\xi,\eta]}}\lambda \\
&= i_{[\widehat{\xi},\widehat{\eta}]}\lambda \\
&= L_{\widehat{\xi}} i_{\widehat{\eta}}\lambda - i_{\widehat{\eta}} L_{\widehat{\xi}}\lambda \\
&= L_{\widehat{\xi}} i_{\widehat{\eta}}\lambda \\
&= L_{\widehat{\xi}}\mu(\eta) \\
&= \widehat{\xi}\mu(\eta) \\
&= X_{\mu(\xi)}\mu(\eta) \\
&= \{\mu(\xi), \mu(\eta)\}.
\end{aligned}
$$

The proof of the momentum lemma is complete. □

For showing existence and uniqueness results for Poisson actions, we recall the basic notions of Lie algebra cohomology [192, Chapter 7]. An additional benefit of studying this homology associated with a Lie algebra is that under certain circumstances one can show that a smooth manifold is a Lie group or not. Or do you know a proof why the sphere $\mathbb{S}^2$ does not admit any Lie group structure at all? Let $\mathfrak{g}$ be a Lie algebra. For all $k \in \mathbb{Z}$ define a nonnegative chain complex by

$$C^k(\mathfrak{g}) := \Lambda^k \mathfrak{g}^*, \qquad \forall k \geq 0,$$

where $C^k(\mathfrak{g}) = 0$ for all $k > \dim \mathfrak{g}$ and $\Lambda^0 \mathfrak{g}^* \cong \mathbb{R}$. Define $d\colon C^k(\mathfrak{g}) \to C^{k+1}(\mathfrak{g})$ by

$$df(X_0, \ldots, X_k) := \sum_{0 \leq i < j \leq k} (-1)^{i+j} f([X_i, X_j], X_0, \ldots, \bar{X}_i, \ldots, \bar{X}_j, \ldots, X_k),$$

whenever $k \geq 1$ and by $d = 0$ if $k = 0$. One checks that $d \circ d = 0$ holds via a tedious computation or using the abstract machinery of derived functors in category theory as in [192, Chapter 7]. The resulting nonnegative chain complex ($C^k(\mathfrak{g})$ vanishes for all $k < 0$ by definition) is called the ***Chevalley–Eilenberg cochain complex***. Then the ***k-th cohomology group of*** $\mathfrak{g}$ is defined by

$$\mathrm{H}^k(\mathfrak{g}; \mathbb{R}) := \frac{\ker d\colon C^k(\mathfrak{g}) \to C^{k+1}(\mathfrak{g})}{\operatorname{im} d\colon C^{k-1}(\mathfrak{g}) \to C^k(\mathfrak{g})}.$$

By [64, Theorem 26.1], we have $\mathrm{H}^*(\mathfrak{g}; \mathbb{R}) \cong \mathrm{H}^*_{\mathrm{dR}}(G)$ for the Lie algebra $\mathfrak{g}$ of a compact and connected Lie group G.

Exercise 2.58 ($\mathrm{H}^3(\mathfrak{sl}_2(\mathbb{R}), \mathbb{R})$) Consider the ***special linear Lie algebra*** $\mathfrak{sl}_2(\mathbb{R}, \mathbb{R})$ with ordered basis

$$e_1 := \begin{pmatrix} 1 & 0 \\ 0 & -1 \end{pmatrix} \qquad e_2 := \begin{pmatrix} 0 & 1 \\ 0 & 0 \end{pmatrix} \qquad e_3 := \begin{pmatrix} 0 & 0 \\ 1 & 0 \end{pmatrix}.$$

Show that $\mathrm{H}^3(\mathfrak{sl}_2(\mathbb{R}), \mathbb{R}) \cong \mathbb{R}$.

Lie algebra cohomology gives obstructions for the existence and uniqueness of momentum maps as in [64, Chapter 26].

Proposition 2.59 (Uniqueness of Momentum Maps for Poisson Actions) *Let μ and $\tilde{\mu}$ be two momentum maps for a Poisson G-action on a connected symplectic manifold. If $\mathrm{H}^1(\mathfrak{g}; \mathbb{R}) = 0$ holds, then $\mu = \tilde{\mu}$.*

Proof By assumption, there exists $\sigma \in \mathfrak{g}^* = \Lambda^1 \mathfrak{g}^*$ such that

$$\mu(\xi) - \tilde{\mu}(\xi) = \sigma(\xi) \qquad \forall \xi \in \mathfrak{g}.$$

Since both μ and $\tilde{\mu}$ are Lie algebra homomorphisms, we have $d\sigma = 0$. Indeed, for $\xi, \eta \in \mathfrak{g}$ we compute

$$d\sigma(\xi, \eta) = \sigma([\eta, \xi]) = \mu([\eta, \xi]) - \tilde{\mu}([\eta, \xi]) = \{\mu(\eta), \mu(\xi)\} - \{\tilde{\mu}(\eta), \tilde{\mu}(\xi)\} = 0.$$

Thus $\sigma \in \ker d = \mathrm{H}^1(\mathfrak{g}; \mathbb{R}) = 0$, implying $\sigma = 0$ and the statement follows. □

Proposition 2.60 (Existence of Poisson Actions) *Suppose we are given a weakly Hamiltonian G-action on a connected symplectic manifold (M, ω). If $\mathrm{H}^2(\mathfrak{g}; \mathbb{R}) = 0$, then the action is Poisson.*

Proof For $\xi, \eta \in \mathfrak{g}$ we compute

$$X_{\mu([\xi,\eta])} = \widehat{[\xi, \eta]} = [\widehat{\xi}, \widehat{\eta}] = [X_{\mu(\xi)}, X_{\mu(\eta)}] = X_{\{\mu(\xi),\mu(\eta)\}}$$

using (2.5). Thus by connectedness of M there exists $\tau \in \Lambda^2\mathfrak{g}^*$ such that

$$\{\mu(\xi), \mu(\eta)\} - \mu([\xi, \eta]) = \tau(\xi, \eta) \qquad \forall \xi, \eta \in \mathfrak{g}.$$

Invoking the Jacobi identity for the Lie as well as the Poisson bracket, yields $d\tau = 0$ and so $[\tau] \in \mathrm{H}^2(\mathfrak{g}; \mathbb{R}) = 0$. Hence there exists $\sigma \in \mathrm{H}^1(\mathfrak{g}; \mathbb{R})$ such that $\tau = d\sigma$. The momentum map

$$\mathfrak{g} \to C^\infty(M), \qquad \xi \mapsto \mu(\xi) - \sigma(\xi)$$

is a Lie algebra homomorphism. □

Recall that a Lie algebra $\mathfrak{g}$ is said to be semisimple if $\mathfrak{g}$ does not admit any nontrivial abelian ideals. Likewise, a Lie group G is called semisimple, if its associated Lie algebra $\mathfrak{g} := T_eG$ is semisimple. By [48, Theorem 16.1], a compact connected Lie group G is semisimple if and only if its fundamental group is finite. In that case

$$\mathrm{H}^1(\mathfrak{g}; \mathbb{R}) = \mathrm{H}^2(\mathfrak{g}; \mathbb{R}) = 0 \qquad \text{and} \qquad \mathrm{H}^3(\mathfrak{g}; \mathbb{R}) \neq 0. \tag{2.9}$$

Corollary 2.61 *Let G be a semisimple Lie group. Then every weakly Hamiltonian G-action on a connected symplectic manifold is Poisson and admits a unique momentum map.*

Proof The statement immediately follows from Propositions 2.59 and 2.60. □

Lie algebra cohomology also gives obstructions for a manifold being a Lie group. In particular, the result for semisimple Lie algebras (2.9) from [48, Theorem 21.1] can be used to show that Lie group structures on spheres are rare!

Exercise 2.62 ([48, Theorem 21.3]) Let $n \geq 0$ be a natural number.

1. Show that $\mathbb{S}^n$ does not admit any Lie group structure for $n \geq 2$ even. *Hint:* Make use of the Hairy Ball Theorem [129, Problem 16-6].
2. Show that only the spheres $\mathbb{S}^0$, $\mathbb{S}^1$ and $\mathbb{S}^3$ are Lie groups.
3. Show that $\mathbb{Z}_2$ is the only nontrivial group that can act freely on the sphere $\mathbb{S}^n$ if n is even.

Let us recapitulate what we have learnt so far in this chapter. It is a good idea to do an analogue exercise at the end of every chapter on your own!

Exercise 2.63 (Summary of the Hamiltonian Formalism)

1. Write a summary of the important concepts of this section about the Hamiltonian formalism and *think* of an answer to each of the following questions:
   - What concepts were new to you?
   - What concepts were you already familiar with?
   - What concepts were difficult to understand and should be further elaborated on?
2. *Pair* up with another student and discuss your answers.
3. *Share* your results with the other students and the lecturer.

Problems

2.64 (Physical Transformation, [77, Section 2.2.1]) Let $\varphi \in \mathrm{Diff}(M)$ be a diffeomorphism of a smooth manifold M. Define the ***cotangent lift of*** φ

$$D_*\varphi \colon T^*M \to T^*M, \qquad D_*\varphi(q, p)(v) := p(D\varphi^{-1}(v))$$

for all $(q, p) \in T^*M$ and $v \in T_{\varphi(q)}M$.

1. Show that $(D_*\varphi)^*\lambda = \lambda$, where λ is the Liouville form on M.
2. Deduce that for any action $\psi \in C^\infty(G \times M, M)$ of a Lie group G the lifted action given by $g \mapsto D_*\psi_g$ is Poisson with corresponding momentum map

$$\mu \colon \mathfrak{g} \to C^\infty(T^*M), \qquad \mu(\xi)(q, p) = p(\widehat{\xi}).$$

3. Let $U \subseteq \mathbb{C}$ be an open subset and suppose that $\varphi \in C^\infty(U, \mathbb{C})$ is holomorphic with $\varphi' \neq 0$ on U for the complex derivative φ' of φ. Then the cotangent lift $D_*\varphi$ of φ is given by

$$D_*\varphi \colon T^*U \to T^*\mathbb{C}, \qquad D_*\varphi(z, w) = \left(\varphi(z), \frac{w}{\overline{\varphi'(z)}}\right).$$

2.65 (The Poincaré Recurrence Theorem, [200, Theorem I.15]) Let $\Sigma = H^{-1}(0)$ be a compact regular energy hypersurface in a Hamiltonian system (M, ω, H) and denote by $\mathcal{B}(\Sigma)$ the Borel σ-algebra generated by all open subsets of Σ.

1. Assume that there exists a probability measure μ_Σ on Σ and let $T : \Sigma \to \Sigma$ be a smooth map satisfying

$$\mu_\Sigma(T^{-1}(U)) = \mu_\Sigma(U) \qquad \forall U \in \mathcal{B}(\Sigma). \tag{2.10}$$

Show that for μ_Σ-almost all points $x \in \Sigma$ there exists a sequence (k_j) of natural numbers with $k_j \to \infty$ as $j \to \infty$ such that

$$\lim_{j\to\infty} T^{k_j}(x) = x.$$

Hint: Show that the set

$$N := \bigcup_{n\in\mathbb{N}} \bigcup_{i\in I_n} \left(B_{1/n}(x_{i,n}) \setminus \left(B_{1/n}(x_{i,n}) \cap \bigcap_{k\geq 0} \bigcup_{l\geq k} T^{-l}\left(B_{1/n}(x_{i,n})\right)\right)\right)$$

is measurable and satisfies $\mu_\Sigma(N) = 0$, where for every $n \in \mathbb{N}$ there exists a finite index set I_n such that $\left(B_{1/n}(x_{i,n})\right)_{i\in I_n}$ is an open cover for Σ.

2. The purpose of this exercise is to construct a probability measure μ_Σ on Σ^{2n-1} in the symplectic manifold (M^{2n}, ω).
   (a) Explain why it is enough to construct a volume form $\Omega_\Sigma \in \Omega^{2n-1}(\Sigma)$ that satisfies $T^*\Omega_\Sigma = \Omega_\Sigma$.
   (b) As $dH \neq 0$ on Σ by the definition of a regular energy hypersurface, we find a neighbourhood U of Σ in M such that $dH \neq 0$ on U as Σ is assumed to be compact. Pick $\alpha \in \Omega^{2n-1}(U)$ such that

$$\omega^n = dH \wedge \alpha\,. \tag{2.11}$$

   Show that there exists a vector field $X \in \mathfrak{X}(U)$ satisfying $dH(X) = 1$. Conclude that (2.11) holds for $\alpha := i_X\omega^n$ and that $\iota_\Sigma^*(\alpha) \in \Omega^{2n-1}(\Sigma)$ is a volume form for the inclusion $\iota_\Sigma : \Sigma \hookrightarrow M$.
   (c) Show that the volume form $\iota_\Sigma^*\alpha \in \Omega^{2n-1}(\Sigma)$ is uniquely determined by the condition (2.11).
   (d) Denote the unique volume form on Σ satisfying (2.11) by Ω_Σ. Show that $T^*\Omega_\Sigma = \Omega_\Sigma$ holds. *Hint:* See the proof of [200, Lemma V.33].

2.66 (Complete Integrability) Consider the Hamiltonian function corresponding to a homogeneous singular potential in Example 2.51 for $m = 2$ and $n = 3$. This is the spatial Kepler problem $(T^*(\mathbb{R}^3 \setminus \{0\}), dp \wedge dq, H)$, with

$$H(q, p) = \frac{1}{2}\|p\|^2 - \frac{1}{\|q\|}$$

The aim of this exercise is to show that the spatial Kepler problem is completely integrable, meaning that there are "enough" conserved quantities to solve the corresponding Hamiltonian equations.

1. By Exercise 2.54 we already know, that the components of the ***angular momentum*** given

$$L\colon \mathbb{R}^3 \times \mathbb{R}^3 \to \mathbb{R}^3, \qquad L(q, p) := q \times p$$

are integral of motions for the spatial Kepler problem. Show that $\|L\|^2$ is an additional integral of motion for the spatial Kepler problem.

2. Show that the components of the ***Runge–Lenz vector***

$$A\colon \mathbb{R}^3 \setminus \{0\} \times \mathbb{R}^3 \to \mathbb{R}^3, \qquad A(q, p) := p \times L(q, p) - \frac{q}{\|q\|}$$

are also integrals of motion for the spatial Kepler problem. Here is where general homogeneous singular potentials for $m \neq 2$ differ from the special case of the Kepler problem!

3. A Hamiltonian system (M^{2n}, ω, H) is said to be ***completely integrable***, if there exist functions $H_1 = H, H_2, \ldots, H_n \in C^\infty(M)$ such that $\{H_i, H_j\} = 0$ for all $i \neq j$. Conclude that the spatial Kepler problem is completely integrable by showing

$$\{L_i, L_j\} = -\varepsilon_{ijk} L_k, \quad \{A_i, A_j\} = 2H\varepsilon_{ijk} L_k \quad \text{and} \quad \{A_i, L_j\} = -\varepsilon_{ijk} A_k,$$

for the Levi–Civita symbol

$$\varepsilon_{ijk} := \begin{cases} 1 & ijk \text{ is an even permutation of } 123, \\ -1 & ijk \text{ is an odd permutation of } 123, \\ 0 & \text{else.} \end{cases}$$

4. Show that the solutions to the Kepler problem are conic sections, that is, curves which are the intersection of a plane and a cone. Moreover, explain these shapes in terms of energy and outly physical implications. *Hint*: See [77, Theorem 3.3.10] for inspiration.

2.67 (Poincaré Continuation Method, [200, Section VIII.4]) Let (M, ω, H) be a Hamiltonian system and let $(\gamma, \tau) \in C^\infty(\mathbb{T}, M) \times (0, +\infty)$ be a parametrised periodic orbit of the Hamiltonian vector field X_H, that is, the loop γ satisfies the ordinary differential equation

$$\dot{\gamma}(t) = \tau X_H(\gamma(t)) \qquad \forall t \in \mathbb{T}$$

and $\gamma(0) \notin \mathrm{Crit}(X_H)$. The eigenvalues of the symplectic endomorphism

$$D\phi_\tau(\gamma(0))\colon T_{\gamma(0)}M \to T_{\gamma(0)}M$$

are called the ***Floquet multipliers*** of the τ-periodic orbit γ.

1. Show that 1 is always a Floquet multiplier of γ.
2. By the canonical form theorem [129, Theorem 9.22], there exists a local embedded hypersurface $\Sigma \subseteq M$ containing $\gamma(0)$ such that X_H is nowhere-tangent to Σ. Define a smooth map

   $$P_\Sigma := \phi(s(x), x), \qquad s(x) := \inf\{t > 0 : \phi_t(x) \in \Sigma\}$$

   locally near $\gamma(0)$. This map is called the ***Poincaré section map of*** γ. Evidently, the fixed points of the Poincaré section map are the initial conditions of all parametrised periodic orbits near $\gamma(0)$. Show that

   $$\det(D\phi_\tau(\gamma(0)) - \lambda) = (1 - \lambda)\det(DP(\gamma(0)) - \lambda).$$

3. By using the implicit function theorem, show that if γ admits precisely two Floquet multipliers equal to 1, then there exists a smooth 1-parameter family of parametrised periodic orbits$(\gamma_\varepsilon, \tau_\varepsilon)$ for $\varepsilon \in (-\varepsilon_0, \varepsilon_0)$ for some $\varepsilon_0 > 0$ satisfying $(\gamma_0, \tau_0) = (\gamma, \tau)$. Thus, any such periodic orbit continues in a whole cylinder of periodic orbits! *Hint:* See [200, Theorem VIII.12] or [9, p. 576].

3 The Lagrangian Formalism

In this chapter, we explore a formalism of classical mechanics that is in some sense dual to the Hamiltonian one introduced in the last chapter. Given a configuration space Q, that is, a smooth finite-dimensional manifold, we can consider the Hamiltonian system $(T^*Q, dp \wedge dq, H)$ for a Hamiltonian function $H \in C^\infty(T^*Q)$. The standard coordinates (q, p) on the cotangent bundle T^*Q can be interpreted as position and momentum. Why do we prefer momentum over velocity? Consider the linear elastic collisions of two masses. Then, by the conservation of energy and momentum, one can compute the velocities after the collision. However, the velocities are usually not conserved! Thus, momentum is conserved, whereas velocity is not, in general. There are still very good reasons to introduce the Lagrangian formalism in classical mechanics, that is, generally speaking, the study of smooth functions on the tangent bundle TQ instead of the cotangent bundle T^*Q. This formalism initially preceded the Hamiltonian one and gave rise to the first description of motions via an extremal action principle of the calculus of variations. This principle was successfully applied for studying the existence of periodic orbits in physical systems as described in the book [135] and the paper [2].

3.1 The Legendre Transform

In analogy to a Hamiltonian system, we introduce the concept of a Lagrangian system.

Definition 3.1 (Lagrangian System) A ***Lagrangian system*** is defined as a tuple (Q, L), where Q is a finite-dimensional smooth manifold, called ***configuration space***, and $L \in C^\infty(TQ)$ is a smooth function, referred to as ***Lagrangian function***. Moreover, the tangent bundle TQ is called the ***state space***.

Y. Bähni, *Lectures on Twisted Rabinowitz-Floer Homology*, Pathways in Mathematics, https://doi.org/10.1007/978-3-032-10673-5_3

Remark 3.2 The physical interpretation of the standard coordinates $(q, \dot{q})$ on the state space TQ of a configuration space Q is that q is the position and $\dot{q}$ is the velocity.

Remark 3.3 Infinite-dimensional configuration spaces are treated in classical field theory [65].

Remark 3.4 Let $\pi : E \to Q$ be a vector bundle and ∇ a connection on E. As in Problem 3.32 denote by $\kappa : TQ \to Q$ the associated connection map. Then

$$(D\pi, \kappa) : TE \to TQ \oplus E$$

is a vector bundle isomorphism along π. In particular

$$TTQ \cong TQ \oplus TQ$$

as vector bundles for every smooth manifold Q. Note that this isomorphism depends on the choice of a particular connection ∇ and is therefore not canonical!

Definition 3.5 (Magnetic Lagrangian Function) Let (Q, m) be a Riemannian manifold with $V \in C^\infty(Q)$ and $\theta \in \Omega^1(Q)$. A ***magnetic Lagrangian function*** is defined to be the function $L \in C^\infty(TQ)$

$$L(q, \dot{q}) := \frac{1}{2}\|\dot{q}\|_m^2 + \theta_q(\dot{q}) - V(q).$$

In contrast to a Hamiltonian system, there is no canonical differential form on the state space TQ. However, given a Lagrangian function $L \in C^\infty(TQ)$, we can construct a differential form depending on the Lagrangian function L.

Definition 3.6 (Legendre Transform) The ***Legendre transform of a Lagrangian system*** (Q, L) is defined to be the mapping $\tau_L \in C^\infty(TQ, T^*Q)$ defined by

$$\tau_L(\dot{q})(v) = \frac{d}{d\varepsilon}\bigg|_{\varepsilon=0} L_q(\dot{q} + \varepsilon v) \qquad \forall \dot{q}, v \in T_qQ, q \in Q. \tag{3.1}$$

The map τ_L is also known as the ***fibre derivative of*** L.

The fibre derivative makes the notion of duality precise as it relates the state space TQ with the phase space T^*Q. We may ask what

$$\lambda_L := \tau_L^*\lambda \in \Omega^1(TQ)$$

looks like, where $\lambda \in \Omega^1(T^*Q)$ denotes the Liouville form (2.3). For $(q, \dot{q}) \in TQ$ and $v \in T_{(q,\dot{q})}TQ$ we compute

$$
\begin{aligned}
(\tau_L^*\lambda)_{(q,\dot{q})}(v) &= \lambda_{(q,\tau_L(\dot{q}))}(D\tau_L(v)) \\
&= \tau_L(\dot{q})\left(D\pi_{T^*Q}\left(D\tau_L(v)\right)\right) \\
&= \tau_L(\dot{q})\left(D\left(\pi_{T^*M}\circ\tau_L\right)(v)\right) \\
&= \tau_L(\dot{q})\left(D\pi_{TQ}(v)\right) \\
&= dL_q\left(\Phi_q\left(D\pi_{TQ}(v)\right)\right),
\end{aligned}
$$

where $\Phi\colon \pi_{TQ}^*TQ \to VTQ$ is the bundle isomorphism given by

$$
\Phi_q(\dot{q}, v_q) := \left.\frac{d}{d\varepsilon}\right|_{\varepsilon=0} (\dot{q} + \varepsilon v_q) \tag{3.2}
$$

as in Problem 3.32 between the pull-back bundle

$$
\pi_{TQ}^*TQ := \{(\dot{q}, v_q) \in TQ \times TQ : \pi_{TQ}(q) = \pi_{TQ}(v_q)\}
$$

and the vertical bundle

$$
VTQ := \coprod_{\dot{q}\in TQ} \ker D\pi_{TQ}|_{\dot{q}}.
$$

Here we consider $D\pi_{TQ}$ as a vector bundle homomorphism

$$
D\pi_{TQ}\colon TTQ \to \pi_{TQ}^*TQ
$$

by defining

$$
D\pi_{TQ}(v) := \left(\dot{q}, D\pi_{TQ}|_{\dot{q}}(v)\right) \qquad \forall v \in T_{(q,\dot{q})}TQ.
$$

Hence, we have found our desired form $\lambda_L \in \Omega^1(TQ)$!

Definition 3.7 (Noether Form) Let (Q, L) be a Lagrangian system. Define the ***Noether form***, written $\lambda_L \in \Omega^1(TQ)$, by

$$
\lambda_L(v) := dL((\Phi \circ D\pi_{TQ})v) \qquad \forall v \in TTQ. \tag{3.3}
$$

Using the Noether form (3.3), we can give an alternative description of the Legendre transformation (Definition 3.6).

Lemma 3.8 *Let $F_L \in C^\infty(TQ, T^*Q)$ be such that*

$$
\pi_{T^*Q} \circ F_L = \pi_{TQ} \qquad \textit{and} \qquad F_L^*\lambda = \lambda_L.
$$

Then $F_L = \tau_L$ is the Legendre transform.

Proof For $(q, \dot{q}) \in TQ$ and $v \in T_{(q,\dot{q})}TQ$ we compute

$$(F_L^*\lambda)_{(q,\dot{q})}(v) = F_L(\dot{q})\left(D(\pi_{T^*Q} \circ F_L)(v)\right) = F_L(\dot{q})\left(D\pi_{TQ}(v)\right)$$

by the first assumption and

$$\begin{aligned}\lambda_L|_{(q,\dot{q})}(v) &= dL((\Phi \circ D\pi_{TQ})(v)) \\ &= dL\left(\frac{d}{d\varepsilon}\bigg|_{\varepsilon=0} \left(\dot{q} + \varepsilon D\pi_{TQ}(v)\right)\right) \\ &= \frac{d}{d\varepsilon}\bigg|_{\varepsilon=0} L_q(\dot{q} + \varepsilon D\pi_{TQ}(v)) \\ &= \tau_L(\dot{q})\left(D\pi_{TQ}(v)\right).\end{aligned}$$

Hence, by the second assumption, we conclude

$$F_L(\dot{q})\left(D\pi_{TQ}(v)\right) = (F_L^*\lambda)_{(q,\dot{q})}(v) = \lambda_L|_{(q,\dot{q})}(v) = \tau_L(\dot{q})\left(D\pi_{TQ}(v)\right).$$

As the projection π_{TQ} is a surjective smooth submersion, that is, the derivative $D\pi_{TQ}$ is surjective at every point, we conclude that $F_L(\dot{q}) = \tau_L(\dot{q})$. □

Exercise 3.9 (Magnetic Lagrangian Function) Compute the Legendre transform for a magnetic Lagrangian function L as in Definition 3.5 and show that it is a diffeomorphism. What is the physical interpretation of L?

If the fibre derivative is a diffeomorphism, we can turn our Lagrangian system on the state space TQ of a configuration space Q into a Hamiltonian system on the phase space T^*Q. Thus, the dualisation is between the tangent and the cotangent bundle, similar to the one in symplectic and Riemannian geometry.

Definition 3.10 (Energy) The ***energy of a Lagrangian system*** (Q, L) is defined to be the function $E_L \in C^\infty(TQ)$ given by

$$E_L(q, \dot{q}) := \tau_L(\dot{q})(\dot{q}) - L(q, \dot{q})$$

for $(q, \dot{q}) \in TQ$.

Definition 3.11 (Hamiltonian Function) Let (Q, L) be a Lagrangian system such that the Legendre transform τ_L is a diffeomorphism. The function $H_L \in C^\infty(T^*Q)$ defined by

$$H_L := E_L \circ \tau_L^{-1}$$

is called the ***Hamiltonian function associated with the Lagrangian function*** L.

Exercise 3.12 (Magnetic Hamiltonian Function) Compute the Hamiltonian function H_L associated with a magnetic Lagrangian function in Exercise 3.9. What is the physical interpretation of H_L and the standard coordinates (q, p) on T^*Q?

There is a characterisation for the Legendre transform being a diffeomorphism. Moreover, by Problem 3.33, the Legendre transform is connected to the classical Legendre–Fenchel transformation from convex analysis.

Remark 3.13 (Tonelli Lagrangians) Let (Q^n, L) be a Lagrangian system and fix a Riemannian metric m on Q. The Lagrangian function L is said to be ***Tonelli*** if the following conditions are satisfied:

1. The fibrewise Hessian of L is positive-definite, that is,
$$\sum_{i,j=1}^{n} \frac{\partial^2 L}{\partial \dot{q}_i \partial \dot{q}_j}(q, \dot{q}) v_i v_j > 0$$
for all $(q, \dot{q}) \in TQ$ and $v := \sum_{i=1}^{n} v_i \partial_i \in T_q Q$ such that $v \neq 0$.
2. The Lagrangian function L is fibrewise supercoersive, that is, for all $(q, \dot{q}) \in TQ$ holds
$$\lim_{\|\dot{q}\|_m \to \infty} \frac{L(q, \dot{q})}{\|\dot{q}\|_m} = +\infty.$$

By [135, Proposition 1.2.1] the fibre derivative $\tau_L \colon TQ \to T^*Q$ of a fibrewise convex Lagrangian function L is a diffeomorphism if and only if L is Tonelli.

Exercise 3.14 Show that every magnetic Lagrangian is Tonelli.

Definition 3.15 (Symmetry Group) A Lie group G is called a ***symmetry group of a Lagrangian system*** (Q, L), if there exists a G-action $\psi \colon G \times Q \to Q$ of G on Q with $L \circ D\psi_g = L$ for all $g \in G$.

If the Legendre transform is a diffeomorphism, the symmetry group of a Lagrangian system is also the symmetry group of the corresponding Hamiltonian system.

Theorem 3.16 (Noether's Theorem, Lagrangian Version [176, Corollary 2.18]) *Let (Q, L) be a Lagrangian system with symmetry group G and such that the Legendre transform τ_L is a diffeomorphism. Then G is a symmetry group of the corresponding Hamiltonian system $(T^*Q, d\lambda, H_L)$ with*

$$\mu(\xi)(q, \tau_L(\dot{q})) = \tau_L(\dot{q}) \left(\frac{d}{dt}\bigg|_{t=0} \psi_{\exp(-t\xi)}(q) \right) \qquad \forall \xi \in \mathfrak{g}, (q, \dot{q}) \in TQ,$$

where ψ denotes the smooth left G-action on the configuration space Q. Moreover, the induced action on the phase space T^*Q is Hamiltonian and Poisson.

Proof Define a smooth left G-action Ψ on the phase space T^*Q by

$$\Psi_g := \tau_L \circ D\psi_g \circ \tau_L^{-1} \qquad \forall g \in G.$$

Applying the Momentum Lemma 2.57 to this action yields the Theorem. We proceed in five steps.
Step 1: $D\psi_g^* \lambda_L = \lambda_L$ for all $g \in G$. For all $v \in T_{(q,\dot{q})}TQ$ we compute

$$\begin{aligned}
((D\psi_g)^* \lambda_L)(v) &= dL((\Phi \circ D\pi_{TQ} \circ DD\psi_g)(v)) \\
&= dL(\Phi(D\psi_g(\dot{q}), D\pi_{TQ} \circ DD\psi_g(v))) \\
&= dL(\Phi(D\psi_g(\dot{q}), D\psi_g(D\pi_{TQ}(v)))) \\
&= dL\left(\frac{d}{d\varepsilon}\bigg|_{\varepsilon=0} D\psi_g(\dot{q} + \varepsilon D\pi_{TQ}(v)) \right) \\
&= dL \circ D\psi_g \left(\frac{d}{d\varepsilon}\bigg|_{\varepsilon=0} (\dot{q} + \varepsilon D\pi_{TQ}(v)) \right) \\
&= dL((\Phi \circ D\pi_{TQ})(v)) \\
&= \lambda_L(v).
\end{aligned}$$

Step 2: The induced action Ψ preserves the Liouville form. For $g \in G$ we compute

$$\begin{aligned}
\Psi_g^* \lambda &= \left(\tau_L \circ D\psi_g \circ \tau_L^{-1} \right)^* \lambda \\
&= \left(\tau_L^{-1} \right)^* (D\psi_g)^* \tau_L^* \lambda \\
&= \left(\tau_L^{-1} \right)^* (D\psi_g)^* \lambda_L \\
&= \left(\tau_L^{-1} \right)^* \lambda_L \\
&= \lambda.
\end{aligned}$$

Step 3: The momentum map is of the stated form. We compute

$$\begin{aligned}
\mu(\xi)(q, \tau_L(\dot{q})) &= i_{\underline{\xi}} \lambda(q, \tau_L(\dot{q})) \\
&= \lambda_{(q, \tau_L(\dot{q}))} \left(\frac{d}{dt}\bigg|_{t=0} \Psi_{\exp(-t\xi)}(q, \tau_L(\dot{q})) \right)
\end{aligned}$$

$$
\begin{aligned}
&= p\left(\left.\frac{d}{dt}\right|_{t=0} \pi_{T^*Q} \circ \Psi_{\exp(-t\xi)}(q, \tau_L(\dot{q}))\right) \\
&= p\left(\left.\frac{d}{dt}\right|_{t=0} \pi_{TQ} \circ D\psi_{\exp(-t\xi)} \circ \tau_L^{-1}(q, \tau_L(\dot{q}))\right) \\
&= p\left(\left.\frac{d}{dt}\right|_{t=0} \psi_{\exp(-t\xi)} \circ \pi_{TQ} \circ \tau_L^{-1}(q, \tau_L(\dot{q}))\right) \\
&= p\left(\left.\frac{d}{dt}\right|_{t=0} \psi_{\exp(-t\xi)} \circ \pi_{T^*Q}(q, \tau_L(\dot{q}))\right) \\
&= \tau_L(\dot{q})\left(\left.\frac{d}{dt}\right|_{t=0} \psi_{\exp(-t\xi)}(q)\right)
\end{aligned}
$$

for all $\xi \in \mathfrak{g}$ and $(q, \dot{q}) \in TQ$.

Step 4: $E_L \circ D\psi_g = E_L$ for all $g \in G$. For all $g \in G$ and $(q, \dot{q}) \in TQ$ we compute

$$
\begin{aligned}
E_L(\psi_g(q), D\theta_g(\dot{q})) &= \tau_L(D\psi_g(\dot{q})(D\psi_g(\dot{q})) - L \circ D\psi_g(q, \dot{q}) \\
&= \left.\frac{d}{d\varepsilon}\right|_{\varepsilon=0} L_{\psi_g(q)} \circ D\psi_g((1+\varepsilon)\dot{q}) - L(q, \dot{q}) \\
&= \left.\frac{d}{d\varepsilon}\right|_{\varepsilon=0} L_q((1+\varepsilon)\dot{q}) - L(q, \dot{q}) \\
&= E_L(q, \dot{q})
\end{aligned}
$$

Step 5: $H_L \circ \Psi_g = H_L$ for all $g \in G$. Using Step 4, we conclude for all $g \in G$ that

$$
H_L \circ \Psi_g = E_L \circ D\psi_g \circ \tau_L^{-1} = E_L \circ \tau_L^{-1} = H_L. \qquad \square
$$

Exercise 3.17 (Conserved Quantities in Lagrangian Systems) Let Q^n be a real vector space and let $L \in C^\infty(TQ)$ be a mechanical Lagrangian function for the potential function $V \in C^\infty(Q)$. Consider the inner product on Q coming from the weighted standard inner product

$$
\langle \cdot, \cdot \rangle_m \colon \mathbb{R}^n \times \mathbb{R}^n \to \mathbb{R}, \qquad \langle x, y \rangle_m := \sum_{j=1}^n m_j x_j y_j,
$$

for masses $m_j > 0$ by means of a coordinate isomorphism $Q \cong \mathbb{R}^n$.

1. Consider the one-parameter subgroup $\mathbb{R} \times Q \to Q$, $s \cdot q := q - sq_0$ for some translation $q_0 \in Q$. Assume that the potential function V is invariant under this one-parameter subgroup. Compute the corresponding moment map and show that

momentum is a conserved quantity. What has this to do with Newton's third law about the equality of action and reaction forces?
2. Consider the one-parameter subgroup $\mathbb{R} \times Q \to Q$, $s \cdot q := e^{-sA}q$ for some rotation $A \in \mathrm{SO}(Q)$. Assume that the potential function V is invariant under this one-parameter subgroup. Compute the corresponding moment map and show that the angular momentum is a conserved quantity.
3. Define a new Lagrangian function

$$\widehat{L}\colon T(Q \times \mathbb{R}) \to \mathbb{R}, \qquad \widehat{L}(q, \tau, \dot{q}, \dot{\tau}) := -L\left(q, \frac{\dot{q}}{\dot{\tau}}\right)\dot{\tau}.$$

on the extended state space $T(Q \times \mathbb{R})$. This Lagrangian function is invariant under time translations $\tau \mapsto \tau - s$ for all times $s \in \mathbb{R}$. Show that

$$\mu(1)(q, \tau, p, 1) = H_L(q, p) \qquad \forall (q, p, \tau) \in T^*Q \times \mathbb{R}$$

is the total energy of the system.

3.2 The Lagrangian and Hamiltonian Action Functional

In this section, we lay the foundation for a variational treatment of classical mechanics via the calculus of variations. Extremal principles are present in many areas of mathematics and physics. For example, using variational methods, one can solve partial differential equations as explained in [43].

Definition 3.18 (Lagrangian Action Functional) For a Lagrangian system (Q, L) the corresponding ***Lagrangian action functional*** is defined by

$$\mathscr{E}_L\colon \mathscr{L}Q \to \mathbb{R}, \qquad \mathscr{E}_L(q) := \int_0^1 L(q(t), \dot{q}(t))dt,$$

where $\mathscr{L}Q := C^\infty(\mathbb{T}, Q)$ denotes the loop space of Q with $\mathbb{T} := \mathbb{R}/\mathbb{Z}$.

Exercise 3.19 Let (Q, L) be a Lagrangian system such that the associated Legendre transform τ_L is a diffeomorphism. Show that for all loops $q \in \mathscr{L}Q$ it holds

$$\mathscr{E}_L(q) = \int_0^1 (\tau_L \circ \dot{q})^*\lambda - \int_0^1 H_L(\tau_L(\dot{q}(t)))dt.$$

Using Exercise 3.19, we immediately get a dual action functional in the Hamiltonian formalism.

Definition 3.20 (Hamiltonian Action Functional) Let (M, λ, H) be an exact Hamiltonian system. Then the corresponding ***Hamiltonian action functional*** is

$$\mathscr{A}_H : \mathscr{L}M \to \mathbb{R}, \qquad \mathscr{A}_H(\gamma) := \int_0^1 \gamma^*\lambda - \int_0^1 H(\gamma(t))dt.$$

We refer to this functional also as the ***symplectic action functional***.

The following principle underlies classical mechanics [120, Chapter 8].

Axiom 3.21 (Hamilton's Principle of Extremal Action) A path $q \in \mathscr{L}Q$ describes a motion of a Lagrangian system (Q, L), if

$$\left.\frac{d}{d\varepsilon}\right|_{\varepsilon=0} \mathscr{E}_L(q_\varepsilon) = 0$$

for all variations q_ε of q, that is, given a smooth vector field $X \in \Gamma(q^*TQ)$ along q, for some Levi–Civita connection ∇ on Q we set

$$q_\varepsilon(t) := \exp^{\nabla}_{q(t)}(\varepsilon X(t)) \qquad \forall t \in \mathbb{T}.$$

Remark 3.22 Axiom 3.21 is just the statement that the motions of Lagrangian systems are critical points $q \in \mathrm{Crit}(\mathscr{E}_L)$, where the differential $d\mathscr{E}_L$ of the Lagrangian action functional $\mathscr{E}_L$ is defined by

$$d\mathscr{E}_L : \mathscr{L}Q \to T^*\mathscr{L}Q, \qquad d\mathscr{E}_L(q)(X) = \left.\frac{d}{d\varepsilon}\right|_{\varepsilon=0} \mathscr{E}_L(q_\varepsilon), \tag{3.4}$$

with the infinitesimal variation $\frac{d}{d\varepsilon}\big|_{\varepsilon=0} q_\varepsilon(t) = X(t)$ for all $t \in \mathbb{T}$.

The next result gives a local description of the critical points of the Lagrangian action functional. For the infinite-dimensional setting, see [9, Theorem 3.5.17].

Lemma 3.23 (Euler–Lagrange Equations) *In standard coordinates $(q, \dot{q})$ on the tangent bundle TQ of a configuration space Q, the differential $d\mathscr{E}_L$ of the Lagrangian action functional $\mathscr{E}_L$ is given by*

$$d\mathscr{E}_L(\gamma)(X) = \int_0^1 \left(\frac{\partial L}{\partial q} - \frac{d}{dt}\frac{\partial L}{\partial \dot{q}}\right) X(t)dt.$$

In particular, $d\mathscr{E}_L$ is well-defined, that is, does not depend on the choice of variation, and any motion of a Lagrangian system locally satisfies the ***Euler–Lagrange equations***

$$\frac{d}{dt}\frac{\partial L}{\partial \dot{q}} = \frac{\partial L}{\partial q}. \tag{3.5}$$

Remark 3.24 A physically relevant example where one can explicitly solve the Euler–Lagrange equations is provided in Problem 3.34 intimately connected to Newton's law of inertia or ***Newton's first law***.

Proof We compute

$$
\begin{aligned}
d\mathscr{E}_L(q)(X) &= \frac{d}{d\varepsilon}\bigg|_{\varepsilon=0} \mathscr{E}_L(q_\varepsilon) \\
&= \int_0^1 \frac{d}{d\varepsilon}\bigg|_{\varepsilon=0} L(\dot{q}_\varepsilon(t))dt \\
&= \int_0^1 dL\left(\frac{d}{d\varepsilon}\bigg|_{\varepsilon=0} \dot{q}_\varepsilon(t)\right) dt \\
&= \int_0^1 \left(\frac{\partial L}{\partial q}\frac{dq_\varepsilon(t)}{d\varepsilon}\bigg|_{\varepsilon=0} + \frac{\partial L}{\partial \dot{q}}\frac{d\dot{q}_\varepsilon(t)}{d\varepsilon}\bigg|_{\varepsilon=0}\right) dt \\
&= \int_0^1 \left(\frac{\partial L}{\partial q}\frac{dq_\varepsilon(t)}{d\varepsilon}\bigg|_{\varepsilon=0} + \frac{\partial L}{\partial \dot{q}}\frac{d}{dt}\frac{dq_\varepsilon(t)}{d\varepsilon}\bigg|_{\varepsilon=0}\right) dt \\
&= \int_0^1 \left(\frac{\partial L}{\partial q}X(t) + \frac{\partial L}{\partial \dot{q}}\frac{d}{dt}X(t)\right) dt \\
&= \int_0^1 \frac{\partial L}{\partial q}X(t)dt + \frac{\partial L}{\partial \dot{q}}X(t)\Big|_0^1 - \int_0^1 \frac{d}{dt}\frac{\partial L}{\partial \dot{q}}X(t)dt \\
&= \int_0^1 \frac{\partial L}{\partial q}X(t)dt - \int_0^1 \frac{d}{dt}\frac{\partial L}{\partial \dot{q}}X(t)dt \\
&= \int_0^1 \left(\frac{\partial L}{\partial q} - \frac{d}{dt}\frac{\partial L}{\partial \dot{q}}\right) X(t)dt
\end{aligned}
$$

using integration by parts. As $X \in \Gamma(q^*TQ)$ is arbitrary, we conclude (3.5) by the fundamental lemma of the calculus of variations in Exercise 3.25. □

Exercise 3.25 (Fundamental Lemma of Calculus of Variations) Let Q be a smooth manifold and let $f \in C^\infty(Q)$ be a smooth function. Prove that

$$
\forall \varphi \in C_c^\infty(Q)\colon \int_Q f\varphi = 0 \qquad \Rightarrow \qquad f = 0.
$$

Exercise 3.26 Let (Q, L) be a Lagrangian system and $f \in C^\infty(Q)$. Show that the Euler–Lagrange equations (3.5) of L and $L + df$ coincide.

A more elegant way of writing the Euler–Lagrange equations (3.5) globally is through the Hamiltonian formalism.

Lemma 3.27 *A path $\gamma \in \mathscr{L}M$ is a motion of an exact Hamiltonian system (M, λ, H) if and only if it is an integral curve of the Hamiltonian vector field X_H, that is, γ satisfies the ordinary differential equation, called* **Hamilton's equation**, *given by*

$$\dot{\gamma}(t) = X_H(\gamma(t)) \qquad \forall t \in \mathbb{T}. \tag{3.6}$$

Proof We compute

$$\begin{aligned}
d\mathscr{A}_H(\gamma)(X) &= \frac{d}{d\varepsilon}\bigg|_{\varepsilon=0} \mathscr{A}_H(\gamma_\varepsilon) \\
&= \int_0^1 \frac{d}{d\varepsilon}\bigg|_{\varepsilon=0} \gamma_\varepsilon^*\lambda - \int_0^1 \frac{d}{d\varepsilon}\bigg|_{\varepsilon=0} H(\gamma_\varepsilon(t))dt \\
&= \int_0^1 \gamma^* L_X\lambda - \int_0^1 dH(X(t))dt \\
&= \int_0^1 \gamma^*(i_X d\lambda + d i_X\lambda) - \int_0^1 dH(X(t))dt \\
&= \int_0^1 d\lambda(X(t), \dot{\gamma}(t))dt + \lambda(X)\big|_0^1 - \int_0^1 dH(X(t))dt \\
&= \int_0^1 d\lambda(X(t), \dot{\gamma}(t))dt + \int_0^1 d\lambda(X_H(\gamma(t)), X(t))dt \\
&= \int_0^1 d\lambda(X(t), \dot{\gamma}(t) - X_H(\gamma(t)))dt
\end{aligned}$$

using Fisherman's formula (Lemma 2.2) and Stoke's theorem. As $X \in \Gamma(\gamma^*TM)$ is arbitrary, we conclude (3.6) again by Exercise 3.25 and nondegeneracy of $d\lambda$. □

Exercise 3.28 Compare the proof of Lemma 3.23 with the proof of Lemma 3.27. What are similarities and differences? Which proof is more elegant?

Exercise 3.29 Let (Q, L) be a Lagrangian system such that the Legendre transform τ_L is a diffeomorphism. Give a proof of Lemma 3.27 in the case (T^*Q, pdq, H_L) by using the local coordinates (q, p) on the phase space T^*Q and the Euler–Lagrange equations in Lemma 3.23.

Remark 3.30 (Lagrangian vs. Hamiltonian Formulation of Classical Mechanics) Given a configuration space in classical mechanics such that the Legendre transform is a diffeomorphism, it does not really matter whether we describe motions of a Lagrangian system or the corresponding Hamiltonian system. So why bother to study both formulations? According to [106], the most significant difference between the Lagrangian and Hamiltonian formulation comes from how the two approaches are applied in other areas of modern physics, such as relativity theory or quantum mechanics. Lagrangian mechanics is used more in relativistic theories to describe the dynamics of particles as well as in relativistic (both classical and quantum) field theories to describe the dynamics of fields. It turns out that using Hamiltonian mechanics in curved spacetime is much more complicated, which is why the Hamiltonian formulation is not so commonly used in general relativity. However, Hamiltonian mechanics is used more in ordinary non-relativistic quantum mechanics to describe the dynamics of particles, as well as in a formulation of quantum field theory called canonical quantisation. The following are just a few reasons why the concept of "momentum" instead of "velocity" is appropriate in those areas:

- Momentum is a conserved quantity, while velocity is generally not. This often makes momentum much more useful in, for example, quantum mechanics or special relativity, where the relationship between momentum and velocity differs from the classical formula $p = m\dot{q}$.
- For describing massless particles like photons, momentum is a much more useful concept than velocity. This is because massless particles always move at a constant velocity, at the speed of light. However, the momentum of a massless particle does not necessarily have to be constant.
- Position and momentum are generally conjugate variables, while position and velocity are not. This is of significant importance in quantum mechanics, particularly because it means that the momentum and position operators do not commute. This then leads to observations such as the Heisenberg uncertainty relation between position and momentum [179, Theorem 8.2].
- In quantum mechanics, we often describe objects in terms of waves, not necessarily particles that are localized somewhere, governed by the time-dependent Schrödinger equation

$$i\hbar\frac{d\psi}{dt} = H\psi,$$

  where H is a self-adjoint operator on an infinite-dimensional separable complex Hilbert space as in [176]. This makes it difficult to generally define the notion of velocity, but the notion of momentum is still perfectly valid by the De-Broglie equation for matter waves [181, Equation (32.13)]. See also [201] for an amusing introduction to the realm of quantum physics!

Later, we will see another reason why we prefer the Lagrangian formalism in some cases to study the existence of periodic orbits! To summarise our observations in Lagrangian and Hamiltonian mechanics, we conclude this chapter by working out a summary.

Table 3.1 Hamiltonian vs. Lagrangian formalism

	Hamiltonian formalism	Lagrangian formalism
Configuration space		
Energy function		
Domain		
Coordinates		
Dimension		
Equations of motion		
Functional		
Hamilton's principle		
Transformation		
Symmetries		
Momentum map		

Exercise 3.31 Let Q be a smooth manifold. Pair up and compare and contrast the Hamiltonian formalism of the cotangent bundle T^*Q with the Lagrangian formalism of the tangent bundle TQ by filling in Table 3.1.

Problems

3.32 (The Connection Map) In this problem we carry out some details regarding the Noether form introduced in this chapter in Definition 3.3. Let $\pi : E \to M$ be an arbitrary vector bundle.

1. Show that

$$\Phi\colon \pi^*E \to VE, \qquad \Phi(u, v) := \frac{d}{d\varepsilon}\bigg|_{\varepsilon=0} (u + \varepsilon v)$$

is a vector bundle isomorphism.
2. Show that the map

$$\mathrm{pr}_2 \circ \Phi^{-1}\colon VE \to E, \qquad (\mathrm{pr}_2 \circ \Phi^{-1})(\Phi(u, v)) = v,$$

is a vector bundle isomorphism and that we have a commutative diagram

$$\begin{array}{ccc} VE & \xrightarrow{\mathrm{pr}_2 \circ \Phi^{-1}} & E \\ {\scriptstyle \pi_{VE}}\downarrow & & \downarrow{\scriptstyle \pi} \\ E & \xrightarrow[\pi]{} & M \end{array}$$

3. Let ∇ be a connection on E. The connection ∇ gives rise to a distribution on E inducing a splitting $TE \cong \nabla \oplus VE$ into a horizontal component ∇ and a vertical

component VE. Define the ***connection map associated with the connection*** ∇ by

$$\kappa\colon TE \to E, \qquad \kappa(\zeta^h + \zeta^v) := (\mathrm{pr}_2 \circ \Phi^{-1})(\zeta^v).$$

Show that $(D\pi, \kappa)$ is a vector bundle isomorphism in the commutative diagram

$$\begin{array}{ccc} TE & \xrightarrow{(D\pi,\kappa)} & TM \oplus E \\ {\scriptstyle \pi_{TE}}\downarrow & & \downarrow{\scriptstyle (\pi_{TM},\pi)} \\ E & \xrightarrow[\pi]{} & M \end{array}$$

In particular, if we consider $E = TQ$ for a configuration space Q, we get the vector bundle isomorphism $TTQ \cong TQ \oplus TQ$ for the double tangent bundle TTQ of Q.

3.33 (Legendre–Fenchel Duality, [150, Section 6.1]) Let $L \in C^\infty(\mathbb{R}^n \times \mathbb{R}^n)$ be such that $L_x \in C^\infty(\mathbb{R}^n)$ defined by $L_x(y) := L(x, y)$ for all $x \in \mathbb{R}^n$ is convex, that is, the Hessian matrix of L_x is positive definite at every point $y \in \mathbb{R}^n$.

1. For every $x \in \mathbb{R}^n$ define the ***Legendre–Fenchel transform of*** L_x to be

$$(L_x)^*\colon \mathbb{R}^n \to \mathbb{R} \cup \{\infty\}, \qquad (L_x)^*(y) := \sup_{z \in \mathbb{R}^n} \{\langle z, y\rangle - L_x(z)\}.$$

   Show that $(L_x)^*$ is also of class C^∞ and convex whenever $(L_x)^*(y)$ is finite. What is the geometrical interpretation of the Legendre–Fenchel transform?
2. Compute the Legendre–Fenchel transform of the function

$$L\colon \mathbb{R} \times \mathbb{R} \to \mathbb{R}, \qquad L(x, y) := x^2 e^y.$$

3. Suppose that the Legendre transform τ_L is a diffeomorphism. Show that

$$H_L(x, y) = (L_x)^*(y) \qquad \forall x, y \in \mathbb{R}^n.$$

For applications, see [150, Section 6.6]. The Legendre–Fenchel transform is used in the theory of Hamilton–Jacobi equations, which are not discussed in this book but in any standard text on classical mechanics and, for example, in [129, p. 585–595].

3.34 (Free Particle on a Pseudo-Riemannian Manifold, [77, p. 13]) Let (Q, m) be a pseudo-Riemannian manifold and denote by $V \in \mathfrak{X}(TQ)$ the geodesic vector field associated with the Levi–Civita connection (see [130, p. 129]). Show that solutions of the Euler–Lagrange equations (3.5) for the mechanical Lagrangian function

$$L \colon TQ \to \mathbb{R}, \qquad L(q, \dot{q}) := \frac{1}{2}\|\dot{q}\|_m^2$$

are precisely integral curves of V and thus geodesics.

3.35 (Mañé Critical Value, [61]) Let (Q, L) be a Lagrangian system with a compact configuration space Q without boundary. Moreover, assume that the Legendre transform $\tau_L \in C^\infty(TQ, T^*Q)$ is a diffeomorphism. The definition of the Lagrangian action functional can be extended to the set of all absolutely continuous curves

$$\mathscr{E}_L \colon W^{1,1}(\mathbb{T}, Q) \to \mathbb{R}, \qquad \mathscr{E}_L(q) = \int_0^1 L(q(t), \dot{q}(t))dt.$$

1. Show that $q \in \operatorname{Crit}(\mathscr{E}_{L+c})$ satisfies $\dot{q}(\mathbb{T}) \subseteq E^{-1}(c)$ for all $k \in \mathbb{R}$.
2. Define the ***Mañé critical value*** by

   $$c(L) := \inf\{c \in \mathbb{R} : \mathscr{E}_{L+c}(q) \geq 0 \, \forall q \in W^{1,1}(\mathbb{T}, Q)\}.$$

   Show that

   $$\min(E) \leq e_0(L) \leq c(L),$$

   where

   $$e_0(L) := \inf\{c \in \mathbb{R} : \pi_{TQ}(E^{-1}(c)) = Q\}.$$

3. Show that the sublevel set

   $$\{H_L \leq c\} := \{(q, p) \in T^*Q : H_L(q, p) \leq c\}$$

   is displaceable in $(T^*Q, dp \wedge dq)$ for all $c < e_0(L)$. *Hint:* [61, Proposition 8.2].

The Limit Set of a Family of Periodic Orbits 4

In this final chapter of the first part, we analyse the bifurcation and limit behaviour of a family of parametrised periodic orbits on a homotopy of stable regular energy surfaces. The case of chords on homotopies of regular energy surfaces of restricted contact type was treated in [35]. In particular, we generalise [35, Theorem A] to the stable case. The significance of such a generalisation to the stable case is illustrated by considering families of regular energy hypersurfaces parametrised by energy and their possible change in topology as the energy passes through Mañé's critical value introduced in Problem 3.35 and studied in [56] or [134] for various hypersurfaces in magnetic Hamiltonian systems. Our main result is that blue sky catastrophes as in [77, Section 7.6] do not occur for families of parametrised periodic orbits on homotopies of stable regular energy surfaces, meaning that the periods of periodic orbits do not explode and thus the periodic orbits will not disappear into the blue sky! This terminology was made popular by Leonid Shilnikov in the mathematical theory of dynamical chaos [91] and was also used in the paper [172] for analysing neuron models.

4.1 Blue Sky Catastrophes

Let (M, ω, H) be a Hamiltonian system and let $\Sigma = H^{-1}(0)$ be a compact regular energy hypersurface as in Definition 2.35. First, we introduce some additional structure on Σ because without any additional structure, such hypersurfaces might not carry any periodic orbits by the counterexamples to the Hamiltonian Seifert conjecture constructed in the papers [88], [89] and [90]. One such convenient structure that guarantees periodic orbits under some favorable circumstances is provided by a stable Hamiltonian structure [57]. As we will see, stable Hamiltonian structures naturally arise on regular energy hypersurfaces in magnetic Hamiltonian systems [56]. Moreover, stable Hamiltonian structures arise in hydrodynamics [57, Corollary 2.5] or constitute boundary conditions for compactness results in

Y. Bähni, *Lectures on Twisted Rabinowitz-Floer Homology*, Pathways in Mathematics, https://doi.org/10.1007/978-3-032-10673-5_4

symplectic field theory as explained in [195]. We start with the underlying notion of a Hamiltonian manifold, the odd-dimensional analogue of a symplectic manifold.

Definition 4.1 (Hamiltonian Manifold) A ***Hamiltonian manifold*** is a tuple (Σ, ω), where Σ is an odd-dimensional smooth manifold and $\omega \in \Omega^2(\Sigma)$ is closed so that $\ker \omega$ is a line distribution. The foliation that induces the line distribution $\ker \omega$ is called ***characteristic foliation***.

Hamiltonian manifolds naturally arise as regular energy hypersurfaces in Hamiltonian systems.

Exercise 4.2 (Regular Energy Hypersurface) Let Σ be a regular energy hypersurface in a Hamiltonian system (M, ω, H). Show that $(\Sigma, \omega|_\Sigma)$ is a Hamiltonian manifold and that the line distribution $\ker \omega|_\Sigma$ is spanned by the Hamiltonian vector field $X_H|_\Sigma \colon \Sigma \to T\Sigma$.

A Hamiltonian structure is not very useful for the existence of periodic orbits in view of the counterexamples to the Hamiltonian Seifert conjecture, as every regular energy hypersurface is a Hamiltonian manifold by Exercise 4.2.

Definition 4.3 (Stable Hamiltonian Manifold, [56, p. 1773]) A Hamiltonian manifold (Σ, ω) is called ***stable***, if there exists $\alpha \in \Omega^1(\Sigma)$ that vanishes nowhere in $\ker \omega$ such that the inclusion $\ker \omega \subseteq \ker(d\alpha)$ holds. We write (Σ, ω, α) for a stable Hamiltonian manifold and call α a ***stabilising form***.

Every stable Hamiltonian manifold is orientable, as the next exercise shows.

Exercise 4.4 (Equivalent Definition of a Stable Hamiltonian Manifold) Let (Σ^{2n-1}, ω) be a Hamiltonian manifold and $\alpha \in \Omega^1(\Sigma)$ such that $\ker \omega \subseteq \ker(d\alpha)$. Show that (Σ, ω, α) is a stable Hamiltonian manifold if and only if $\alpha \wedge \omega^{n-1}$ is a volume form on Σ.

Again, a natural example of a stable Hamiltonian manifold is a regular energy hypersurface in a Hamiltonian system satisfying an additional property.

Exercise 4.5 (Regular Energy Hypersurface) Let Σ be a regular energy hypersurface in a Hamiltonian system (M, ω, H) such that there exists a vector field X in a neighbourhood of Σ with X being nowhere tangent to Σ and $\ker \omega|_\Sigma \subseteq \ker L_X\omega|_\Sigma$. Show that $(\Sigma, \omega|_\Sigma, i_X\omega|_\Sigma)$ is a stable Hamiltonian manifold.

We provide further examples of a stable Hamiltonian manifold.

Example 4.6 (Magnetic Torus) Let $\mathbb{T}^n$ be the standard flat torus for $n \geq 2$ and let $J \colon \mathbb{R}^n \to \mathbb{R}^n$ be an antisymmetric nonzero linear map as in [56, Section 6.1].

Define $\rho \in \Omega^2(\mathbb{T}^n)$ by setting $\rho(\cdot, \cdot) := \langle \cdot, J \cdot \rangle$ and denote by $\omega_\rho = dp \wedge dq + \pi^* \rho$ the magnetic symplectic form on $T^*\mathbb{T}^n \cong \mathbb{T}^n \times \mathbb{R}^n$. For an energy value $k \in \mathbb{R}$ set $\Sigma_k := H^{-1}(k)$ for the mechanical Hamiltonian function

$$H(q, p) := \frac{1}{2}\|p\|^2 \qquad \forall (q, p) \in \mathbb{T}^n \times \mathbb{R}^n.$$

Define $A := (J|_{\operatorname{im} J})^{-1}$ and $\beta \in \Omega^1(\operatorname{im} J)$ by

$$\beta_x(v) := \frac{1}{2}\langle x, Av \rangle.$$

By [56, Proposition 6.3], the energy hypersurface Σ_k is stable and displaceable for every $k > 0$. The stabilising form λ on Σ_k is given by

$$\alpha = f^*(pdq) + (\mathrm{pr}_{\|} \circ \mathrm{pr})^* \beta, \tag{4.1}$$

where

$$\mathrm{pr}_\perp \colon \mathbb{R}^n \to \ker J, \qquad \mathrm{pr}_{\|} \colon \mathbb{R}^n \to \operatorname{im} J, \qquad \mathrm{pr} \colon \mathbb{T}^n \times \mathbb{R}^n \to \mathbb{R}^n$$

denote the projections with respect to the orthogonal splitting

$$\mathbb{R}^n = \ker J \oplus \operatorname{im} J,$$

and

$$f \colon \mathbb{T}^n \times \mathbb{R}^n \to \mathbb{T}^n \times \mathbb{R}^n, \qquad f(q, p) := \big(q, \mathrm{pr}_\perp(p)\big).$$

It is illustrative to consider the special case $n = 2$ and $\sigma = dq_1 \wedge dq_2$. Then

$$\alpha = -\frac{1}{2}(p_1 dp_2 - p_2 dp_1).$$

Example 4.7 (Mapping Torus of a Symplectomorphism) Let (M, ω) be a compact symplectic manifold and let $\varphi \in \operatorname{Symp}(M, \omega)$ be a symplectomorphism. Then the quotient manifold

$$\Sigma_\varphi := \frac{M \times \mathbb{R}}{(x, t) \sim (\varphi(x), t + 1)}$$

is a stable Hamiltonian manifold with stabilising form $\alpha = dt$. The stable Hamiltonian manifold $(\Sigma_\varphi, \omega|_{\Sigma_\varphi}, \alpha)$ is called the ***mapping torus of*** φ.

A very important example of a stable Hamiltonian manifold is a contact manifold. It is actually a special case, where ω is exact.

Definition 4.8 (Contact Manifold) A ***contact manifold*** is defined to be a stable Hamiltonian manifold $(\Sigma, d\alpha, \alpha)$. We simply write (Σ, α) for a contact manifold and call $\alpha \in \Omega^1(\Sigma)$ a ***contact form***.

Example 4.9 (Regular Energy Surface) Let Σ be a compact regular energy surface in a mechanical Hamiltonian system $(T^*M, dp \wedge dq, H)$. Then by [1, Theorem 1.2.2] there exists $\alpha \in \Omega^1(\Sigma)$ such that $d\alpha = \omega|_\Sigma$ and (Σ, α) is a contact manifold.

Example 4.10 (Star-Shaped Hypersurfaces) Let $\Sigma = \Sigma_\rho \subseteq \mathbb{R}^{2n}$ be a star-shaped hypersurface for some positive function $\rho \in C^\infty(\mathbb{S}^{2n-1})$. The vector field

$$X = \frac{1}{2}\sum_{j=1}^{n}\left(x_j \frac{\partial}{\partial x_j} + y_j \frac{\partial}{\partial y_j}\right) = \frac{1}{2}\sum_{j=1}\left(z_j \frac{\partial}{\partial z_j} + \bar{z}_j \frac{\partial}{\partial \bar{z}_j}\right) \in \mathfrak{X}(\mathbb{R}^{2n})$$

satisfies $i_X d\lambda_0 = \lambda_0$ and is nowhere tangent to Σ because

$$dH_\Sigma(X)|_\Sigma = d\lambda_0(X, X_{H_\Sigma})|_\Sigma = \lambda_0(X_{H_\Sigma})|_\Sigma = 1.$$

Thus $(\Sigma, d\lambda_0|_\Sigma, \lambda_0|_\Sigma)$ is a contact manifold by Exercise 4.5 since

$$\ker(L_X d\lambda_0) = \ker(i_X dd\lambda_0 + di_X d\lambda_0) = \ker(d\lambda_0)$$

using Cartan's Magic Formula.

Every stable Hamiltonian manifold admits a distinguished vector field.

Definition 4.11 (Reeb Vector Field) Let (Σ, ω, α) be a stable Hamiltonian manifold. The vector field $R = R_{\omega,\alpha} \in \mathfrak{X}(\Sigma)$ implicitly defined by

$$i_R\omega = 0 \qquad \text{and} \qquad i_R\alpha = 1$$

is called the ***Reeb vector field of*** (Σ, ω, α).

Exercise 4.12 Let (Σ, ω, α) be a compact stable Hamiltonian manifold.

1. Explain why the Reeb vector field is unique.
2. Show that the flow $\phi_t \colon \Sigma \times \mathbb{R} \to \Sigma$ of the Reeb vector field $R \in \mathfrak{X}(\Sigma)$ conserves the stabilising form α, that is, it holds

$$\phi_t^*\alpha = \alpha \qquad \forall t \in \mathbb{R}.$$

3. Explain why stable Hamiltonian manifolds are defined as in Definition 4.3.

4. Conclude, that in the case where α is a contact form, the linearised flow $D\phi_t$ preserves the splitting

$$T\Sigma = \ker\alpha \oplus \langle R\rangle.$$

Example 4.13 (Star-Shaped Hypersurfaces) Let $\Sigma = H_\Sigma^{-1} \subseteq \mathbb{R}^{2n}$ be a star-shaped hypersurface. Then the restriction $R_\Sigma := X_{H_\Sigma}|_\Sigma \in \mathfrak{X}(\Sigma)$ of the Hamiltonian vector field $X_{H_\Sigma} \in \mathfrak{X}(\mathbb{R}^{2n})$ satisfies

$$R_\Sigma \in \ker d\lambda_0|_\Sigma \qquad \text{and} \qquad \lambda_0(R_\Sigma)|_\Sigma = 1,$$

by (1.3). Hence R_Σ is the Reeb vector field of the contact manifold $(\Sigma, \lambda_0|_\Sigma)$.

Example 4.14 (Magnetic Torus) The flow ϕ_t of the magnetic Hamiltonian system in Example 4.6 is given by

$$\phi_t(q,p) = \left(\int_0^t e^{sJ} p ds + q, e^{tJ} p\right),$$

as one can explicitly compute this flow using (2.4), and $(q,p) \in \Sigma_k$ gives rise to a contractible closed orbit of period τ if and only if

$$\mathrm{pr}_\perp(p) = 0, \quad e^{\tau J}\,\mathrm{pr}_\parallel(p) = \mathrm{pr}_\parallel(p), \quad \text{and} \quad \left\|\mathrm{pr}_\parallel(p)\right\|^2 = 2k. \tag{4.2}$$

The projection of such a contractible periodic orbit to $\mathbb{T}^2$ is depicted in Fig. 4.1.

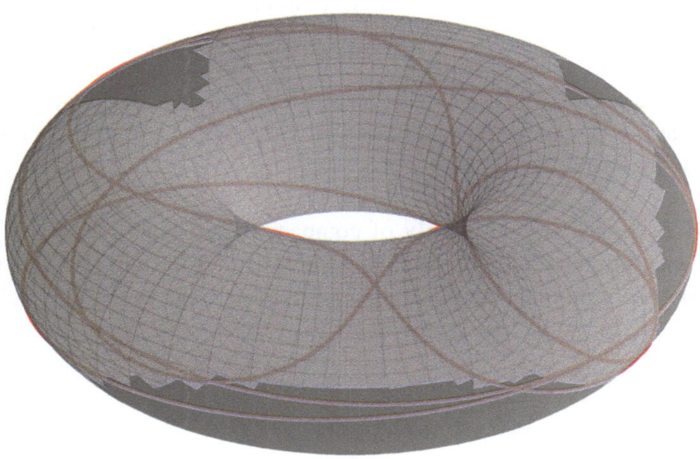

Fig. 4.1 A contractible periodic Reeb orbit on a magnetic torus $\mathbb{T}^2$

Definition 4.15 (Parametrised Periodic Reeb Orbit) Let (Σ, ω, α) be a stable Hamiltonian manifold. A ***parametrised periodic Reeb orbit*** in Σ is defined to be a tuple $(\gamma, \tau) \in C^\infty(\mathbb{T}, \Sigma) \times (0, +\infty)$ satisfying

$$\dot{\gamma}(t) = \tau R(\gamma(t)) \qquad \forall t \in \mathbb{T},$$

where $R = R_{\omega,\alpha} \in \mathfrak{X}(\Sigma)$ denotes the Reeb vector field of (Σ, ω, α). We denote the set of all parametrised periodic Reeb orbits in Σ by $\mathscr{P}(\Sigma, \omega, \alpha)$.

Remark 4.16 (Stable Weinstein Conjecture) If $(\Sigma^3, \omega, \alpha)$ is a compact stable Hamiltonian manifold, then we have $\mathscr{P}(\Sigma, \omega, \alpha) \neq \varnothing$ except in the case where Σ is a $\mathbb{T}^2$-bundle over $\mathbb{S}^1$ by [112, Theorem 1.1]. If (Σ^3, α) is a compact contact manifold without boundary, then $\mathscr{P}(\Sigma, \alpha) \neq \varnothing$ holds by [177]. More existence results on periodic orbits in stable Hamiltonian manifolds in special cases are given in [134], [56, Theorem 4.9] and [2].

In the remaining part of this section, we would like to give a dynamical motivation for defining a stable Hamiltonian manifold. A geometrical one was already provided in Exercise 4.12. We start by a variational characterisation of the parametrised periodic Reeb orbits in a stable Hamiltonian manifold.

Definition 4.17 (Maupertuis Functional) Let (Σ, ω, α) be a stable Hamiltonian manifold. Define the ***Maupertuis action functional*** by

$$\mathscr{A}_0 \colon \mathscr{L}\Sigma \to \mathbb{R}, \qquad \mathscr{A}_0(\gamma) := \int_0^1 \gamma^*\alpha.$$

Exercise 4.18 Let (Σ, ω, α) be a stable Hamiltonian manifold.

1. Show that $\mathscr{P}(\Sigma, \omega, \alpha) \subseteq \operatorname{Crit}(\mathscr{A}_0)$.
2. Assume that $\alpha \in \Omega^1(\Sigma)$ is a contact form. Show that then $\operatorname{Crit}(\mathscr{A}_0) = \mathscr{P}(\Sigma, \alpha)$.
3. Show that it is not necessarily the case that $\operatorname{Crit}(\mathscr{A}_0) \subseteq \mathscr{P}(\Sigma, \omega, \alpha)$ if α is only a stabilising form.

Let (Σ, ω_σ) be a smooth family of compact Hamiltonian manifolds for $\sigma \in [0, 1]$. Given a smooth family of stabilising forms $(\alpha_\sigma)_{\sigma \in [0,1]} \subseteq \Omega^1(\Sigma)$ on (Σ, ω_σ), we can consider a smooth family of parametrised periodic orbits $(\gamma_\sigma, \tau_\sigma)_{\sigma \in [0, \sigma_\infty)}$ for some $0 < \sigma_\infty \leq 1$, that is, the maps $(\gamma_\sigma, \tau_\sigma) \in \mathscr{L}\Sigma \times (0, +\infty)$ solve the equation

$$\dot{\gamma}_\sigma(t) = \tau_\sigma R_\sigma(\gamma_\sigma(t)) \qquad \forall t \in \mathbb{T},$$

for all $\sigma \in [0, \sigma_\infty)$, where $R_\sigma := R_{\alpha_\sigma} \in \mathfrak{X}(\Sigma)$ denotes the Reeb vector field associated with the stabilising form α_σ on (Σ, ω_σ) for all $\sigma \in [0, 1]$. In order to systematically study the limit behaviour of the smooth family of parametrised periodic Reeb orbits $(\gamma_\sigma, \tau_\sigma)_{\sigma\in[0,\sigma_\infty)}$, it is convenient to introduce the ***limit set*** consisting of all

$$(\gamma_{\sigma_\infty}, \tau_{\sigma_\infty}) \in \mathscr{L}\Sigma \times (0, +\infty)$$

such that there exists a sequence $(\sigma_k) \subseteq [0, \sigma_\infty)$ with

$$\sigma_k \to \sigma_\infty \qquad \text{and} \qquad (\gamma_{\sigma_k}, \tau_{\sigma_k}) \xrightarrow{C^\infty} (\gamma_{\sigma_\infty}, \tau_{\sigma_\infty})$$

as $k \to \infty$. As we shall see, if the periods τ_σ are uniformly bounded from above and below for all $\sigma \in [0, \sigma_\infty)$, the limit set is not empty by Ascoli's theorem and bootstrapping (4.4). However, it could be $\tau_\sigma \to +\infty$ as $\sigma \to \sigma_\infty$. This scenario is referred to as ***blue sky catastrophe***. For a homotopy of stable Hamiltonian manifolds, no blue sky catastrophe does occur, giving a motivation for the notion of stability!

Lemma 4.19 *Let $(\gamma_\sigma, \tau_\sigma)_{\sigma\in[0,\sigma_\infty)}$ be a smooth family of parametrised periodic Reeb orbits for some $0 < \sigma_\infty \le 1$ on a homotopy of stable Hamiltonian manifolds $(\Sigma, \omega_\sigma, \alpha_\sigma)_{\sigma\in[0,1]}$. Then there exists a constant $C > 0$ such that*

$$\frac{1}{C} \le \tau_\sigma \le C \qquad \forall \sigma \in [0, \sigma_\infty).$$

Proof Consider the smooth family

$$\mathscr{A}_0 \colon \mathscr{L}\Sigma \times [0,1] \to \mathbb{R}, \qquad \mathscr{A}_0(\gamma, \sigma) := \int_0^1 \gamma^* \alpha_\sigma$$

of Maupertuis functionals from Definition 4.17. Let $\sigma \in [0, \sigma_\infty)$. Observe, that we have the period-action equality

$$\mathscr{A}_0(\gamma_\sigma, \sigma) = \int_0^1 \alpha_\sigma(\dot{\gamma}_\sigma(t))dt = \tau_\sigma \int_0^1 \alpha_\sigma(R_\sigma(\gamma_\sigma(t)))dt = \tau_\sigma.$$

Thus we compute

$$\begin{aligned}
\partial_\sigma \tau_\sigma &= \partial_\sigma(\mathscr{A}_0(\gamma_\sigma, \sigma)) \\
&= \int_0^1 (\partial_\sigma \alpha_\sigma)(\dot{\gamma}_\sigma) + d\mathscr{A}_0|_{(\gamma_\sigma,\sigma)}(\partial_\sigma \gamma_\sigma, \sigma) \\
&= \int_0^1 (\partial_\sigma \alpha_\sigma)(\dot{\gamma}_\sigma) \\
&= \tau_\sigma \int_0^1 (\partial_\sigma \alpha_\sigma)(R_\sigma \circ \gamma_\sigma)
\end{aligned}$$

since $\mathscr{P}(\Sigma, \omega_\sigma, \alpha_\sigma) \subseteq \mathrm{Crit}(\mathscr{A}_0(\cdot, \sigma))$ by Exercise 4.18. As Σ is compact by assumption, there exists $\kappa > 0$ such that

$$|(\partial_\sigma \alpha_\sigma)(R_\sigma(x))| \leq \kappa \qquad \forall (x, \sigma) \in \Sigma \times [0, 1],$$

and so

$$0 < \tau_{\sigma_0} e^{-\kappa(\sigma_1 - \sigma_0)} \leq \tau_{\sigma_1} \leq \tau_{\sigma_0} e^{\kappa(\sigma_1 - \sigma_0)} < +\infty \tag{4.3}$$

for all $0 \leq \sigma_0 < \sigma_1 < \sigma_\infty$ by Gronwall's Lemma [200, Lemma IV.4], implying

$$\frac{1}{C} \leq \tau_\sigma \leq C, \qquad C := \tau_0 e^{\kappa \sigma_\infty}$$

for all $\sigma \in [0, \sigma_\infty)$. □

The following exercise provides a scenario where the setting of Lemma 4.19 occurs in practice.

Exercise 4.20 Let $\Sigma = \Sigma_\rho \subseteq \mathbb{R}^{2n}$ be a star-shaped hypersurface for a positive function $\rho \in C^\infty(\mathbb{S}^{2n-1})$. Consider the diffeomorphism

$$f : \mathbb{S}^{2n-1} \to \Sigma_\rho, \qquad f(z) := \rho(z)z.$$

1. Show that $f^*\lambda_0|_{\Sigma_\rho} = \rho^2\lambda_0|_{\mathbb{S}^{2n-1}}$ and conclude $\ker(f^*\lambda_0|_{\Sigma_\rho}) = \ker(\lambda_0|_{\mathbb{S}^{2n-1}})$.
2. Compute the Reeb vector field associated with the contact form $\rho^2\lambda_0|_{\mathbb{S}^{2n-1}}$ and conclude that the dynamics does not generally coincide with the dynamics of the Reeb vector field associated with the contact form $\lambda_0|_{\mathbb{S}^{2n-1}}$.

Lemma 4.19 enables us to describe the topological properties of the limit set of a family of parametrised periodic Reeb orbits on a smooth homotopy of stable Hamiltonian manifolds.

Theorem 4.21 *The limit set of a smooth family of parametrised periodic Reeb orbits on a homotopy of stable Hamiltonian manifolds is nonempty, compact, and connected. Moreover, the Maupertuis functional $\mathscr{A}_0$ is constant on the limit set.*

Remark 4.22 Theorem 4.21 is a standard result in the theory of dynamical systems, see [200, Proposition IV.8]. For example, this result also applies to finite energy planes, as proven in [77, Section 13.3].

Proof It suffices to show that the limit set is nonempty. In fact, compactness and connectedness follow from that along the lines of the proof of [35, Theorem 5.2]. That the limit set is nonempty follows by Ascoli's theorem [145, Theorem 45.4] and bootstrapping the defining equation of a parametrised periodic Reeb orbit. Let us denote by $C > 0$ the uniform bound of Lemma 4.19. Define a family

$$\mathscr{F} \subseteq C^0(\mathbb{T}, \Sigma \times [1/C, C]) \qquad \text{by} \qquad \mathscr{F} := \{(\gamma_\sigma, \tau_\sigma) : \sigma \in [0, \sigma_\infty)\}.$$

Fix a Riemannian metric m on Σ. Then $\mathscr{F}$ is pointwise bounded since

$$\mathscr{F}_t := \{(\gamma_\sigma(t), \tau_\sigma) : \sigma \in [0, \sigma_\infty)\} \qquad \forall t \in \mathbb{T}$$

is contained in the compact, thus bounded, space $\Sigma \times [1/C, C]$. Moreover, the family $\mathscr{F}$ is equicontinuous. Indeed, for $t, t_0 \in \mathbb{T}$ we estimate

$$\begin{aligned} d_m(\gamma_\sigma(t), \gamma_\sigma(t_0)) &\leq \int_{t_0}^{t} |\dot{\gamma}_\sigma(s)| ds \\ &= \tau_\sigma \int_{t_0}^{t} |R_\sigma(\gamma_\sigma(s))| ds \\ &\leq C \int_{t_0}^{t} |R_\sigma(\gamma_\sigma(s))| ds \\ &\leq C_0 |t - t_0|, \end{aligned}$$

by definition of the metric d_m induced by the Riemannian metric and since

$$\Sigma \times [0, 1] \to T\Sigma, \qquad (x, \sigma) \mapsto R_\sigma(x)$$

is bounded. By Ascoli's Theorem, the closure $\bar{\mathscr{F}}$ of $\mathscr{F}$ is compact. Choose a sequence $(\sigma_k) \subseteq [0, \sigma_\infty)$ with $\sigma_k \to \sigma_\infty$ as $k \to \infty$. Then by compactness of $\bar{\mathscr{F}}$ there exists a subsequence of $(\gamma_{\sigma_k}, \tau_{\sigma_k})$, which can without loss of generality be assumed to be indexed by $(\sigma_k) \subseteq [0, \sigma_\infty)$ as well, such that

$$\lim_{k \to \infty} (\gamma_{\sigma_k}, \tau_{\sigma_k}) = (\gamma_{\sigma_\infty}, \tau_{\sigma_\infty}) \in C^0(\mathbb{T}, \Sigma \times [1/C, C])$$

in the uniform C^0-topology. In fact $\gamma_{\sigma_\infty} \in C^0(\mathbb{T}, \Sigma)$ is a periodic Reeb orbit of R_{σ_∞}, since

$$\dot{\gamma}_{\sigma_\infty} = \lim_{k \to \infty} \dot{\gamma}_{\sigma_k} = \lim_{k \to \infty} \tau_{\sigma_k}(R_\sigma \circ \gamma_{\sigma_k}) = \tau_{\sigma_\infty}(R_{\sigma_\infty} \circ \gamma_{\sigma_\infty})$$

uniformly and conclude that $\gamma_{\sigma_\infty} \in C^1(\mathbb{T}, \Sigma_{\sigma_\infty})$. We can inductively improve the regularity of γ_{σ_∞}. This method is referred to as ***bootstrapping***. Indeed, for all k we have that

$$\ddot{\gamma}_{\sigma_k} = \partial_t \tau_{\sigma_k}(R_{\sigma_k} \circ \gamma_{\sigma_k}) = \tau_{\sigma_k} DX_{R_{\sigma_k}}(\dot{\gamma}_{\sigma_k}) = \tau_{\sigma_k}^2 DR_{\sigma_k}(R_{\sigma_k} \circ \gamma_{\sigma_k})$$

and so $\gamma_{\sigma_\infty} \in C^2(\mathbb{T}, \Sigma)$. Repeating this process yields $\gamma_{\sigma_\infty} \in C^\infty(\mathbb{T}, \Sigma)$. Consequently, the limit set is not empty, as $\tau_{\sigma_\infty} \geq \frac{1}{C} > 0$. Finally, that $\mathscr{A}_0$ is action-constant on the limit set follows from (4.3). □

Consider a Hamiltonian system (M, ω, H). If $\Sigma = H^{-1}(0)$ is a regular energy hypersurface in (M, ω, H), there is another variational characterisation of parametrised periodic Reeb orbits in $\mathscr{P}(\Sigma, \omega|_\Sigma, \alpha)$ if Σ admits a stabilising form $\alpha \in \Omega^1(\Sigma)$. Recall that the symplectic action functional in Definition 3.20 is suitable for detecting periodic orbits of a fixed period in the phase space of arbitrary energy. Another approach is to look at periodic orbits of fixed energy but of arbitrary periods.

Definition 4.23 (Rabinowitz Action Functional) Let (M, λ, H) be an exact Hamiltonian system. Then the corresponding ***Rabinowitz action functional*** is given by

$$\mathscr{A}^H \colon \mathscr{L}M \times \mathbb{R} \to \mathbb{R}, \qquad \mathscr{A}^H(\gamma, \tau) := \int_0^1 \gamma^*\lambda - \tau \int_0^1 H(\gamma(t))dt.$$

Exercise 4.24 Let (M, λ, H) be an exact Hamiltonian system. What are the differences and similarities of the symplectic action functional $\mathscr{A}_H$ and the Rabinowitz action functional $\mathscr{A}^H$? Sketch the corresponding critical points!

Exercise 4.25 Work through [50, Section 4] to explore some additional occurrences of Lagrange multiplier functionals in mathematics and physics.

Definition 4.26 (Homotopy of Stable Energy Hypersurfaces) Let (M, ω) be a connected symplectic manifold. A ***homotopy of stable energy hypersurfaces*** is defined to be a time-dependent Hamiltonian function $H \in C^\infty(M \times [0, 1])$ such that

1. 0 is a regular value of $H_\sigma := H(\cdot, \sigma) \in C^\infty(M)$ for all $\sigma \in [0, 1]$.
2. $\Sigma_\sigma := H_\sigma^{-1}(0)$ is connected for all $\sigma \in [0, 1]$.
3. $H^{-1}(0)$ is compact.
4. There exists a smooth time-dependent vector field $(X_\sigma)_{\sigma \in [0,1]}$ on a neighbourhood of $H^{-1}(0)$ such that

$$dH_\sigma(X_\sigma)|_{\Sigma_\sigma} > 0 \qquad \text{and} \qquad \ker(\omega|_{\Sigma_\sigma}) \subseteq \ker(L_{X_\sigma}\omega|_{\Sigma_\sigma})$$

for all $\sigma \in [0, 1]$.

Remark 4.27 By the Ehresmann fibration theorem [77, Corollary 5.7.7], all hypersurfaces Σ_σ of a homotopy of stable regular energy hypersurfaces are diffeomorphic to each other.

Example 4.28 (Star-Shaped Hypersurfaces) Let $\rho_\sigma \in C^\infty(\mathbb{S}^{2n-1})$ be a smooth family of positive functions for $\sigma \in [0, 1]$. Then

$$H_\sigma \colon \mathbb{C}^n \to \mathbb{R}, \qquad H_\sigma(z) := \begin{cases} h\left(\|z\|^2 \rho_\sigma(\frac{z}{\|z\|})^{-2} - 1\right), & z \neq 0, \\ -\frac{1}{2}, & z = 0, \end{cases}$$

is a homotopy of star-shaped hypersurfaces $\Sigma_\sigma = \Sigma_{\rho_\sigma}$.

In the restricted three-body problem, families of periodic orbits are known to exist, as shown in [34]. As Problem 4.45 about hyperbolic space shows, a family of periodic orbits cannot always be extended indefinitely among a homotopy of regular energy hypersurfaces. As we have already seen, this phenomenon is known as a blue sky catastrophe as in [77, Section 7.6]. In practical applications, this can sometimes be achieved by a numerical simulation as described in [173]. Let $(\Sigma_\sigma, \omega_\sigma, \alpha_\sigma)_{\sigma \in [0,1]}$ with $\omega_\sigma := \omega|_{\Sigma_\sigma}$ and $\alpha_\sigma := i_{X_\sigma}\omega_\sigma$ be a homotopy of stable regular energy hypersurfaces for some $H \in C^\infty(M \times [0, 1])$. Consider a smooth family of parametrised periodic orbits $(\gamma_\sigma, \tau_\sigma)_{\sigma \in [0,\sigma_\infty)}$ for $0 < \sigma_\infty \leq 1$, that is, we have that $(\gamma_\sigma, \tau_\sigma) \in \mathscr{L}\Sigma_\sigma \times (0, +\infty)$ solves the Hamiltonian equation

$$\dot{\gamma}_\sigma(t) = \tau_\sigma X_{H_\sigma}(\gamma_\sigma(t)) \qquad \forall t \in \mathbb{T}, \sigma \in [0, \sigma_\infty). \tag{4.4}$$

Example 4.29 By Problem 2.67, every periodic orbit with precisely two Floquet multipliers equal to one extends to a whole cylinder of periodic orbits. If $\sigma_\infty < 1$ is maximal with the property that every orbit of the cylinder has precisely two Floquet multipliers equal to one, then bifurcation phenomena can occur at σ_∞ as analytically explored in [117, Theorem 2.1] or [9, Chapter 8].

We now state the analogue of Lemma 4.19 for a homotopy of stable regular energy hypersurfaces. The main difficulty is constructing a suitable smooth family of Rabinowitz action functionals since we cannot use the properties of the symplectic area functional in Exercise 4.18 anymore. The crucial idea is to extend every stabilising form to the whole symplectic manifold. This and related techniques will be crucial throughout the book and are based on [56, Section 4.2.2].

Theorem 4.30 *Let $(\gamma_\sigma, \tau_\sigma)_{\sigma \in [0,\sigma_\infty)}$ be a smooth family of parametrised periodic orbits for some $0 < \sigma_\infty \leq 1$ on a homotopy of stable regular energy hypersurfaces $(\Sigma_\sigma, \omega_\sigma, \alpha_\sigma)_{\sigma \in [0,1]}$. Then there exists a constant $C > 0$ such that*

$$\frac{1}{C} \leq \tau_\sigma \leq C \qquad \forall \sigma \in [0, \sigma_\infty).$$

In particular, the limit set of a family of periodic orbits on a homotopy of stable regular-energy hypersurfaces is nonempty, compact, and connected.

Proof The result immediately follows from Theorem 4.21, if we can prove an analogue of Lemma 4.19. In order to prove this result, we need to show that the smooth family of parametrised periodic orbits $(\gamma_\sigma, \tau_\sigma)_{\sigma\in[0,\sigma_\infty)}$ can be reparametrised to a smooth family of Reeb orbits which are critical points of a suitable smooth family of Rabinowitz action functionals. Define a smooth function

$$\chi\colon H^{-1}(0) \to (0, +\infty), \qquad \chi(x,\sigma) := dH_\sigma(X_\sigma)(x).$$

Then we claim that $X_{H_\sigma}(x) = \chi(x,\sigma)R_\sigma(x)$ holds for every $\sigma \in [0, 1]$ and $x \in \Sigma_\sigma$. Indeed, both sides belong to the characteristic line bundle $\ker \omega_\sigma$ and

$$\alpha_\sigma(X_{H_\sigma}) = i_{X_\sigma}\omega_\sigma(X_{H_\sigma}) = \omega_\sigma(X_\sigma, X_{H_\sigma}) = dH_\sigma(X_\sigma)|_{\Sigma_\sigma} = \chi.$$

Thus we may assume that the smooth family $(\gamma_\sigma, \tau_\sigma)_{\sigma\in[0,\sigma_\infty)}$ is a smooth family of parametrised periodic Reeb orbits. As $H^{-1}(0)$ is compact by assumption, the reparametrisation does not affect the convergence of the periods τ_σ. Next, we construct the family of Rabinowitz action functionals. By Problem 4.46, there exists a smooth family $(\psi_\sigma)_{\sigma\in[0,1]}$ of stable tubular neighbourhoods of $(\Sigma_\sigma, \omega_\sigma, \alpha_\sigma)$. More precisely, for $\varepsilon > 0$ fixed sufficiently small, for all $\sigma \in [0, 1]$ the smooth embeddings $\psi_\sigma\colon (-\varepsilon, \varepsilon) \times \Sigma_\sigma \hookrightarrow M$ satisfy for all $r \in (-\varepsilon, \varepsilon)$ the conditions

$$\psi_\sigma|_{\{0\}\times\Sigma_\sigma} = \iota_{\Sigma_\sigma}\colon \Sigma_\sigma \hookrightarrow M \qquad \text{and} \qquad \psi_\sigma^*\omega = \omega_\sigma + d(r\alpha_\sigma),$$

The embeddings ψ_σ are defined via following the flow of the vector fields X_σ which exists locally. Next, we choose some sufficiently small mollifications $f \in C^\infty(\mathbb{R})$ and $h \in C^\infty(\mathbb{R})$ as depicted in Fig. 4.2, satisfying the following properties:

- $f(r) = r + 1$ for $r \in \left[-\frac{\varepsilon}{2}, \frac{\varepsilon}{2}\right]$.
- $\operatorname{supp} f \subseteq (-\varepsilon, \varepsilon)$.

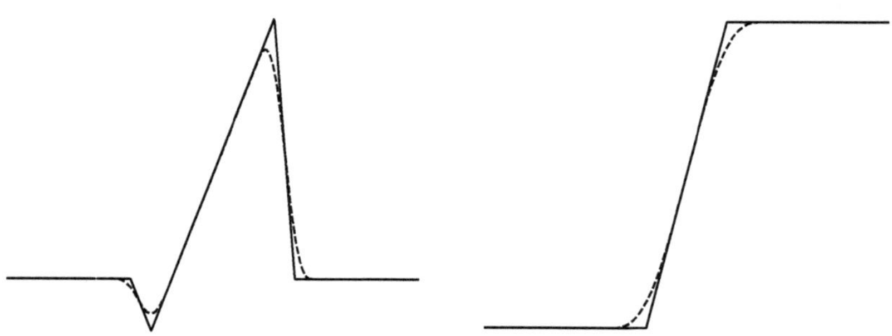

(a) (b)

Fig. 4.2 (**a**) Possible choice of extension function f and its mollification. (**b**) Mollification of the piecewise-linear function h

- h satisfies

$$h(r) = \begin{cases} r & r \in \left[-\frac{\varepsilon}{3}, \frac{\varepsilon}{3}\right] \\ \frac{\varepsilon}{3} & r \in \left[\frac{\varepsilon}{3}, +\infty\right) \\ -\frac{\varepsilon}{3} & r \in \left(-\infty, -\frac{\varepsilon}{3}\right] \end{cases}.$$

For all $\sigma \in [0, 1]$ define extensions

$$\bar{\alpha}_\sigma(p) := \begin{cases} f(r)\alpha_\sigma(x) & p = \psi_\sigma(r, x), \\ 0 & p \notin U_\sigma, \end{cases}$$

for $U_\sigma := \psi_\sigma((-\varepsilon, \varepsilon) \times \Sigma_\sigma)$, and modifications

$$\bar{H}_\sigma(p) := \begin{cases} h(r) & p = \psi_\sigma(r, x), \\ \frac{\varepsilon}{3} & p \in U_\sigma^c \cap H_\sigma^{-1}(0, +\infty), \\ -\frac{\varepsilon}{3} & p \in U_\sigma^c \cap H_\sigma^{-1}(-\infty, 0), \end{cases}$$

Note that $\bar{H} \in C^\infty(M \times [0, 1])$ is well-defined, as by Definition 4.26 we have that the vector fields X_σ satisfy $dH_\sigma(X_\sigma)|_{\Sigma_\sigma} > 0$ by assumption. Consider the smooth family of Rabinowitz action functionals $\mathscr{A}_\sigma : \mathscr{L}M \times (0, +\infty) \to \mathbb{R}$ defined by

$$\mathscr{A}_\sigma(\gamma, \tau) := \int_0^1 \gamma^* \bar{\alpha}_\sigma - \tau \int_0^1 \bar{H}_\sigma \circ \gamma.$$

We claim that

$$d\mathscr{A}_\sigma|_{(\gamma,\tau)}(X, \eta) = \bar{m}_\sigma \left(\operatorname{grad} \mathscr{A}_\sigma|_{(\gamma,\tau)}, (X, \eta)\right) \tag{4.5}$$

holds for all $(X, \eta) \in T_\gamma \mathscr{L}M \times \mathbb{R}$ and $(\gamma, \tau) \in \mathscr{L}M \times (0, +\infty)$, where

$$\bar{m}_\sigma((X, \eta), (Y, \xi)) := \int_0^1 d\bar{\alpha}_\sigma(JX(t), Y(t))dt + \eta\xi,$$

for all $(Y, \xi) \in \mathscr{L}M \times \mathbb{R}$ with J being an ω-compatible almost complex structure on the symplectic manifold (M, ω), is a bilinear form and

$$\operatorname{grad} \mathscr{A}_\sigma|_{(\gamma,\tau)}(t) := \begin{pmatrix} J(\dot{\gamma}(t) - \tau X_{\bar{H}_\sigma}(\gamma(t))) \\ -\int_0^1 \bar{H}_\sigma \circ \gamma \end{pmatrix} \qquad \forall t \in \mathbb{T}.$$

In order to show (4.5), we show first that $X_{\bar{H}_\sigma}|_{\Sigma_\sigma} = R_\sigma$ for all $\sigma \in [0, 1]$ for the Reeb vector field $R_\sigma = R_{\omega_\sigma,\alpha_\sigma} \in \mathfrak{X}(\Sigma_\sigma)$. We compute

$$\begin{aligned} i_{h'(r)R_\sigma}\psi_\sigma^*\omega &= i_{h'(r)R_\sigma}(\omega_\sigma + d(r\alpha_\sigma)) \\ &= i_{h'(r)R_\sigma}(\omega_\sigma + dr \wedge \alpha_\sigma + r d\alpha_\sigma) \\ &= h'(r) i_{R_\sigma}\omega_\sigma + h'(r) i_{R_\sigma} dr\alpha_\sigma - h'(r) i_{R_\sigma}\alpha_\sigma dr + r h'(r) i_{R_\sigma} d\alpha_\sigma \\ &= -h'(r)dr \\ &= -d\bar{H}_\sigma, \end{aligned}$$

as R_σ belongs to $\ker\omega_\sigma \subseteq \ker d\alpha_\sigma$ by stability. Moreover, we compute

$$\begin{aligned} i_{X_{\bar{H}_\sigma}} d\bar{\alpha}_\sigma &= i_{X_{\bar{H}_\sigma}}(f'(r)dr \wedge \alpha_\sigma + f(r)d\alpha_\sigma) \\ &= f'(r) i_{X_{\bar{H}_\sigma}} dr\alpha_\sigma - f'(r) i_{X_{\bar{H}_\sigma}}\alpha_\sigma dr + f(r) i_{X_{\bar{H}_\sigma}} d\alpha_\sigma \\ &= -f'(r)h'(r)dr \\ &= -h'(r)dr \\ &= -d\bar{H}_\sigma. \end{aligned}$$

on U_σ, as by definition $X_{\bar{H}_\sigma} = 0$ else. Hence

$$\begin{aligned} d\mathscr{A}_\sigma|_{(\gamma,\tau)}(X,\eta) &= \int_0^1 d\bar{\alpha}_\sigma(X,\dot{\gamma}) - \tau\int_0^1 d\bar{H}_\sigma(X) - \eta\int_0^1 \bar{H}_\sigma\circ\gamma \\ &= \int_0^1 d\bar{\alpha}_\sigma(X, \dot{\gamma} - \tau X_{\bar{H}_\sigma}\circ\gamma) - \eta\int_0^1 \bar{H}_\sigma\circ\gamma \\ &= \bar{m}\left(\operatorname{grad}\mathscr{A}_\sigma|_{(\gamma,\tau)}, (X,\eta)\right) \end{aligned}$$

by a routine computation of the differential of the Rabinowitz action functional similar to Lemma 3.27. This shows $\mathscr{P}(\Sigma_\sigma, \omega_\sigma, \alpha_\sigma) \subseteq \mathrm{Crit}(\mathscr{A}_\sigma)$. □

4.2 Noncontractible Periodic Orbits on Lens Spaces

Let $\Sigma = \Sigma_\rho \subseteq \mathbb{R}^{2n}$, $n \geq 2$, be a star-shaped hypersurface for some positive radius function $\rho \in C^\infty(\mathbb{S}^{2n-1})$. As we have seen, $\Sigma = H_\Sigma^{-1}(0)$ is the regular energy hypersurface of the Hamiltonian function H_Σ defined in (1.2). We have encountered a way of variationally describing periodic orbits as critical points of the Rabinowitz action functional $\mathscr{A}^{H_\Sigma}$. In order to show that any star-shaped hypersurface Σ admits a periodic orbit, we can therefore prove that $\mathrm{Crit}(\mathscr{A}^{H_\Sigma}) \neq \varnothing$. As Σ is diffeomorphic to $\mathbb{S}^{2n-1}$ and the flow on $\mathbb{S}^{2n-1}$ is periodic by Exercise 1.6, the idea is to infer the existence of periodic orbits on Σ from periodic Reeb orbits on $\mathbb{S}^{2n-1}$. But, as the example of an ellipsoid $E(a_1, \ldots, a_n)$ already shows, if we deform the sphere into an ellipsoid with rationally independent weights a_j, many periodic orbits do not survive! Assume that Σ_σ is a smooth homotopy of star-shaped hypersurfaces as in Example 4.28 for $\sigma \in [0, 1]$ and that there exists a periodic Reeb orbit on Σ_0 with precisely two Floquet multipliers equal to one. This is, for example, the case for an ellipsoid with rationally independent weights. Then by Poincaré's continuation method described in Problem 2.67, there always exists a whole family of periodic orbits on Σ_σ for all $\sigma \in [0, \sigma_\infty)$ for some $0 < \sigma_\infty \leq 1$. Thus, by Theorem 4.30, there always exists a periodic Reeb orbit on the limiting hypersurface Σ_{σ_∞}. But for an arbitrary star-shaped hypersurface we cannot tell if such a family extends over the limiting hypersurface Σ_{σ_∞} to the hypersurface Σ_1, in general. How far the family extends is hard to quantify by analytical means, although one can often numerically verify the existence of such a family for a wide range of parameters. There is a large body of literature on numerical continuation methods as in [20]. We will not pursue the numerical approach further in this book. Instead, we will use the following guiding idea. As we have seen, the critical points of the Rabinowitz action functional on different star-shaped hypersurfaces do not necessarily coincide. But if we can define a homology theory with generators some suitable equivalence classes of critical points, then the resulting homologies might be the same for different star-shaped hypersurfaces! This is a much weaker requirement, and we are going to explain this idea for lens spaces in this section.

Definition 4.31 (Lens Space) Let $m \geq 1$ be an integer and let $k_1, \ldots, k_n \in \mathbb{Z}$ be coprime to m. Consider the $\mathbb{Z}_m$-action

$$\mathbb{Z}_m \times \mathbb{S}^{2n-1} \to \mathbb{S}^{2n-1}, \qquad [k] \cdot (z_1, \ldots, z_n) := \varphi^k(z_1, \ldots, z_n),$$

induced by the rotation

$$\varphi \colon \mathbb{S}^{2n-1} \to \mathbb{S}^{2n-1}, \qquad \varphi(z_1, \ldots, z_n) := \left(e^{2\pi i k_1/m} z_1, \ldots, e^{2\pi i k_n/m} z_n\right).$$

A ***lens space*** is defined to be the smooth quotient manifold $L_m(k_1, \ldots, k_n) := \Sigma_\rho/\mathbb{Z}_m$ for a star-shaped hypersurface $\Sigma_\rho \subseteq \mathbb{C}^n$ satisfying $\rho \circ \varphi = \rho$.

Exercise 4.32 Show that $L_m(k_1, \ldots, k_n)$ is a contact manifold with the contact form given by the restriction $\lambda_0|_{\Sigma/\mathbb{Z}_m}$.

The following examples show an abundance of lens spaces throughout celestial mechanics.

Example 4.33 (The Hénon–Heiles Hamiltonian) As in [168, Section 2], consider the mechanical Hamiltonian function

$$H\colon \mathbb{R}^4 \to \mathbb{R}, \qquad H(q_1, q_2, p_1, p_2) := \frac{1}{2}\left(\|p\|^2 + \|q\|^2\right) + q_1^2 q_2 - \frac{1}{3}q_2^3.$$

This Hamiltonian function is known as the ***Hénon–Heiles Hamiltonian*** and exhibits very complex chaotic dynamics with a long history in the numerical simulation of stellar motion [102, Section I.3]. On $\mathbb{R}^4 \cong \mathbb{C}^2$ consider the coordinates

$$z := q_1 + iq_2 \qquad \text{and} \qquad w := p_1 + ip_2.$$

Define $\varphi\colon \mathbb{C}^2 \to \mathbb{C}^2$, $\varphi(z, w) := e^{2\pi i/3}(z, w)$. Then $\varphi^*\lambda = \lambda$ for

$$\lambda = \frac{1}{2}(p_1 dq_1 - q_1 dp_1) + \frac{1}{2}(p_2 dq_2 - q_2 dp_2).$$

For every $0 < k < \frac{1}{6}$, the regular energy surface $H^{-1}(k)$ contains a strictly convex sphere-like component $\Sigma_k \cong \mathbb{S}^3$ by [168, Theorem 2.2]. The resulting quotient $\Sigma_k/\mathbb{Z}_3$ is diffeomorphic to the lens space $L_3(1, 1)$, but not contactomorphic to it with the standard contact distribution. Instead, the quotient $\Sigma_k/\mathbb{Z}_3$ is contactomorphic to $L_3(1, 2)$ with its standard contact distribution. This is mainly due to the use of different coordinates.

Example 4.34 (Hill's Lunar Problem) The mechanical Hamiltonian function

$$H(q, p) := \frac{1}{2}\|p\|^2 - \frac{1}{\|q\|} + q_1 p_2 - q_2 p_1 - q_1^2 + \frac{1}{2}q_2^2$$

on the phase space $T^*(\mathbb{R}^2 \setminus \{0\})$ is called ***Hill's lunar Hamiltonian*** following [77, Section 5.8]. After Levi–Civita regularisation, the regularised Hill's lunar Hamiltonian $K\colon T^*\mathbb{R}^2 \to \mathbb{R}$ is given by

$$K(q, p) = \frac{1}{2}\left(\|p\|^2 + \|q\|^2\right) + 2\|q\|^2(q_2 p_1 - q_1 p_2) - 4(q_1^6 - 3q_1^4 q_2^2 - 3q_1^2 q_2^4 + q_2^6).$$

For $k > 0$ sufficiently small, the energy hypersurface $K^{-1}(k)$ contains a strictly convex sphere-like component $\Sigma_k \cong \mathbb{S}^3$. On $T^*\mathbb{R}^2 \cong \mathbb{C}^2$ consider the coordinates

$$z := q_1 + iq_2 \qquad \text{and} \qquad w := p_1 + ip_2$$

and the rotation

$$\varphi\colon \mathbb{C}^2 \to \mathbb{C}^2, \qquad \varphi(z, w) := e^{\pi i/2}(z, w).$$

Then K is invariant under the rotation φ and thus $\Sigma/\mathbb{Z}_4$ is diffeomorphic to the lens space $L_4(1, 1)$, but again due to the choice of nonstandard coordinates, not contactomorphic to it. It is a delicate question in Contact Topology to decide the correct value of k_1 and k_2, such that the obtained lens space $L_4(1, 1)$ in Hill's lunar problem is contactomorphic to $L_4(k_1, k_2)$. The tight contact structures on the lens spaces $L(m, k_2)$ are classified up to isotopy by [110, Theorem 2.1], so in principle it should be possible to obtain the correct value of k_2.

Example 4.35 (Stark–Zeeman Systems) Planar Stark–Zeeman systems as in [58] and [59] generalise many important physical systems, including the diamagnetic Kepler problem and the restricted three-body problem [141]. By [58, Corollary 1], for energy values below the first critical value, the Moser regularised energy hypersurfaces are diffeomorphic to the unit cotangent bundles $S^*\mathbb{S}^n$. In particular, for $n = 2$ we obtain $S^*\mathbb{S}^2 \cong \mathbb{RP}^3$, a real projective space by Exercise 1.2.

Let $L_m(k_1, \ldots, k_n) = \Sigma/\mathbb{Z}_m$ be a lens space. By examining the long exact sequence of homotopy groups of a fibration [104, Theorem 4.41], there is a short exact sequence

$$0 \longrightarrow \pi_1(\Sigma, x) \longrightarrow \pi_1(\Sigma/\mathbb{Z}_m, \pi(x)) \longrightarrow \pi_0(\mathbb{Z}_m) \longrightarrow 0,$$

implying $\pi_1(\Sigma/\mathbb{Z}_m) \cong \mathbb{Z}_m$ by [128, Corollary 12.9]. Thus we may look for noncontractible periodic Reeb orbits on lens spaces! What about homology? For computing the singular homology $\mathrm{H}_*(L_m; \mathbb{Z})$, it is most convenient to use cellular homology [104, p. 137]. Any cellular decomposition of a compact topological space Q induces a finite cellular filtration $\varnothing =: U_{-1} \subseteq U_0 \subseteq \cdots \subseteq U_n = Q$ for open subsets $U_1, \ldots, U_n \subseteq Q$. One can check that in that case

$$\mathrm{H}_k(U_l, U_{l-1}; \mathbb{Z}) = 0 \qquad \forall k \neq l,$$

so that the relative singular homology of the pair (U_l, U_{l-1}) is concentrated in degree k. One can build a chain complex with any finite cellular filtration of Q.

Exercise 4.36 Let Q be a topological space with cellular filtration $(U_1, \ldots, U_n)$. For all $k \geq 0$ set $C_k := \mathrm{H}_k(U_k, U_{k-1}; \mathbb{Z})$ and define $\partial_k \colon C_k \to C_{k-1}$ to be the composition of

$$\mathrm{H}_k(U_k, U_{k-1}) \to \mathrm{H}_{k-1}(U_{k-1}) \quad \text{and} \quad \mathrm{H}_{k-1}(U_k - 1) \to \mathrm{H}_{k-1}(U_{k-1}, U_{k-2}),$$

where the first map is the connecting homomorphism for the long exact sequence of the pair (U_k, U_{k-1}) and the latter is induced by inclusion.

1. Show that (C_*, ∂_*) is a chain complex.
2. Show that the homology of the chain complex (C_*, ∂_*) is isomorphic to singular homology $\mathrm{H}_*(Q; \mathbb{Z})$.

By [104, Example 2.43], any lens space $L_m(k_1, \ldots, k_n)$ admits a cellular decomposition with precisely one cell e^k for every $0 \leq k \leq 2n-1$. The corresponding boundary maps are given by

$$0 \longrightarrow \mathbb{Z} \xrightarrow{0} \mathbb{Z} \xrightarrow{m} \mathbb{Z} \xrightarrow{0} \ldots \xrightarrow{0} \mathbb{Z} \xrightarrow{m} \mathbb{Z} \xrightarrow{0} \mathbb{Z} \longrightarrow 0$$

using the Cellular Boundary Formula [104, p. 140], and thus the resulting homology is given by

$$\mathrm{H}_k(L_m(k_1, \ldots, k_n); \mathbb{Z}) \cong \begin{cases} \mathbb{Z} & k = 0, 2n-1, \\ \mathbb{Z}_m & 0 < k < 2n-1, k \text{ odd}, \\ 0 & \text{else}. \end{cases} \tag{4.6}$$

There is a more dynamical interpretation of cellular homology using Morse functions. Let (M^n, m) be a compact Riemannian manifold and $f \in C^\infty(M)$ a Morse function. Consider the negative gradient flow $\phi \colon \mathbb{R} \times M \to M$ defined by

$$\frac{d}{dt}\phi_t = -\operatorname{grad}_m f \circ \phi_t \qquad \forall t \in \mathbb{R}.$$

For any $x \in \operatorname{Crit}(M)$ we can consider the ***stable manifold***

$$W^s(x) := \left\{ y \in M : \lim_{t \to +\infty} \phi_t(y) = x \right\} \cong \mathbb{R}^{n - \operatorname{ind}_f(x)},$$

and ***unstable manifold***

$$W^u(x) := \left\{ y \in M : \lim_{t \to -\infty} \phi_t(y) = x \right\} \cong \mathbb{R}^{\operatorname{ind}_f(x)},$$

where $\operatorname{ind}_f(x)$ denotes the Morse index of x, that is, the number of negative eigenvalues of the Hessian matrix at a critical point counted with multiplicity. We say that the negative gradient flow of f satisfies the ***Morse–Smale condition***, if for all critical points $a, b \in \operatorname{Crit}(f)$ the condition

$$T_y W^u(a) + T_y W^s(b) = T_y M \qquad \forall y \in W^u(a) \cap W^s(b)$$

holds, implying that the stable and unstable manifolds intersect transversally. A consequence of the Morse–Smale condition is that the moduli space $W^u(a) \cap W^s(b)$ of all parametrised negative gradient flow lines from a to b is a smooth manifold of dimension $\operatorname{ind}_f(a) - \operatorname{ind}_f(b)$. It is now a deep result that M has the homotopy type of a finite cell complex with a cell of dimension $\operatorname{ind}_f(x)$ for each critical point $x \in \operatorname{Crit}(f)$ as in [140, Theorem 3.5]. A preliminary step in its proof is provided in Problem 4.48. Thus, any Morse function f on M gives rise to a finite cellular

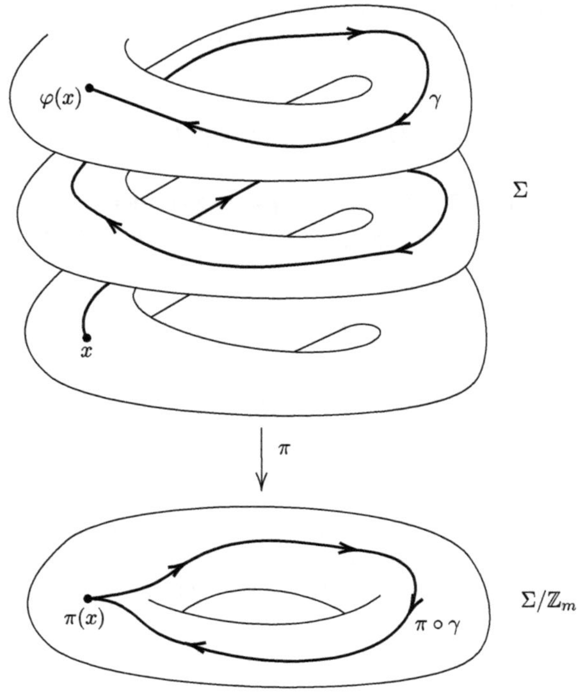

Fig. 4.3 The projection $\pi \circ \gamma \in \mathscr{L}(\Sigma/\mathbb{Z}_m, \pi(x))$ of $\gamma \in \mathscr{L}_\varphi(\Sigma, x)$ is not contractible for the deck transformation $\varphi \neq \mathrm{id}_\Sigma$

filtration $U_1, \dots, U_n$ of M, and in particular to a homology $\mathrm{HM}_*(f; \mathbb{Z})$ isomorphic to singular homology. The idea of the following two parts of the book is to build a homology theory which can detect noncontractible periodic Reeb orbits on lens spaces. For this, it is reasonable to use a suitable generalisation of the Rabinowitz action functional. First of all, since the $\mathbb{Z}_m$-action is not free at the origin, it should not be possible to represent every lens space as a regular energy hypersurface in a suitable symplectic manifold. However, by [129, Theorem 21.13], the canonical projection $\Sigma \to \Sigma/\mathbb{Z}_m$ is a smooth covering map (Fig. 4.3).

Exercise 4.37 Let $\Sigma/\mathbb{Z}_m$ be a lens space with $m \neq 1$. For $x \in \Sigma$ define the ***based twisted loop space of*** Σ ***and*** φ by

$$\mathscr{L}_\varphi(\Sigma, x) := \{\gamma \in C^0([0,1], \Sigma) : \gamma(0) = x \text{ and } \gamma(1) = \varphi(x)\}.$$

Show that if $\gamma \in \mathscr{L}_\varphi(\Sigma, x)$ for some $x \in \Sigma$, then $\pi \circ \gamma \in \mathscr{L}(\Sigma/\mathbb{Z}_m, \pi(x))$ is not contractible. In contrast, if $\gamma \in \mathscr{L}(\Sigma/\mathbb{Z}_m, \pi(x))$ is not contractible, then there exists $1 \leq k < m$ such that $\widetilde{\gamma}_x \in \mathscr{L}_{\varphi^k}(\Sigma, x)$ for the unique lift $\widetilde{\gamma}_x$ of γ with $\widetilde{\gamma}_x(0) = x$.

Motivated by Exercise 4.37, we make the following definition in the smooth case based on [137, p. 452–453].

Definition 4.38 (Twisted Loop Space) Let (M, ω) be a symplectic manifold and let $\varphi \in \mathrm{Symp}(M, \omega)$ be a symplectomorphism of finite order. Define the ***free twisted loop space of*** M ***and*** φ by

$$\mathscr{L}_\varphi M := \{\gamma \in C^\infty(\mathbb{R}, M) : \gamma(t+1) = \varphi(\gamma(t))\ \forall t \in \mathbb{R}\}.$$

Remark 4.39 (Twisted Loops are Symmetric) Let (M, ω) be a symplectic manifold, and let $\varphi \in \mathrm{Symp}(M, \omega)$ be a symplectomorphism of finite order $m := \operatorname{ord}\varphi$. Then we can define the inclusion

$$\iota\colon \mathscr{L}_\varphi M \hookrightarrow \mathscr{L}M, \qquad \iota(\gamma)(t) := \gamma(mt) \qquad \forall t \in \mathbb{T}.$$

Any such loop $\iota(\gamma)$ is symmetric with respect to φ, that is, we have

$$\varphi(\iota(\gamma)(\mathbb{T})) = \iota(\gamma)(\mathbb{T}).$$

Let (M, ω) be a symplectic manifold and let $\varphi \in \mathrm{Symp}(M, \omega)$ be a symplectomorphism of finite order. Given a twisted loop $\gamma \in \mathscr{L}_\varphi M$ and $\varepsilon_0 > 0$, we say that a curve $(-\varepsilon_0, \varepsilon_0) \to \mathscr{L}_\varphi M, \varepsilon \mapsto \gamma_\varepsilon$ starting at γ is ***smooth***, if the induced variation

$$\mathbb{R} \times (-\varepsilon_0, \varepsilon_0) \to M, \qquad (t, \varepsilon) \mapsto \gamma_\varepsilon(t)$$

is smooth. Since $\gamma_\varepsilon(t+1) = \varphi(\gamma_\varepsilon(t))$ holds for all $\varepsilon \in (-\varepsilon_0, \varepsilon_0)$ and $t \in \mathbb{R}$, we call such a variation a ***twisted variation***. Then the infinitesimal variation

$$\delta\gamma := \left.\frac{\partial \gamma_\varepsilon}{\partial \varepsilon}\right|_{\varepsilon=0} \in \Gamma(\gamma^* TM),$$

satisfies $\delta\gamma(t+1) = D\varphi(\delta\gamma(t))$ for all $t \in \mathbb{R}$.

Lemma 4.40 *Let (M, ω) be a symplectic manifold. Suppose that $\gamma \in \mathscr{L}_\varphi M$ and assume that $X \in \Gamma(\gamma^* TM)$ satisfies $X(t+1) = D\varphi(X(t))$ for all $t \in \mathbb{R}$. Then there exists a twisted variation of γ such that $\delta\gamma = X$.*

Proof As φ is assumed to be of finite order, there exists a φ-invariant ω-compatible almost complex structure J on M by [137, Lemma 5.5.6]. With respect to the induced Riemannian metric $m_J := \omega(J\cdot, \cdot)$ the symplectomorphism φ is an isometry. Define the exponential variation $\mathbb{R} \times (-\varepsilon_0, \varepsilon_0) \to M, \gamma_\varepsilon(t) := \exp^{\nabla_J}_{\gamma(t)}(\varepsilon X(t))$, for $\varepsilon_0 > 0$ sufficiently small and ∇_J denoting the Levi–Civita connection associated with m_J. This $\varepsilon_0 > 0$ exists because of the naturality of geodesics [130, Corollary 5.14]. Then we compute

$$\begin{aligned}\gamma_\varepsilon(t+1) &= \exp^{\nabla_J}_{\gamma(t+1)}(\varepsilon X(t+1)) \\ &= \exp^{\nabla_J}_{\varphi(\gamma(t))}(D\varphi(\varepsilon X(t)))\end{aligned}$$

$$= \varphi\left(\exp^{\nabla_J}_{\gamma(t)}(\varepsilon X(t))\right)$$
$$= \varphi(\gamma_\varepsilon(t))$$

by naturality of the exponential map [130, Proposition 5.20]. □

This discussion, together with Lemma 4.40, motivates the following definition of the tangent space to the free twisted loop space.

Definition 4.41 (Tangent Space to the Free Twisted Loop Space) Let (M, ω) be a symplectic manifold. For $\gamma \in \mathscr{L}_\varphi M$ define the ***tangent space to the free twisted loop space at*** γ by

$$T_\gamma \mathscr{L}_\varphi M := \{X \in \mathfrak{X}(\gamma) : X(t+1) = D\varphi(X(t))\ \forall t \in \mathbb{R}\}.$$

Exercise 4.42 Let M be a smooth manifold and let $\varphi \in \mathrm{Diff}(M)$ be a diffeomorphism. Show that the continuous free twisted loop space

$$\mathscr{L}^0_\varphi M := \{\gamma \in C^0([0,1], M) : \gamma(1) = \varphi(\gamma(0))\},$$

is the pullback of

$$(\mathrm{ev}_0, \mathrm{ev}_1)\colon \mathscr{P}M \to M \times M, \qquad \gamma \mapsto (\gamma(0), \gamma(1)),$$

where we abbreviate $\mathscr{P}M := C^0([0,1], M)$, along the graph of φ

$$\Gamma_\varphi\colon M \to M \times M, \qquad \Gamma_\varphi(x) := (x, \varphi(x)),$$

in the category of smooth Banach manifolds. Conclude that

$$T_\gamma \mathscr{L}^0_\varphi M = \{X \in \Gamma^0(\gamma^* TM) : X(1) = D\varphi(X(0))\}$$

holds for all $\gamma \in \mathscr{L}^0_\varphi M$.

Exercise 4.43 ([67]) Let (M, λ) be an exact symplectic manifold. For $\varphi \in \mathrm{Diff}(M)$ such that $\varphi^*\lambda = \lambda$, consider the ***twisted symplectic area functional***

$$\mathscr{A}^0_\varphi\colon \mathscr{L}_\varphi M \to \mathbb{R}, \qquad \mathscr{A}^0_\varphi(\gamma) := \int_0^1 \gamma^*\lambda.$$

1. Show that the differential of $\mathscr{A}^0_\varphi$ is given by

$$d\mathscr{A}^0_\varphi|_\gamma\colon T_\gamma \mathscr{L}_\varphi M \to \mathbb{R}, \qquad d\mathscr{A}^0_\varphi|_\gamma(X) = \int_0^1 d\lambda(X(t), \dot\gamma(t))dt$$

at $\gamma \in \mathscr{L}_\varphi M$. Conclude that $\gamma \in \mathrm{Crit}(\mathscr{A}_\varphi^0)$ if and only if $\gamma(0)$ is a fixed point of the symplectomorphism φ.
2. Can the twisted symplectic area functional $\mathscr{A}_\varphi^0$ be a Morse function in general?
3. How could one modify the twisted symplectic area functional $\mathscr{A}_\varphi^0$ to variationally detect noncontractible periodic Reeb orbits on lens spaces?

Remark 4.44 Let G be a symmetry group on an exact Hamiltonian system (M, λ, H). Then one can attempt to define a "moment" Rabinowitz action functional similar to the one in [73, Section 4.2]

$$\mathscr{A}^{H,\mu}\colon C^\infty(\mathbb{T}, M \times \mathfrak{g}) \times \mathbb{R} \to \mathbb{R}$$

by

$$\mathscr{A}^{H,\mu}(\gamma, \xi, \tau) := \mathscr{A}^H(\gamma, \tau) - \int_0^1 \mu(\xi(t))(\gamma(t))dt$$

for the corresponding moment map $\mu\colon \mathfrak{g} \to C^\infty(M)$. If G is not discrete, the critical points of the moment Rabinowitz action functional are given by

$$(\gamma, \xi, \tau) \in \mathrm{Crit}(\mathscr{A}^{H,\mu}) \qquad \Leftrightarrow \qquad \begin{cases} \gamma \in \mathscr{L}\Sigma, \\ \dot{\gamma}(t) = \tau X_H(\gamma(t)) - \widehat{\xi}(t)\ \forall t \in \mathbb{T}, \\ \mu^*(\gamma(t)) = 0\ \forall t \in \mathbb{T}, \end{cases}$$

where $\mu^*\colon M \to \mathfrak{g}^*$ denotes the comoment map defined by $\mu^*(x)(\xi) := \mu(\xi)(x)$. Thus, if we pass to the quotient, we consider periodic orbits on

$$\left(H^{-1}(0) \cap (\mu^*)^{-1}(0)\right)/G \cong H^{-1}(0)/G \cap (\mu^*)^{-1}(0)/G.$$

Under additional assumptions on the moment map μ, the quotient $(\mu^*)^{-1}(0)/G$ is the well-known Marsden–Weinstein quotient [137, Section 5.4], a compact symplectic manifold. This symplectic form cannot be exact because of compactness. Therefore, it seems difficult to investigate this scenario with the pictures from Exercise 4.24 in mind.

Problems

4.45 ([56, Section 5.2]) The aim of this problem is to show that without the stability assumption, Theorem 4.30 is false. Consider the path of matrices $(X_s)_{s\in\mathbb{R}}$ in $\mathfrak{sl}(2, \mathbb{R})$ given by

$$X_s := \begin{pmatrix} 1/2 & 0 \\ 0 & -1/2 \end{pmatrix} + s\begin{pmatrix} 0 & 1/2 \\ -1/2 & 0 \end{pmatrix}.$$

Let Γ be a cocompact lattice of $\mathrm{PSL}(2, \mathbb{R})$ acting without fixed points on the hyperbolic plane $\mathbb{H}$. Then $M := \Gamma \setminus \mathrm{PSL}(2, \mathbb{R})$ is a compact hyperbolic surface, and the unit cotangent bundle SM can be identified with $\Gamma \setminus \mathrm{PSL}(2, \mathbb{R})$.

1. Show that the flow

$$\phi^s : SM \to SM, \qquad \phi^s_t(\Gamma g) = \Gamma g e^{tX_s}$$

   is the Hamiltonian flow of the mechanical Hamiltonian

$$H(x, v) := \frac{1}{2} \|v\|_x^2$$

   with respect to the magnetic symplectic form ω_s for all $s \in \mathbb{R}$ on TM given by

$$\omega_s := d\lambda - s\pi^*\sigma.$$

2. Show that there exists $c \in \mathrm{PSL}(2, \mathbb{R})$ such that

$$c^{-1}X_1 c = \begin{pmatrix} 0 & 1 \\ 0 & 0 \end{pmatrix}$$

   holds. Define a map

$$f : SM \to SM, \qquad f(\Gamma g) := \Gamma g c.$$

   Show that this map satisfies

$$f \circ \phi^1_t = h_t \circ f \qquad \forall t \in \mathbb{R},$$

   where

$$h : SM \to SM, \qquad h_t(\Gamma g) := \Gamma g \begin{pmatrix} 1 & t \\ 0 & 1 \end{pmatrix}$$

   is the *horocycle flow*.
3. For every $k > 0$ set $\Sigma_k := H^{-1}(k)$. Show that the following statements are true:
   (a) For $k > 1/2$, the dynamics are after rescaling conjugate to the underlying geodesic flow and all level sets Σ_k are contact manifolds at the level of the universal covering.
   (b) At $k = 1/2$ we hit the horocycle flow.
   (c) For $k < 1/2$ all orbits are closed and contractible. Again, all level sets Σ_k are contact manifolds but with opposite orientation.

4. Show that the horocycle flow does not admit any periodic orbits and that $\Sigma_{1/2}$ is unstable. *Hint:* Look at [68, Lemma 11.28] if you get stuck.
5. Compute *Mañé's critical value* defined by

$$c(\widetilde{H}) := \inf_{\theta} \sup_{q \in \widetilde{M}} \widetilde{H}(q, \theta_q),$$

where the infimum is taken over all 1-forms θ on the universal covering manifold $\Pi\colon \widetilde{M} \to M$ such that $d\theta = \Pi^*\sigma$.

4.46 ([56, Proposition 2.6 (a)]) Let $(\Sigma_\sigma, \omega_\sigma, \alpha_\sigma)_{\sigma\in[0,1]}$ be a homotopy of stable regular energy hypersurfaces. Show that there exists a smooth family $(\psi_\sigma)_{\sigma\in[0,1]}$ of smooth embeddings $\psi_\sigma\colon (-\varepsilon, \varepsilon) \times \Sigma_\sigma \hookrightarrow M$ satisfying

$$\psi_\sigma|_{\{0\}\times\Sigma_\sigma} = \iota_{\Sigma_\sigma}\colon \Sigma_\sigma \hookrightarrow M \qquad \text{and} \qquad \psi_\sigma^*\omega = \omega_\sigma + d(r\alpha_\sigma)$$

for some $\varepsilon > 0$ sufficiently small. Such a smooth family is called a ***stable tubular neighbourhood of*** $(\Sigma_\sigma, \omega_\sigma, \alpha_\sigma)_{\sigma\in[0,1]}$.

4.47 Give a simplified proof of Theorem 4.30 in the case of a homotopy of contact hypersurfaces. You may want to review the proofs given in [77, Section 7.6] and in [35]. What remains unchanged and what becomes easier in the case of a contact form?

4.48 (Morse Theory) Let M be a compact smooth manifold without boundary. A ***Morse function on*** M is a smooth function $h \in C^\infty(M)$ such that every critical point $x_0 \in \mathrm{Crit}(h)$ is nondegenerate, that is, $dh(x_0) = 0$ for the differential $dh \in \Omega^1(M)$ of h, and the locally defined Hessian

$$\mathrm{Hess}_h(x_0) = \left(\frac{\partial^2 h}{\partial x_i \partial x_j}(x_0)\right)_{i,j=1,\dots,n}$$

satisfies $\det \mathrm{Hess}_h(x_0) \neq 0$.

1. Show that there exists a coordinate chart about $x_0 \in \mathrm{Crit}(f)$ such that locally

$$h(x) = h(x_0) + \frac{1}{2}\mathrm{Hess}_h(x_0)\|x\|^2.$$

Conclude that nondegenerate critical points are isolated and thus there are only finitely many critical points of a Morse function h on the compact manifold M. *Hint:* This result can be proven using Moser's Trick for proving the Darboux Theorem [129, Theorem 22.13], see [9, Lemma 3.2.3].

2. Fix a Riemannian metric m on M and consider the vector field

$$X := \operatorname{grad} h / \| \operatorname{grad} h \|_m^2 \in \mathfrak{X}(M \setminus \{x_1, \ldots, x_k\}),$$

where $x_1, \ldots, x_k$ denote the critical points of h. Show that $h(\phi_t(x)) = h(x) + t$ holds whenever $\phi_t(x)$ is defined for the flow ϕ of X. *Hint:* See [63, Section XI.3].
3. Let $[a, b] \subseteq \mathbb{R}$ be a compact interval containing no critical values of h. Show that ϕ restricts to a diffeomorphism from $[0, b-a] \times h^{-1}(a)$ to $h^{-1}([a, b])$. *Hint:* See [63, Section XI.3].
4. Assume that h admits precisely two critical points. Show that then M is diffeomorphic to the n-sphere $\mathbb{S}^n$. Is it true that this homeomorphism can also be chosen to be a diffeomorphism? Explain your answer!

Part II

Hamiltonian Floer Homology

Hamiltonian Floer homology is the semi-infinite dimensional Morse homology associated with the symplectic action functional introduced by Andreas Floer in the mid eighties in order to prove the Arnold conjecture. For an introduction and recent developments see the survey article [5]. For a review of Morse homology with a view towards Floer homology either see the lecture notes [75] or the book [170]. For more technical details of the construction of Floer homology we suggest [24] or [136]. Floer homology and its flavours are an active area of research. The aim of this part is to introduce the relevant technicalities in order to construct twisted Rabinowitz–Floer homology. Hence we introduce Floer homology in the Morse–Bott setting as opposed to the more standard Morse setting. Moreover, we introduce the basic language of Polyfold Theory developed by Hofer–Wysocki–Zehnder [109] to deal with abstract transversality, compactness and gluing problems for moduli spaces. This enables the reader to get in contact with a new viewpoint of nonlinear functional analysis and is a preparation for diving deeper into recent literature.

5 Morse–Bott Homology

Morse–Bott homology is a generalisation of Morse homology to functions with degenerate critical points [111]. See [51, Appendix A] for a brief introduction via the cascade approach and [73, Appendix A] for a more extensive treatment. Morse–Bott functions occur in the presence of symmetries. More precisely, let G be a Lie group acting on a smooth manifold M. If $f \in C^\infty(M)$ is G-invariant, that is, $f(g \cdot x) = f(x)$ holds for all $g \in G$ and $x \in M$, then $\mathrm{Crit}(f)$ is also G-invariant. In particular, f does not have isolated critical points, in general. However, the presence of symmetries usually simplifies the explicit computation of Morse homology and occurs throughout Physics, as we have already seen in Part I. For an explicit example, consider the embedded torus $\mathbb{T}^2 \subseteq \mathbb{R}^3$ lying on the (x, y)-plane with the standard height function $h(x, y, z) := z$ as illustrated in Fig. 5.1. The function h is invariant under rotations of $\mathbb{T}^2$ around the z-axis.

Exercise 5.1 (Morse Homology Warm-Up, [75]) Let M be a smooth manifold without boundary and let $h \in C^\infty(M)$ be a smooth function. We write

$$\mathrm{Crit}(h) := \{x \in M : dh(x) = 0\}$$

for the critical set of h, where $dh \in \Omega^1(M)$ denotes the differential of h. Pick a Riemannian metric m on M and denote by ∇ the corresponding Levi–Civita connection [130, Chapter 5]. Define the Hessian of h with respect to ∇ by

$$\mathrm{Hess}_h^\nabla(X, Y) := \nabla_X(dh)(Y) \qquad \forall X, Y \in \mathfrak{X}(M).$$

For a critical point $x \in \mathrm{Crit}(h)$, the pointwise Hessian $\mathrm{Hess}_h(x) \colon T_xM \times T_xM \to \mathbb{R}$ is a well-defined symmetric bilinear form independent of the choice of torsion-free connection ∇. A function $h \in C^\infty(M)$ is said to be a ***Morse function***, if for every critical point the pointwise Hessian is nondegenerate. The number of negative eigenvalues of the pointwise Hessian at a critical point $x \in \mathrm{Crit}(h)$ counted with

Y. Bähni, *Lectures on Twisted Rabinowitz-Floer Homology*, Pathways in Mathematics, https://doi.org/10.1007/978-3-032-10673-5_5

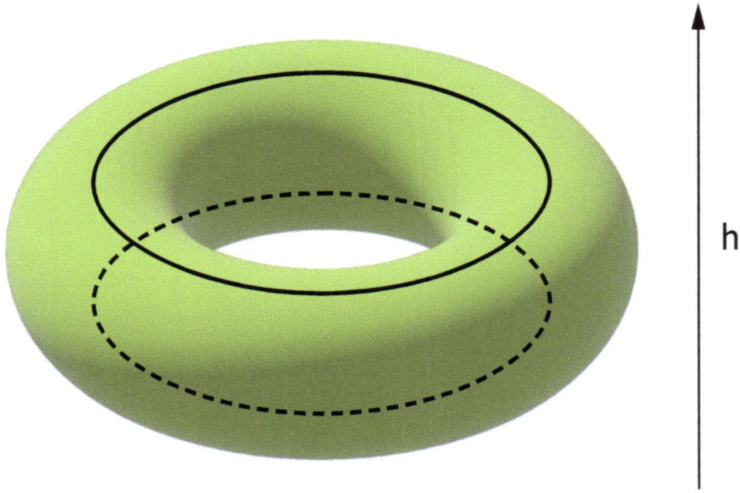

Fig. 5.1 The rotationally invariant embedded torus $\mathbb{T}^2 \subseteq \mathbb{R}^3$

multiplicity is defined to be the Morse index $\text{ind}_h(x)$ of x. Set

$$\text{Crit}_k(h) := \{x \in \text{Crit}(h) : \text{ind}_h(x) = k\} \qquad \forall k \in \mathbb{Z}.$$

Any Morse function h gives rise to a nonnegative chain complex, called the ***Morse–Smale–Witten chain complex*** $(\text{CM}_*(h), \partial)$, where

$$\text{CM}_k(h) := \text{Crit}_k(h) \otimes \mathbb{Z}_2$$

is the free $\mathbb{Z}_2$-vector space generated by the critical points of Morse index k, together with the boundary map $\partial_k\colon\ \text{CM}_k(h) \to \text{CM}_{k-1}(h)$ defined on generators by

$$\partial x^- := \sum_{x^+ \in \text{Crit}_{k-1}(h)} \#_2 \mathscr{M}(x^-, x^+) x^+,$$

where $\mathscr{M}(x^-, x^+)$ denotes the moduli space of all unparametrised gradient flow lines from x^- to x^+, that is, all equivalence classes $[u]$ of maps $u \in C^\infty(\mathbb{R}, M)$ satisfying

$$\partial_s u(s) = -\operatorname{grad}_m h(u(s)) \qquad \text{and} \qquad \lim_{s \to \pm\infty} u(s) = x^\pm,$$

for all $s \in \mathbb{R}$ and the equivalence class $[u] \in \mathscr{M}(x^-, x^+)$ is given by reparametrisation

$$[u] := \{s \mapsto u(s + r) : r \in \mathbb{R}\} \subseteq C^\infty(\mathbb{R}, M).$$

Here $\operatorname{grad}_m h \in \mathfrak{X}(M)$ denotes the gradient of h implicitly defined by

$$m(\operatorname{grad}_m h, \cdot) = dh(\cdot).$$

Finally, define the ***Morse homology*** $\mathrm{HM}_*(h)$ ***of*** h ***with coefficients in*** $\mathbb{Z}_2$ by setting

$$\mathrm{HM}_k(h) := \frac{\ker \partial_k}{\operatorname{im} \partial_{k+1}} \qquad \forall k \in \mathbb{Z}.$$

1. Explain why we only count unparametrised gradient flow lines.
2. Show that the pointwise Hessian Hess_h of a smooth function h is a well-defined symmetric bilinear form for any torsion-free connection ∇ in TM.
3. On the sphere $\mathbb{S}^2 \subseteq \mathbb{R}^3$ depicted in Fig. 5.2a and the heart in Fig. 5.2b consider the Morse function h given by the height. Compute the Morse homology $\mathrm{HM}_*(h)$ by first drawing every critical point of h and then all negative gradient flow lines connecting critical points with Morse index difference equal to one.
4. Consider the upright torus $\mathbb{T}^2 = \mathbb{S}^1 \times \mathbb{S}^1 \subseteq \mathbb{R}^3$ as in Fig. 5.2c together with the standard height function h. What are analytical issues in computing $\mathrm{HM}_*(h)$?
5. Compute the Morse homology $\mathrm{HM}_*(h)$ of the real projective plane $\mathbb{RP}^2 = \mathbb{S}^2/\mathbb{Z}_2$. *Hint:* Find a Morse function on $\mathbb{S}^2$ identifying antipodal points z and $-z$ as in Fig. 5.2d.
6. Let $m \geq 2$ be an integer and $k_1, \ldots, k_n \in \mathbb{Z}$ coprime to m. Compute the Morse homology of the lens spaces $\mathbb{S}^{2n-1}/\mathbb{Z}_m$, where the $\mathbb{Z}_m$-action is induced by the rotation $\varphi\colon \mathbb{S}^{2n-1} \to \mathbb{S}^{2n-1}$, $\varphi(z_1, \ldots, z_n) := \left(e^{2\pi i k_1/m} z_1, \ldots, e^{2\pi i k_n/m} z_n\right)$.

5.1 $\mathbb{Z}_2$-Coefficients

Let (M, ω) be a symplectic manifold without boundary and denote by $\Lambda M \subseteq \mathscr{L}M$ the component of all contractible free loops in M. For $H \in C^\infty(M \times \mathbb{T})$ define by

$$\mathscr{A}_H\colon \Lambda M \to \mathbb{R}, \qquad \mathscr{A}_H(\gamma) := \int_{\mathbb{D}} \bar{\gamma}^*\omega - \int_0^1 H_t(\gamma(t))dt \tag{5.1}$$

the ***symplectic action functional***, where $\bar{\gamma} \in C^\infty(\mathbb{D}, M)$ is a filling disc for γ, that is, we have that

$$\bar{\gamma}(e^{2\pi i t}) = \gamma(t) \qquad \forall t \in \mathbb{T}$$

with $\mathbb{D} := \{z \in \mathbb{C} : |z| = 1\}$ denoting the closed unit disc.

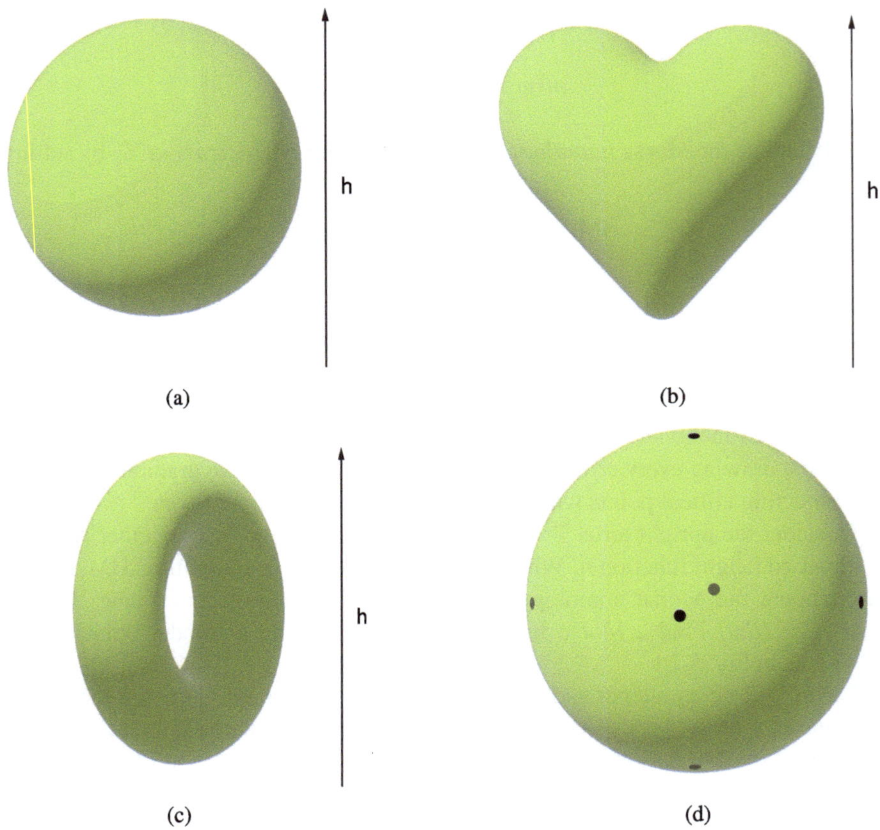

Fig. 5.2 (**a**) A sphere $\mathbb{S}^2 \subseteq \mathbb{R}^3$. (**b**) A deformed sphere resembling a heart. (**c**) An upright torus $\mathbb{T}^2 \subseteq \mathbb{R}^3$. (**d**) A $\mathbb{Z}_2$-symmetric sphere

Exercise 5.2 (Symplectic Asphericity) A symplectic manifold (M, ω) is said to be ***symplectically aspherical***, if

$$\int_{\mathbb{S}^2} f^*\omega = 0 \qquad \forall f \in C^\infty(\mathbb{S}^2, M).$$

1. Show that a connected symplectic manifold is symplectically aspherical if and only if $[\omega]|_{\pi_2(M)} = 0$, where $[\omega] \in \mathrm{H}^2_{\mathrm{dR}}(M; \mathbb{R})$ denotes the cohomology class of the closed form ω.
2. Assume that (M, ω) is symplectically aspherical. Show that then the symplectic action functional (5.1) is well-defined.

Example 5.3 Every exact symplectic manifold is symplectically aspherical by Stokes' theorem, and the magnetic torus (Example 4.6) is also symplectically aspherical since we have $\pi_2(T^*\mathbb{T}^n) = 0$ or the magnetic symplectic form admits a primitive.

The next exercise motivates us to be interested in a generalisation of Morse homology.

Exercise 5.4 Consider the symplectic action functional $\mathscr{A}_H$ on a symplectically aspherical symplectic manifold (M, ω).

1. Show that the critical points $\mathrm{Crit}(\mathscr{A}_H)$ consist precisely of all contractible loops $\gamma \in C^\infty(\mathbb{T}, M)$ that satisfy the time-dependent Hamiltonian equation

$$\dot{\gamma}(t) = X_{H_t}(\gamma(t)) \qquad \forall t \in \mathbb{T}. \tag{5.2}$$

2. Explain why in the time-independent case, that is, the case of an autonomous Hamiltonian function $H \in C^\infty(M)$, critical points of the symplectic action functional $\mathscr{A}_H$ are never isolated if there exists a nonconstant solution to (5.2).

Since we are interested in periodic orbits on regular energy surfaces and they are nonisolated critical points of the symplectic action functional $\mathscr{A}_H$, it is hopeless to construct a corresponding Morse homology as done by Andreas Floer in the mid-80s. We start by describing the construction of the Morse–Bott homology with coefficients in $\mathbb{Z}_2$ in a finite-dimensional setting.

Definition 5.5 (Morse–Bott Function, [42, Lecture 3]) Let M be a smooth manifold. A ***Morse–Bott function on*** M is defined to be a function $f \in C^\infty(M)$ with

1. $\mathrm{Crit}(f) \subseteq M$ is an embedded submanifold.
2. $T_x \mathrm{Crit}(f) = \ker \mathrm{Hess}_f(x)$ for all $x \in \mathrm{Crit}(f)$.

Remark 5.6 The second assumption on a Morse–Bott function is crucial for proving the Morse–Bott lemma [29, Lemma 3.51] about a local normal form, an analogue of the Morse lemma. The Morse–Bott Lemma will be a key technical ingredient in establishing the exponential decay of gradient flow lines in Sect. 5.2. Generally speaking, the critical points of a Morse–Bott function f are non-degenerate in the normal direction to $\mathrm{Crit}(f)$.

Exercise 5.7 (Morse Function) Explain why every Morse function on a compact smooth manifold is a Morse–Bott function.

Example 5.8 (Atijah–Guillemin–Sternberg) Let (M, ω) be a compact and connected symplectic manifold, and let $\mathbb{T}^m \to \mathrm{Symp}(M, \omega), \theta \mapsto \psi_\theta$ be a Hamiltonian torus action generated by a comoment map $\mu^* \colon M \to \mathbb{R}^m$ defined in Remark 4.44.

Then by [137, Theorem 5.5.1] the image $\mu^*(M) \subseteq \mathbb{R}^m$ of the comoment map is a convex polytope. More precisely, the set of fixed points of the action is a finite union of pairwise disjoint connected symplectic submanifolds $C_1, \ldots, C_N$, the moment map is constant on each connected component C_j of the set of fixed points, and the image of the comoment map μ^* is the convex hull of the points $\mu^*(C_j) \in \mathbb{R}^m$. The proof is based on the following observation. By [137, Lemma 5.5.8] for every $\theta \in \mathbb{R}^m$ the function

$$H_\theta \colon M \to \mathbb{R}, \qquad H_\theta(x) := \mu^*(x)(\theta)$$

is a Morse–Bott function. The resulting convex polytope is studied in the theory of symplectic toric manifolds [137, Exercise 5.4.18], an interesting class of symplectic manifolds with a view toward algebraic geometry [64, Part XI]. The toric domains also play an important role in the mathematical formulation of physics, as in the recent papers [74] and [52].

Exercise 5.9 (Bott's Perfect Morse Function) On the odd-dimensional sphere

$$\mathbb{S}^{2n-1} := \left\{ (z_1, \ldots, z_n) \in \mathbb{C}^n : \sum_{j=1}^{n} |z_j|^2 = 1 \right\}$$

consider ***Bott's perfect Morse function*** defined in [29, Example 3.7] by

$$f \colon \mathbb{S}^{2n-1} \to \mathbb{R}, \qquad f(z_1, \ldots, z_n) := \sum_{j=1}^{n} j |z_j|^2.$$

This function is invariant under the natural $\mathbb{S}^1$-action on $\mathbb{S}^{2n-1}$ given by

$$e^{i\theta} \cdot (z_1, \ldots, z_n) := \left(e^{i\theta} z_1, \ldots, e^{i\theta} z_n \right),$$

and thus descends to the quotient $\mathbb{CP}^{n-1} := \mathbb{S}^{2n-1}/\mathbb{S}^1$. In particular, f is invariant under the rotation

$$\varphi \colon \mathbb{S}^{2n-1} \to \mathbb{S}^{2n-1}, \qquad \varphi(z_1, \ldots, z_n) := \left(e^{2\pi i k_1/m} z_1, \ldots, e^{2\pi i k_n/m} z_n \right),$$

for $m \geq 2$ an integer and $k_1, \ldots, k_n \in \mathbb{Z}$ coprime to m. Since f is invariant under the $\mathbb{S}^1$-action on $\mathbb{S}^{2n-1}$, it is also invariant under the induced $\mathbb{Z}_m$-action of φ and thus descends to the lens space $\mathbb{S}^{2n-1}/\mathbb{Z}_m$. Show that f is a Morse–Bott function on the sphere $\mathbb{S}^{2n-1}$ by explicitly computing the critical manifold $\operatorname{Crit}(f)$.

Next, we describe the construction of Morse–Bott homology associated with a Morse–Bott function generalising the construction of Morse homology reviewed in

Exercise 5.1. Let (M, m) be a compact Riemannian manifold and $f \in C^\infty(M)$ be a Morse–Bott function. Choose an additional Morse function $h \in C^\infty(\mathrm{Crit}(f))$ and a Riemannian metric m_0 on $\mathrm{Crit}(f)$ such that (h, m_0) is a Morse–Smale pair, that is, the stable and unstable manifolds intersect transversally. This is actually generically true by perturbing the metric m_0. Using Theorem 5.41 below, one can define a nonnegative $\mathbb{Z}$-graded chain complex $(\mathrm{CM}_*(f), \partial_*)$ of $\mathbb{Z}_2$-vector spaces by

$$\mathrm{CM}_k(f) := \mathrm{Crit}_k(h) \otimes \mathbb{Z}_2, \quad \mathrm{Crit}_k(h) := \left\{x \in \mathrm{Crit}(h) : \mathrm{ind}_f(x) + \mathrm{ind}_h(x) = k\right\},$$

for all $k \in \mathbb{Z}$, where ind denotes the Morse index, with boundary operator

$$\partial_k\colon \mathrm{CM}_k(f) \to \mathrm{CM}_{k-1}(f), \qquad \partial_k x^- := \sum_{x^+ \in \mathrm{Crit}_{k-1}(h)} \#_2 \mathcal{M}(x^-, x^+) x^+,$$

where $\mathcal{M}(x^-, x^+)$ denotes the abstractly perturbed moduli space $\mathcal{M}(x^-, x^+)$ of unparametrised negative gradient flow lines with cascades from x^- to x^+. A ***gradient flow line with cascades*** starts with a negative gradient flow line of h on $\mathrm{Crit}(f)$ that converges at its negative asymptotic end to a critical point of h. In finite time, this gradient flow line meets the asymptotic end of a gradient flow line of the Morse–Bott function f. We refer to this gradient flow line of f as the first cascade. The cascade converges again at its positive end to a point in $\mathrm{Crit}(f)$. There, the flow continues with the gradient flow of h on $\mathrm{Crit}(f)$. After finite time, a second cascade might appear, but having passed through finitely many cascades, we finally end up with a gradient flow line of h, which we follow until it converges asymptotically to a critical point of h. Gradient flow lines with zero cascades correspond to ordinary Morse flow lines of h on the manifold $\mathrm{Crit}(f)$. We compute

$$\begin{aligned}
(\partial_k \circ \partial_{k+1}) x^- &= \sum_{x^+ \in \mathrm{Crit}_{k-1}(h)} \sum_{x \in \mathrm{Crit}_k(h)} \#_2 \mathcal{M}(x^-, x) \#_2 \mathcal{M}(x, x^+) x^+ \\
&= \sum_{x^+ \in \mathrm{Crit}_{k-1}(h)} \#_2 \partial \mathcal{M}(x^-, x^+) x^+ \\
&= 0
\end{aligned}$$

for all $x^- \in \mathrm{Crit}_{k+1}(h)$ and $k \in \mathbb{Z}$ using the identification

$$\partial \mathcal{M}(x^-, x^+) \cong \coprod_{x \in \mathrm{Crit}(h)} \mathcal{M}(x^-, x) \times \mathcal{M}(x, x^+).$$

In fact, the 1-dimensional moduli space of all unparametrised negative gradient flow lines with cascades $\mathcal{M}(x^-, x^+)$ is a compact smooth manifold and thus by the classification of all 1-dimensional smooth manifolds [129, Problems 15-13 and 15-14], we find that $\mathcal{M}(x^-, x^+)$ is a disjoint union of copies of the closed interval $[0, 1]$ or the circle $\mathbb{S}^1$. Consequently, the boundary $\partial \mathcal{M}(x^-, x^+)$ counted modulo two is zero as there are always an even number of boundary points. Hence we can define the ***Morse–Bott homology of*** f by

$$\mathrm{HM}_k(f) := \frac{\ker \partial_k}{\operatorname{im} \partial_{k+1}}, \qquad \forall k \in \mathbb{Z}.$$

As our notation suggests, $\mathrm{HM}_*(f)$ is independent of any auxiliary data up to natural isomorphisms. In particular, as every Morse function is a Morse–Bott function by Exercise 5.7, Morse–Bott homology is canonically isomorphic to the ordinary Morse homology of M, and thus to the singular homology of M with coefficients in $\mathbb{Z}_2$, that is, $\mathrm{HM}_*(f) \cong \mathrm{H}_*(M;\mathbb{Z}_2)$ by [73, Corollary A.19] combined with Problem 5.46.

Exercise 5.10 (Morse Homology) Show that Morse–Bott homology does generalise Morse homology. More precisely, let M be a compact smooth manifold. Can you pick a Morse–Bott function f on M such that $\mathrm{HM}_*(f)$ is just the Morse homology?

Exercise 5.11 Show that the natural isomorphism $\mathrm{HM}_*(f) \cong \mathrm{H}_*(M;\mathbb{Z}_2)$ is, in general, wrong if the manifold is not compact.

The following example illustrates the computation of Morse–Bott homology in a very geometric setting, similar to the deformed sphere resembling a heart in Morse homology.

Example 5.12 (The Teapot) Consider the surface of genus one that resembles a teapot as depicted in Fig. 5.3. Then the standard height function f is a Morse–Bott function on the teapot with critical manifold being the disjoint union of a circle $\mathbb{S}^1$ and eight nondegenerate critical points. On the critical manifold $\mathbb{S}^1$ we consider the standard height function h. We denote the three maxima by x_1, x_2 and x_3 with Morse–Bott index equal to two. Next we denote by y_1, y_2, y_3, y_4 and y_5 the critical points of Morse–Bott index equal to one. Lastly, we denote the two minima by z_1 and z_2 with index zero, as shown in Fig. 5.4. By drawing the negative gradient flow lines with cascades as in Fig. 5.5, we get

$$\partial_2 x_1 = y_1, \qquad \partial_2 x_2 = y_1 + y_3, \qquad \partial_2 x_3 = y_3,$$

as we only count modulo two. Similarly, we compute

$$\partial_1 y_2 = z_1 + z_2 \qquad \text{and} \qquad \partial_1 y_1 = \partial_1 y_3 = \partial_1 y_4 = \partial_1 y_5 = 0.$$

One checks that $\partial_1 \circ \partial_2 = 0$ holds and thus we get the chain complex

$$0 \xrightarrow{\;0\;} \mathbb{Z}_2^3 \xrightarrow{\begin{pmatrix}1&1&0\\0&0&0\\0&1&1\\0&0&0\\0&0&0\end{pmatrix}} \mathbb{Z}_2^5 \xrightarrow{\begin{pmatrix}0&1&0&0&0\\0&1&0&0&0\end{pmatrix}} \mathbb{Z}_2^2 \xrightarrow{\;0\;} 0$$

Fig. 5.3 The deformed surface of genus one resembling a teapot with the standard height function f

Fig. 5.4 The critical manifold $\mathrm{Crit}(f)$ for the standard height function f on the teapot as well as the standard height function on the circle

Hence we compute for the homology

$$\mathrm{HM}_0(f) = \frac{\ker \partial_0}{\operatorname{im} \partial_1} = \frac{\mathbb{Z}_2^2}{\mathbb{Z}_2} = \mathbb{Z}_2,$$

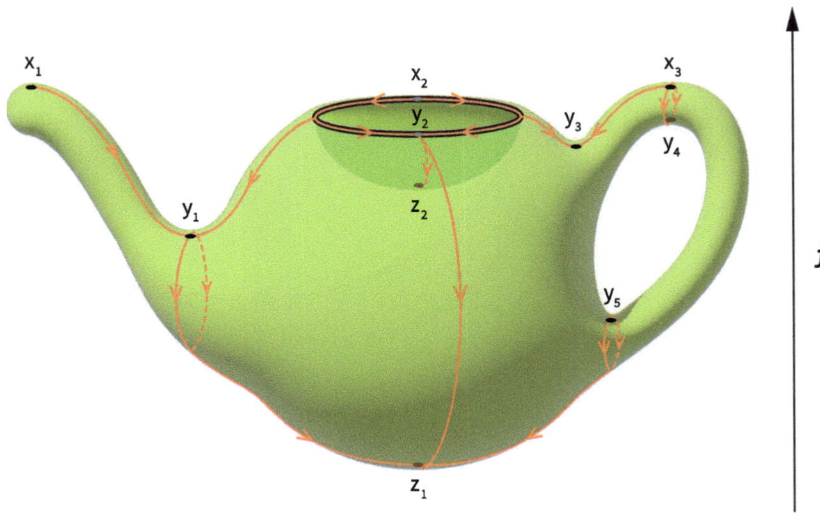

Fig. 5.5 Negative gradient flow lines for the teapot

$$\mathrm{HM}_1(f) = \frac{\ker \partial_1}{\operatorname{im} \partial_2} = \frac{\mathbb{Z}_2^4}{\mathbb{Z}_2^2} = \mathbb{Z}_2^2,$$

$$\mathrm{HM}_2(f) = \frac{\ker \partial_2}{\operatorname{im} \partial_3} = \mathbb{Z}_2.$$

This is precisely the singular homology of the torus $\mathbb{T}^2$ with coefficients in $\mathbb{Z}_2$!

Exercise 5.13 (Broken Teapot) Assume that the teapot from Example 5.12 fell on the floor and the handle broke off as depicted in Fig. 5.6. Compute the corresponding Morse–Bott homology of the standard height function f.

Exercise 5.14 (Lens Spaces) Let $f \in C^\infty(\mathbb{S}^{2n-1})$ be Bott's perfect Morse function from Exercise 5.9 with critical manifold being the disjoint union of n copies of $\mathbb{S}^1$. Consider the free $\mathbb{Z}_m$-action on the odd-dimensional sphere $\mathbb{S}^{2n-1}$ induced by

$$\varphi\colon \mathbb{S}^{2n-1} \to \mathbb{S}^{2n-1}, \qquad \varphi(z_1, \dots, z_n) := \left(e^{2\pi i k_1/m} z_1, \dots, e^{2\pi i k_n/m} z_n\right)$$

for some integer $m \geq 2$ and $k_1, \dots, k_n \in \mathbb{Z}$ coprime to m.

1. Choose the standard height function h on every copy of $\mathbb{S}^1$ and determine the index of every critical point in $\mathrm{Crit}(h)$.
2. Compute the Morse–Bott homology $\mathrm{HM}_*(f)$ of f. What do you expect?
3. Compute the singular homologies $\mathrm{H}_*(\mathbb{S}^{2n-1}/\mathbb{Z}_m; \mathbb{Z})$ and $\mathrm{H}_*(\mathbb{S}^{2n-1}/\mathbb{Z}_m; \mathbb{Z}_2)$.

Fig. 5.6 A broken teapot with the standard height function f

4. Compute the Morse–Bott homology $\mathrm{HM}_*(f/\mathbb{Z}_m)$ for Bott's perfect Morse function in Example 5.9. What do you expect?

Why did we define the Morse–Bott homology only with coefficients in $\mathbb{Z}_2$? As is well known, Morse homology can be defined with integer coefficients via *coherent orientations* as in [72], [170] or [66]. See also Problem 5.46. By [30, Corollary 1.2] the chain complex induced by f via cascades and the Morse–Smale–Witten chain complex induced by a sufficiently small perturbation of f coincide up to a choice of sign. Thus, the Morse homology Theorem immediately gives the desired isomorphism. However, by perturbing the Morse–Bott function f one loses the symmetry properties. In some situations those symmetries facilitate the computation of the homology, as is illustrated by the case of lens spaces in Exercise 5.14. In what follows, we make the **wrong** assumption that the Morse–Bott homology $\mathrm{HM}_*(f;\mathbb{Z})$ with coefficients in $\mathbb{Z}$ can always be constructed for a Morse–Bott function f on a compact manifold M and is independent of the choice of the Morse–Bott function. In particular, we would have that

$$\text{“}\mathrm{HM}_*(f;\mathbb{Z})\text{"} \cong \mathrm{H}_*(M;\mathbb{Z}).$$

As in [161], consider the Morse–Bott function

$$f\colon \mathbb{RP}^5 \to \mathbb{R}, \qquad f([x_0,x_1,x_2,x_3,x_4,x_5]) := \frac{-x_4^2+x_5^2}{x_1^2+\cdots+x_5^2}.$$

The Morse–Bott function f admits an isolated minimum at $[0,0,0,0,1,0]$, a maximum at $[0,0,0,0,0,1]$, and a critical submanifold $\{[x_0,x_1,x_2,x_3,0,0]\}$ of index one diffeomorphic to $\mathbb{RP}^3$. Choose the Morse function

$$h\colon \mathbb{RP}^3 \to \mathbb{R}, \qquad h([x_0, x_1, x_2, x_3]) := x_0^2 + 2x_1^2 + 3x_2^2 + 4x_3^2$$

with a single critical point of each index similar to Exercise 5.1. Using this Morse function, we get $\mathrm{H}_2(f; \mathbb{Z}) = \mathbb{Z}_2 \neq 0 = \mathrm{H}_2(\mathbb{RP}^5; \mathbb{Z})$. Therefore, we cannot define Morse–Bott homology $\mathrm{HM}_*(f; \mathbb{Z})$ with coefficients in $\mathbb{Z}$ in general without further modifications! But what could be the right assumptions?

Definition 5.15 (Poincaré Polynomial and Morse–Bott Polynomial) Let $R = \mathbb{Z}_2$ or $R = \mathbb{Z}$ and let M be a compact smooth manifold. Given a Morse–Bott function $f \in C^\infty(M)$, define the ***Morse–Bott polynomial***

$$P_f(z; R) := \sum_{C \subseteq M \text{ critical submanifold}} P_C(z; R) z^{\mathrm{ind}_f(C)},$$

where the ***Poincaré polynomial*** is given by

$$P_C(z; R) := \sum_{k=0}^{\dim C} \operatorname{rank} \mathrm{H}_k(C; R) z^k.$$

Proposition 5.16 (Morse–Bott Inequalities, [149, Corollary 2.46]) *Let $R = \mathbb{Z}_2$ or $R = \mathbb{Z}$ and let $f \in C^\infty(M)$ be a Morse–Bott function on a compact smooth manifold such that the negative normal bundle $E^-(C)$ is R-orientable for every critical submanifold $C \subseteq \mathrm{Crit}(f)$. Then there exists $Q \in \mathbb{Z}[z]$ with nonnegative coefficients such that*

$$P_f(z; R) = P_M(z; R) + (1 + z) Q(z). \tag{5.3}$$

Remark 5.17 The Morse–Bott inequalities (5.3) stated in [29, Theorem 3.53] are unfortunately stated with a wrong orientability assumption, later clarified in [161]. It is not enough that all critical submanifolds are orientable! Either the negative normal bundles are orientable or one uses local coefficients [31] in the not necessarily orientable negative normal bundles for the general case.

Exercise 5.18 State the Morse–Bott inequalities for a Morse function. What can you say about the number of critical points of a Morse function?

Exercise 5.19 From Proposition 5.16, deduce a statement about the ***Euler characteristic of*** M defined by

$$\chi(M) := \sum_{k=0}^{\dim M} (-1)^k \operatorname{rank} \mathrm{H}_k(M; R).$$

Exercise 5.20 Let $f \in C^\infty(\mathbb{S}^{2n-1})$ be Bott's perfect Morse from Exercise 5.9.

1. Show that the negative normal bundles are $\mathbb{Z}$-orientable and explicitly compute the polynomial Q from Proposition 5.16.
2. Calculate Q from Proposition 5.16 for $f/\mathbb{Z}_m \in C^\infty(\mathbb{S}^{2n-1}/\mathbb{Z}_m)$.
3. Can you sketch the computation of the Morse–Bott homology with integer coefficients for $f/\mathbb{Z}_m$ such that it coincides with the singular homology of the lens space $\mathbb{S}^{2n-1}/\mathbb{Z}_m$ with coefficients in $\mathbb{Z}$?

In the next section, we discuss some analytical technicalities concerning the definition of Morse–Bott homology with coefficients in $\mathbb{Z}_2$. We will highlight a global approach for studying moduli spaces also suited for the semi-infinite-dimensional case as in Floer theory via polyfold theory [69], complementing [73, Appendix A].

5.2 A Polyfold Setup for Morse–Bott Homology

The classical approach to establishing generic transversality results for Floer theories is via the Sard–Smale Theorem [136, Theorem A.5.1]. The idea is to represent the moduli space of negative gradient flow lines as the zero set of a suitable Fredholm section. This is the geometric approach to regularisation, as explained in [69, Section 3.1]. You can explore this approach yourself in Problem 5.47. Another more algebraic approach to abstract perturbations, called the virtual approach via virtual fundamental cycles, is by Kuranishi structures of Fukaya–Oh–Ohta–Ono [81], [82] and Pardon [153]. The main motivation for the introduction of a new notion of differentiability in an infinite-dimensional setting is due to the following observation, as described in [78, Introduction]. In the construction of Morse and Floer homology, negative gradient flow lines play a crucial role. These locally lie in the mapping space $\mathrm{Map}(\mathbb{R}, V)$ for some suitable vector space V. As the differential equation is invariant under the translation

$$\tau\colon \mathbb{R} \times \mathrm{Map}(\mathbb{R}, V) \to \mathrm{Map}(\mathbb{R}, V), \qquad \tau(t, u)(s) := u(s + t), \tag{5.4}$$

we must quotient out this reparametrisation action in order to achieve compactness modulo breaking of the relevant moduli spaces. However, after equipping the mapping space with some topology, such as the compact-open topology or the norm topology, the above ***shift map*** (5.4) has terrible differentiability properties.

Exercise 5.21 Explain why the shift map (5.4) is not differentiable in the classical sense. In order to do that, pick some explicit vector space V.

By [78, Theorem 6.2] and Problem 5.44 the shift map is smooth in a new sense.

Definition 5.22 (Scale Structure, [109, Definition 1.1.1]) A ***scale structure*** on a Banach space E is a decreasing sequence

$$E =: E_0 \supseteq E_1 \supseteq E_2 \supseteq \dots$$

of Banach spaces such that the inclusion $E_{k+1} \hookrightarrow E_k$ is compact for every $k \in \mathbb{N}_0$ and such that $E_\infty := \bigcap_{k=0}^{\infty} E_k$ is dense in every E_k. A Banach space E together with a scale structure (E_k) is called a ***scale Banach space***.

The following fundamental example underlies Morse and Floer homology.

Exercise 5.23 (Weighted Sobolev Spaces, [78, Example 3.9], [69, Lemma 4.10]) Fix an odd monotone cutoff function $\beta \in C^\infty(\mathbb{R}, [-1, 1])$ with

$$\beta(s) = \begin{cases} 1 & s \geq 1, \\ -1 & s \leq -1. \end{cases}$$

Fix $\delta > 0$ and define

$$\gamma_\delta \colon \mathbb{R} \to \mathbb{R}, \qquad \gamma_\delta(s) := e^{\delta \beta(s) s}$$

as depicted in Fig. 5.7. For $p \in [1, +\infty)$ and $k \in \mathbb{N}_0$ define the ***Sobolev spaces with weight*** δ by

$$W^{k,p}_\delta(\mathbb{R}, \mathbb{R}^n) := \{u \in W^{k,p}(\mathbb{R}, \mathbb{R}^n) : \gamma_\delta u \in W^{k,p}(\mathbb{R}, \mathbb{R}^n)\}.$$

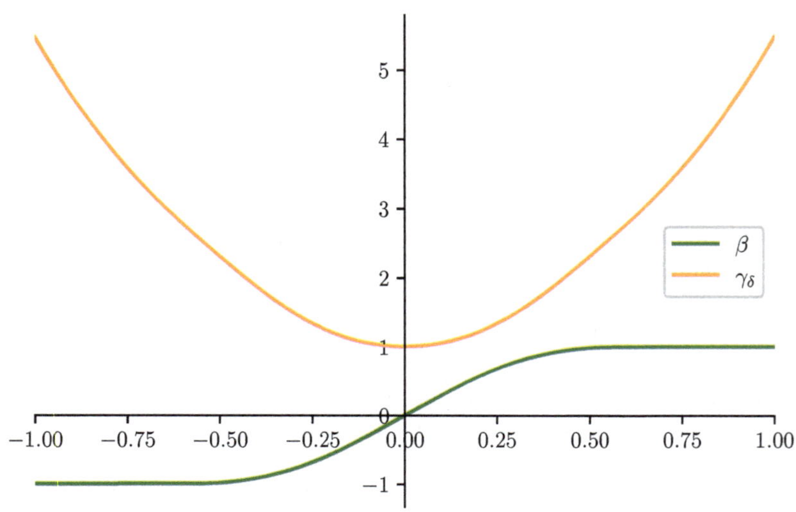

Fig. 5.7 The functions β and γ_δ

Choose a strictly increasing sequence (δ_k) with $\delta_0 = 0$. Show that $(E, (E_k))$ is a scale Banach space where $E := L^p(\mathbb{R}, \mathbb{R}^n)$ with scales

$$E_k := W^{k,p}_{\delta_k}(\mathbb{R}, \mathbb{R}^n), \qquad \forall k \in \mathbb{N}_0,$$

where the scales E_k are equipped with the norm

$$\|u\|_{E_k} := \|\gamma_{\delta_k} u\|_{W^{k,p}} \qquad \forall u \in E_k.$$

Can you suggest a terminology for a function belonging to a weighted Sobolev space?

Given two scale Banach spaces, we can define properties for maps between them in analogy to the classical setting of Banach spaces. All we have to do is to respect the scale structures!

Definition 5.24 (Scale Continuity, [78, Definition 4.1]) Let $(E, (E_k))$ and $(F, (F_k))$ be two scale Banach spaces. A map $f: E \to F$ is called ***scale continuous***, if we have $f(E_k) \subseteq F_k$ for all $k \in \mathbb{N}_0$ and the restriction $f|_{E_k}: E_k \to F_k$ is continuous.

Next we introduce two constructions for creating new scale Banach spaces from old ones in order to make precise the notion of differentiability with respect to scales.

Example 5.25 (Shifted Scale Banach Space, [78, Definition 3.3]) Let $(E, (E_k))$ be a scale Banach space and $l \in \mathbb{N}_0$. Then $(E^l, (E^l_k))$ is a scale Banach space for the scales $E^l_k := E_{l+k}$ for all $k \in \mathbb{N}_0$.

Example 5.26 (Scale Direct Sum, [78, Definition 3.4]) Let $(E, (E_k))$ and $(F, (F_k))$ be scale Banach spaces. Then $(E \oplus F, (E_k \oplus F_k))$ is also a scale Banach space.

Definition 5.27 (Tangent Bundle, [109, Definition 1.1.14]) For a scale Banach space $(E, (E_k))$ define its ***tangent bundle*** by $TE := E^1 \oplus E$.

We are now in the position to state the new notion of differentiability in scale calculus. The main underlying example is provided in Problem 5.44.

Definition 5.28 (Scale Differentiability, [78, Definition 4.2]) A scale continuous map $f: E \to F$ between scale Banach spaces $(E, (E_k))$ and $(F, (F_k))$ is called ***scale differentiable***, if for every $x \in E_1$ there exists a bounded linear operator

$$Df(x): E_0 \to F_0,$$

called the ***scale derivative of*** f, such that the restriction $f|_{E_1}: E_1 \to F_0$ is Fréchet differentiable with derivative $Df|_{E_1}$ and the ***tangent map***

$$Tf: TE \to TF, \qquad Tf(x, h) := (f(x), Df(x)h)$$

is scale continuous.

Using the iterated notion of scale differentiability, one can define higher scale regularity. We say that a scale differentiable map is ***scale smooth***, if its tangent map is infinitely scale differentiable. The following example motivated the development of scale calculus.

Example 5.29 (Weighted Sobolev Spaces, [78, Theorem 6.2]) Let E_k be the scale of weighted Sobolev spaces on the scale Banach space $E := L^p(\mathbb{R}, \mathbb{R}^n)$ introduced in Example 5.23. Then the shift map

$$\tau: \mathbb{R} \oplus E \to E, \qquad \tau(t, u)(s) = u(s + t)$$

is scale smooth.

One important property of scale differentiability is that, surprisingly, the chain rule remains valid in this more general setting.

Proposition 5.30 (Chain Rule, [109, Theorem 1.3.1]) *Suppose that $f: E \to F$ and $g: F \to G$ are scale differentiable maps for scale Banach spaces E, F, and G. Then the composition $g \circ f$ is scale differentiable with tangent map*

$$T(g \circ f) = Tg \circ Tf: TE \to TG.$$

Proof See Problem 5.45. □

The proof of the chain rule in Problem 5.45 is heavily based on the compactness of the embeddings of the scales. Using the chain rule (Proposition 5.30) one can define the notion of scale manifolds and scale differential geometry in analogy to the finite-dimensional case. For details, see [49, Chapter 5]. Thus, with the theory developed so far, one can make sense of the smooth moduli space of unparametrised negative gradient flow lines. For an explicit example, see [69, Section 4.3]. However, for broken negative gradient flow lines, one needs an even more general notion.

Definition 5.31 (Retraction, [109, Definition 2.1.1]) Let E be a scale Banach space. A ***retraction on*** E is a scale smooth map $r: E \to E$ such that $r \circ r = r$.

Remark 5.32 If X is a smooth Banach manifold, then $\operatorname{Fix}(r) = r(M)$ is a smooth submanifold for every smooth retraction $r: X \to X$ by a result of Cartan [109, Proposition 2.1.2].

It is in general not true that the fixed point set of a scale smooth retraction of a scale manifold is a scale submanifold. This led Hofer–Wysocki–Zehnder to the generalised notion of an *M-polyfold*, where "M" stands for "manifold flavoured". Heuristically, an M-polyfold is locally the fixed-point set of a scale smooth retraction of a scale smooth manifold. The main aspect for us is that Fredholm theory is still valid in M-polyfolds in some sense. For an extensive treatment, see [109, Part I]. Next, we partially carry out the roadmap for regularisation approaches via the polyfold approach as described in [69, Section 3.3]. The polyfold approach aims to associate to a compactified moduli space a smooth compact manifold with coherent boundary unique up to a suitable notion of cobordism. In order to achieve this and to eliminate a lot of repetitive work in applications, this approach first establishes compactness, then gluing, and finally transversality. In our exposition, we will focus on the main ingredient for showing that the moduli problem in Morse–Bott homology fits into the polyfold framework, namely the exponential decay of negative gradient flow lines of a Morse–Bott function showing the importance of weighted Sobolev spaces as introduced in Exercise 5.23.

Proposition 5.33 (Exponential Decay, [73, Theorem A.3]) *Let $u \in C^\infty(\mathbb{R}, M)$ be a negative gradient flow line for a Morse–Bott function $f \in C^\infty(M)$ with respect to a Riemannian metric m on a smooth compact manifold M, that is, u satisfies*

$$\partial_s u(s) = -\operatorname{grad}_m f(u(s)) \qquad \forall s \in \mathbb{R}.$$

Then there exist $x^\pm \in \operatorname{Crit}(f)$ with

$$\lim_{s\to\pm\infty} u(s) = x^\pm,$$

and there exists a constant $\delta > 0$, as well as for each $k \in \mathbb{N}_0$ a constant $C_k > 0$ with

$$\|\partial_s^k u(s)\|_m \leq C_k e^{-\delta|s|} \qquad \forall s \in \mathbb{R}, k \in \mathbb{N}_0.$$

Exercise 5.34 Consider a negative gradient flow line $u \in C^\infty(\mathbb{R}, M)$ of a Morse–Bott function $f \in C^\infty(M)$ on a compact smooth manifold M. What could be the significance of the exponential decay established in Proposition 5.33?

Remark 5.35 For a map $u \in C^\infty(\mathbb{R}, M)$ we can consider its ***energy*** defined by

$$E_m(u) := \int_{-\infty}^{+\infty} \|\partial_s u(s)\|_m^2 ds \in [0, +\infty]. \tag{5.5}$$

If u is a negative gradient flow line of a Morse–Bott function $f \in C^\infty(M)$ with asymptotics $x^\pm \in \operatorname{Crit}(f)$, then the energy (5.5) is given by

$$E_m(u) = f(x^-) - f(x^+) \in [0, +\infty).$$

Indeed, using the negative gradient flow equation we compute

$$\begin{aligned}
E_m(u) &= \int_{-\infty}^{+\infty} \|\partial_s u(s)\|_m^2 ds \\
&= \int_{-\infty}^{+\infty} \| \operatorname{grad}_m f(u(s))\|_m^2 ds \\
&= \int_{-\infty}^{+\infty} m\left(\operatorname{grad}_m f(u(s)), \operatorname{grad}_m f(u(s))\right) ds \\
&= \int_{-\infty}^{+\infty} df\left(\operatorname{grad}_m f(u(s)))\right) ds \\
&= -\int_{-\infty}^{+\infty} df\left(\partial_s u(s))\right) ds \\
&= -\int_{-\infty}^{+\infty} \frac{d}{ds} f(u(s)) ds \\
&= \lim_{s\to-\infty} f(u(s)) - \lim_{s\to+\infty} f(u(s)) \\
&= f(x^-) - f(x^+).
\end{aligned}$$

We give a detailed proof of Proposition 5.33 that incorporates fundamental ideas concerning Morse–Bott functions.

Exercise 5.36 Let $f \in C^\infty(M)$ be a Morse–Bott function on a compact smooth manifold M^n. By the Morse–Bott lemma [29, Lemma 3.51], for every critical point $x_0 \in \operatorname{Crit}(f)$ there exists a chart about x_0 such that locally we can write

$$f(x) = \frac{1}{2}\sum_{j=1}^{n} a_j x_j^2, \tag{5.6}$$

where a_j are the eigenvalues of the Hessian matrix $\operatorname{Hess}_f(x_0)$ with respect to the standard Euclidean metric.

1. Show that the gradient flow lines of (5.6) decay exponentially as claimed in Proposition 5.33.
2. Explain why one cannot directly conclude exponential decay if the metric is not the standard Euclidean one.

By Exercise 5.36 we conclude that Proposition 5.33 is true if the Riemannian metric around the critical points is the standard Euclidean one. However, this is not the case in general as in contrast to symplectic manifolds, Riemannian manifolds do carry local invariants such as curvature [130, Chapter 7]. So, the question is how can we transfer the result of Exercise 5.36 to the general case. The key ingredient is the so-called action-energy inequality.

Exercise 5.37 An ***action-energy inequality*** for a Morse–Bott function $f \in C^\infty(M)$ is a local inequality of the form

$$|f(x)| \leq \kappa \| \operatorname{grad}_m f(x) \|_m^2, \tag{5.7}$$

for some Riemannian metric m and constant $\kappa > 0$.

1. Explain why (5.7) is called action-energy inequality.
2. Show that (5.7) holds for the standard Euclidean metric.
3. Suppose that m_1 and m_2 are two Riemannian metrics such that locally the inequality $\| \cdot \|_{m_1} \leq \kappa \| \cdot \|_{m_2}$ holds for a constant $C > 0$. Show that

   $$\| \operatorname{grad}_{m_2} f \|_{m_2} \leq C \| \operatorname{grad}_{m_1} f \|_{m_1}$$

   holds and conclude that f satisfies the action-energy inequality (5.7) with respect to any Riemannian metric.

Next, we explain how exponential decay follows from an action-energy inequality.

Lemma 5.38 ([75, Lemma 6.5]) *Suppose that f satisfies locally the action-energy inequality* (5.7) *on a chart domain U with respect to some Riemannian metric m and let $u\colon [s_-, s_+] \to U$ be a gradient flow line of $\operatorname{grad}_m f$ satisfying $f(u(s)) > 0$ for every $s \in [s_-, s_+]$. Then*

$$d_m\left(u(s_-), u(s_+)\right) \leq \frac{2}{\sqrt{\kappa}} \left(\sqrt{f(u(s_-))} - \sqrt{f(u(s_+))} \right),$$

holds, where d_m denotes the metric induced by m.

Proof Note that by the assumption that $f(x) > 0$ it follows that the gradient never vanishes by the action-energy inequality (5.7). Therefore, we estimate

$$\begin{aligned} d_m\left(x(s_-), x(s_+)\right) &\leq \int_{s_-}^{s_+} \|\partial_s u(s)\|_m ds \\ &= \int_{s_-}^{s_+} \| \operatorname{grad}_m f(u(s)) \|_m ds \\ &= \int_{s_-}^{s_+} \frac{\| \operatorname{grad}_m f(u(s)) \|_m^2}{\| \operatorname{grad}_m f(u(s)) \|_m} ds \end{aligned}$$

$$
\begin{aligned}
&\leq \frac{1}{\sqrt{\kappa}} \int_{s_-}^{s_+} \frac{\| \operatorname{grad}_m f(u(s)) \|^2}{\sqrt{f(u(s))}} ds \\
&= -\frac{1}{\sqrt{\kappa}} \int_{s_-}^{s_+} \frac{df(\partial_s u(s))}{\sqrt{f(x)}} ds \\
&= -\frac{1}{\sqrt{\kappa}} \int_{s_-}^{s_+} \frac{\partial_s f(u(s))}{\sqrt{f(u(s))}} ds \\
&= -\frac{2}{\sqrt{\kappa}} \int_{s_-}^{s_+} \partial_s \sqrt{f(u(s))} ds \\
&= \frac{2}{\sqrt{\kappa}} \left(\sqrt{f(u(s_-))} - \sqrt{f(u(s_+))} \right).
\end{aligned}
$$

This finishes the proof of the lemma. □

Lemma 5.39 ([75, Lemma 6.6]) *Suppose that f satisfies locally the action-energy inequality* (5.7) *on a chart domain U with respect to some Riemannian metric m, and let $u\colon [s_0, \infty) \to U$ be a gradient flow line of* $\operatorname{grad}_m f$ *satisfying $f(u(s)) > 0$ for every $s \geq s_0$. Then*

$$
f(u(s)) \leq f(u(s_0)) e^{\frac{1}{\kappa}(s_0 - s)} \qquad \forall s \geq s_0. \tag{5.8}
$$

Proof The action-energy inequality combined with the gradient flow equation gives rise to the following inequality

$$
f(u(s)) \leq \kappa \| \operatorname{grad}_m f(u(s)) \|_m^2 = -\kappa \frac{d}{ds} f(u(s)).
$$

This implies

$$
\frac{d}{ds} \ln f(u(s)) = \frac{\frac{d}{ds} f(u(s))}{f(u(s))} \leq -\frac{1}{\kappa}.
$$

Integrating this inequality, we obtain

$$
\ln f(u(s)) = \int_{s_0}^{s} \frac{d}{ds} \ln f(u(s)) ds + \ln f(u(s_0)) \leq -\frac{s - s_0}{\kappa} + \ln f(u(s_0)).
$$

Taking exponents, we get the stated estimate. □

Proof of Proposition 5.33 We first show that every negative gradient flow line of a Morse–Bott function asymptotically converges to critical manifolds $C^{\pm} \subseteq \operatorname{Crit}(f)$ at its ends. The finitely many connected critical manifolds of f are all normally hyperbolic with respect to the negative gradient flow of f, meaning that we have the decomposition

$$T_xM = T_xC \oplus E_x^+ \oplus E_x^- \qquad \forall x \in C,$$

where $E_x^\pm$ are spanned by the positive and negative eigenspaces of the Hessian of f at x, respectively. This follows directly from the Morse–Bott lemma (Exercise 5.36). Since f is a Lyapunov function and thus decreases along negative gradient flow lines, any negative gradient flow line must converge to a critical manifold at its ends by applying the La Salle principle [200, Theorem IV.10] and [200, Proposition IV.8].

Next we show that each negative gradient flow line converges in fact exponentially to the critical manifolds in forward and backward time. By Exercise 5.37 we can assume without loss of generality that there exists a time $T \in \mathbb{R}$ such that $u(s) \in U$ for all $s \geq T$, where on U an action-energy inequality holds. Using Lemmas 5.38 and 5.39 we estimate for $T < s_1 < s_2$

$$\begin{aligned} d_m\left(u(s_1), u(s_2)\right) &\leq \frac{2}{\sqrt{\kappa}}\sqrt{f(u(s_1))} \\ &\leq \frac{2}{\sqrt{\kappa}}\sqrt{f(u(s_0))}e^{\frac{1}{2\kappa}(s_0-s_1)} \\ &= \frac{2}{\sqrt{\kappa}}\sqrt{f(u(s_0))}e^{\frac{s_0}{2\kappa}}e^{-\frac{1}{2\kappa}s_1}. \end{aligned}$$

Set

$$\mu_0 := \frac{2}{\sqrt{\kappa}}\sqrt{f(u(s_0))}e^{\frac{s_0}{2\kappa}}, \qquad c := \frac{1}{2\kappa}.$$

With this notation the above inequality becomes

$$d_m\big(u(s_1), u(s_2)\big) \leq \mu_0 e^{-cs_1}.$$

Note that the right-hand side does not depend on s_2. Because $s_2 > s_1$ is arbitrary and u converges to 0 as s goes to infinity, we obtain from this

$$d_m\big(u(s_1), 0\big) \leq \mu_0 e^{-cs_1}.$$

This shows that u converges to the critical point exponentially. That the derivatives converge to zero and also exponentially follows from the exponential convergence of u by bootstrapping the gradient flow equation. This establishes exponential decay and thus finishes the proof of Proposition 5.33. □

As we have established exponential decay in Proposition 5.33 for gradient flow lines of a Morse–Bott function to its critical points, we can now explain how an M-polyfold Fredholm setup for Morse–Bott homology can be defined. We follow

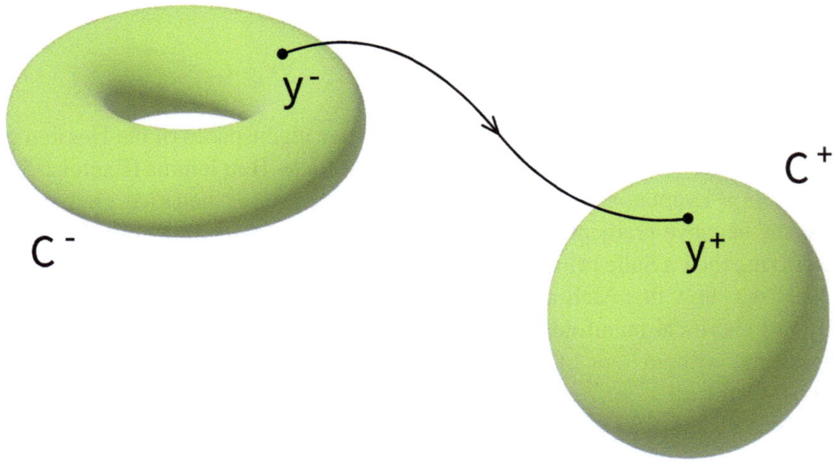

Fig. 5.8 An asymptotically exponential Sobolev path connecting two critical components $C^{\pm}$ of a Morse–Bott function

[49, Section 8.4]. The essential arguments are contained in [73, Appendix A]. We assume the following setup. Let (M, m) be a compact Riemannian manifold and $f \in C^{\infty}(M)$ a Morse–Bott function. Choose an additional Morse function $h \in C^{\infty}(\mathrm{Crit}(f))$ and a Riemannian metric m_0 on $\mathrm{Crit}(f)$ such that (h, m_0) is a Morse–Smale pair, that is, the stable and unstable manifolds intersect transversally. Pick two connected critical components $C^{\pm} \subseteq \mathrm{Crit}(f)$. Fix

$$0 < \delta < \min\left\{|\lambda| : \lambda \in \mathrm{Spec}(\mathrm{Hess}_m(f)(x)) \setminus \{0\}, x \in \mathrm{Crit}(f)\right\},$$

where $\mathrm{Spec}(\mathrm{Hess}(f)(x))) \subseteq \mathbb{R}$ denotes the spectrum of the Hessian of f at a critical point x. For a positive strictly increasing sequence $0 < \delta_0 < \delta_1 < \cdots < \delta$, consider the Banach manifold $\mathscr{E}_k(C^-, C^+)$, $k \geq 0$, consisting of all maps $u \in H^{k+2}_{\mathrm{loc}}(\mathbb{R}, M)$ converging exponentially to $C^{\pm}$, that is, there exist $y^{\pm} \in C^{\pm}$ as well as

$$\xi^- \in H^{k+2}_{\delta_k}((-\infty, -T], T_{y^-}M) \quad \text{and} \quad \xi^+ \in H^{k+2}_{\delta_k}([T, +\infty), T_{y^+}M)$$

for some $T \in \mathbb{R}$ with $u(s) = \exp^{\nabla}_{y^{\pm}}(\xi^{\pm}(s))$ for all $\pm s \geq T$, as in Fig. 5.8, where ∇ denotes the Levi–Civita connection associated with the Riemannian metric m.

The local charts on $\mathscr{E}_k(C^-, C^+)$ are constructed by exponential neighbourhoods around smooth paths [170, Appendix A]. As explained in [69, Remark 4.23], it is important that the charts are based at a smooth map. Similarly to [17] or [69], one can construct scale smooth charts using those exponential neighbourhoods, equipping the Banach manifold $\mathscr{E}(C^-, C^+) := \mathscr{E}_0(C^-, C^+)$ with a scale smooth structure. Moreover, there are two natural scale smooth evaluation maps

$$\mathrm{ev}^{\pm} \colon \mathscr{E}(C^-, C^+) \to \mathrm{Crit}(f), \qquad \mathrm{ev}^{\pm}(u) = y^{\pm}.$$

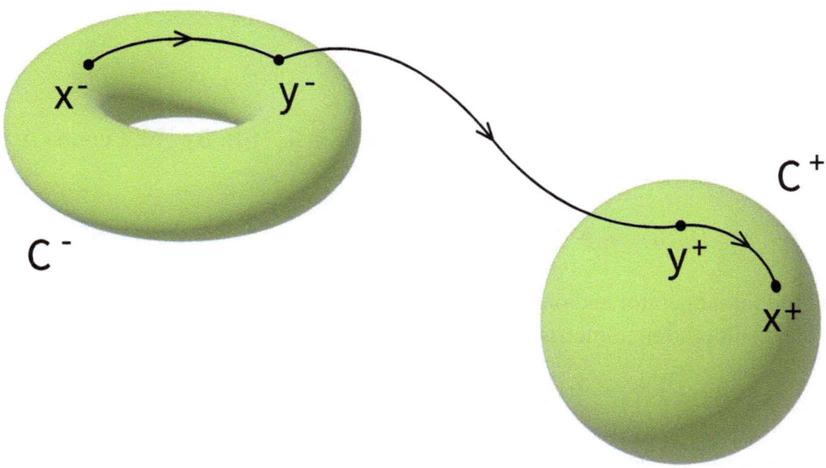

Fig. 5.9 A Sobolev path with one cascade connecting two critical points $x^{\pm}$ of the Morse function on $C^{\pm}$

Fix $x^{\pm} \in C^{\pm} \cap \mathrm{Crit}(h)$ and define the ***stable and unstable manifolds*** by

$$W^{\pm}(x^{\pm}) := \left\{ y \in \mathrm{Crit}(f) : \lim_{t \to \pm\infty} \phi_t(y) = x^{\pm} \right\},$$

where $\phi \colon \mathbb{R} \times \mathrm{Crit}(f) \to \mathrm{Crit}(f)$ denotes the negative gradient flow of h with respect to the Morse–Smale metric m_0. Define

$$\mathscr{E}^1_k(x^-, x^+) := \left\{ u \in \mathscr{E}_k(C^-, C^+) : \mathrm{ev}^{\pm}(u) \in W^{\pm}(x^{\pm}) \right\}$$

for all $k \geq 0$ as in Fig. 5.9. Again, $\mathscr{E}^1(x^-, x^+) := \mathscr{E}^1_0(x^-, x^+)$ is a scale manifold. For details, see the proof of [73, Theorem A.12] and [73, Theorem A.14].

There exists a canonical scale smooth bundle $\mathscr{F}^1(x^-, x^+) \to \mathscr{E}^1(x^-, x^+)$, where the fibre over $u \in \mathscr{E}^1_k(C^-, C^+)$ is given by

$$\mathscr{F}^1_k(x^-, x^+)|_u = H^{k+1}_{\delta_k}(\mathbb{R}, u^* TM).$$

Consider the section

$$\partial \colon \mathscr{E}^1(x^-, x^+) \to \mathscr{F}^1(x^-, x^+), \qquad \partial(u) := \dot{u} + \mathrm{grad}_m f(u).$$

Its zero section $\partial^{-1}(0)$ consists of gradient flow lines with one cascade and belongs to weighted Sobolev spaces according to Proposition 5.33. If we identify $\mathscr{E}^1(x^-, x^+)$ with the zero section of $\mathscr{F}^1(x^-, x^+)$, then for every $u \in \mathscr{E}^1(x^-, x^+)$ we have a canonical splitting

$$T_u \mathscr{F}^1(x^-, x^+) = \mathscr{F}^1(x^-, x^+)|_u \oplus T_u \mathscr{E}^1(x^-, x^+)$$

into horizontal and vertical subspaces. Define the projection

$$\pi: T_u\mathscr{F}^1(x^-, x^+) \to \mathscr{F}^1(x^-, x^+)|_u$$

to be the projection along $T_u\mathscr{E}^1(x^-, x^+)$ for all $u \in \partial^{-1}(0)$. By [73, Lemma A.13], the ***vertical differential***

$$\pi \circ D\partial(u)\colon T_u\mathscr{E}^1(x^-, x^+) \to \mathscr{F}^1(x^-, x^+)|_u, \qquad \forall u \in \partial^{-1}(0)$$

is a Fredholm operator and one can explicitly compute its index locally. If $x^- \neq x^+$, the scale smooth reparametrisation action

$$\tau\colon \mathbb{R} \oplus \mathscr{E}^1(x^-, x^+) \to \mathscr{E}^1(x^-, x^+), \qquad \tau(t, u)(s) = u(s+t),$$

as in Example 5.29 is free, giving rise to a scale smooth bundle

$$\mathscr{F}^1(x^-, x^+)/\mathbb{R} \to \mathscr{E}^1(x^-, x^+)/\mathbb{R}$$

to which ∂ descends to a scale Fredholm section which we also denote by ∂. This construction can be generalised to m cascades and thus yields the scale smooth bundle

$$\mathscr{F}^m(x^-, x^+)/\mathbb{R}^m \to \mathscr{E}^m(x^-, x^+)/\mathbb{R}^m$$

of unparametrised paths from x^- to x^+ with m cascades.

Exercise 5.40 Explain why for the case of m cascades we quotient out an appropriate action by $\mathbb{R}^m$. Do we also reparametrise the negative gradient flow lines of the Morse function h on $\operatorname{Crit}(f)$?

Now we consider broken paths. For $x^0, \dots, x^l \in \operatorname{Crit}(h)$ with

$$f(x^0) \geq f(x^1) \geq \cdots \geq f(x^{l-1}) \geq f(x^l),$$

we define the space of broken paths from x^0 to x^l by

$$\mathscr{E}(x^0, \dots, x^l) := \coprod_{(\alpha_0, \dots, \alpha_l) \in \mathbb{N}_0^{l+1}} \mathscr{E}^{\alpha_0}(x^0, x^1)/\mathbb{R}^{\alpha_0} \times \cdots \times \mathscr{E}^{\alpha_l}(x^{l-1}, x^l)/\mathbb{R}^{\alpha_l}.$$

Then $\mathscr{E}(x^0, \dots, x^l)$ is a scale manifold over which we have the scale bundle

$$\mathscr{F}(x^0, \dots, x^l) := \coprod_{(\alpha_0, \dots, \alpha_l) \in \mathbb{N}_0^{l+1}} \mathscr{F}^{\alpha_0}(x^0, x^1)/\mathbb{R}^{\alpha_0} \times \cdots \times \mathscr{F}^{\alpha_l}(x^{l-1}, x^l)/\mathbb{R}^{\alpha_l}$$

and the scale Fredholm section

$$\underbrace{\partial \times \cdots \times \partial}_{l \text{ times}} \colon \mathscr{E}(x^0, \ldots, x^l) \to \mathscr{F}(x^0, \ldots, x^l).$$

For $x^{\pm} \in \mathrm{Crit}(h)$ with $f(x^-) \geq f(x^+)$, we define the space of broken paths from x^- to x^+ by

$$X(x^-, x^+) := \coprod_{l \in \mathbb{N}} \coprod_{\substack{x^1, \ldots, x^{l-1} \in \mathrm{Crit}(h) \\ f(x^-) \geq f(x^1) \geq \cdots \geq f(x^{l-1}) \geq f(x^+)}} \mathscr{E}(x^-, x^1, \ldots, x^{l-1}, x^+). \tag{5.9}$$

Similarly, one defines

$$Y(x^-, x^+) := \coprod_{l \in \mathbb{N}} \coprod_{\substack{x^1, \ldots, x^{l-1} \in \mathrm{Crit}(h) \\ f(x^-) \geq f(x^1) \geq \cdots \geq f(x^{l-1}) \geq f(x^+)}} \mathscr{F}(x^-, x^1, \ldots, x^{l-1}, x^+).$$

One can show that $X(x^-, x^+)$ carries the natural structure of an M-polyfold and

$$Y(x^-, x^+) \to X(x^-, x^+)$$

is a bundle. Moreover, the scale Fredholm section above induces a scale Fredholm section $\partial \colon X(x^-, x^+) \to Y(x^-, x^+)$ of index

$$\operatorname{ind} \partial = \operatorname{ind}(x^-) - \operatorname{ind}(x^+) - 1,$$

where either

$$\operatorname{ind} = \operatorname{ind}_f + \operatorname{ind}_h \qquad \text{or} \qquad \operatorname{ind} = -\frac{1}{2} \operatorname{sgn} \operatorname{Hess} f - \frac{1}{2} \operatorname{sgn} \operatorname{Hess} h.$$

For details see [51, Appendix A]. One can show that the level set $\partial^{-1}(0)$ of unparametrised broken negative gradient flow lines is compact and that there exists an *abstract perturbation* of ∂, that is, a scale smooth section $\sigma : X \to Y$, where

$$X := \coprod_{\substack{x^{\pm} \in \mathrm{Crit}(h) \\ f(x^-) \geq f(x^+)}} X(x^-, x^+) \qquad \text{and} \qquad Y := \coprod_{\substack{x^{\pm} \in \mathrm{Crit}(h) \\ f(x^-) \geq f(x^+)}} Y(x^-, x^+),$$

with the property that for all $x^{\pm} \in \mathrm{Crit}(h)$ the moduli space

$$\mathscr{M}(x^-, x^+) := \{u \in X(x^-, x^+) : (\partial + \sigma)(u) = 0\}$$

of ***all flow lines with cascades from*** x^- ***to*** x^+ is compact and $\partial + \sigma$ is transverse to the zero section. Hence we arrive at a prototypical result for Morse–Bott homology, compare [49, Corollary 8.9].

Theorem 5.41 (M-Polyfold Setup for Morse–Bott Homology) *Let* (M, m) *be a compact Riemannian manifold without boundary and* $f \in C^\infty(M)$ *a Morse–Bott function. Choose an additional Morse function* $h \in C^\infty(\mathrm{Crit}(f))$ *and an additional Riemannian metric* m_0 *on* $\mathrm{Crit}(f)$ *such that* (h, m_0) *is a Morse–Smale pair. For all critical points* $x^\pm \in \mathrm{Crit}(h)$*, the moduli space*

$$\mathcal{M}(x^-, x^+) := \{u \in X(x^-, x^+) : (\partial + \sigma)(u) = 0\}$$

of abstractly perturbed negative gradient flow lines with cascades from x^- *to* x^+ *is a smooth compact manifold with corners of dimension*

$$\dim \mathcal{M}(x^-, x^+) = \mathrm{ind}(x^-) - \mathrm{ind}(x^+) - 1.$$

Moreover, there is a canonical diffeomorphism

$$\partial \mathcal{M}(x^-, x^+) = \mathcal{M}(x^-, x^+) \cap \partial X \cong \coprod_{x \in \mathrm{Crit}(h)} \mathcal{M}(x^-, x) \times \mathcal{M}(x, x^+).$$

Proof This is the adapted statement of [49, Corollary 8.9]. The proof is an exercise in applying the abstract theory provided by polyfold theory. There are two ways of proving such a result: Either one relies on it as a kind of black box and dives deeper into the technical details when needed later on, or one fully tries to understand the new subject. In the latter case, we suggest starting with the overview paper [69] and trying to solve Problem 6.19. More technical details on the matter of gluing can, for example, be found in the analytical paper [79]. □

Remark 5.42 (Mathematical Folklore, [160]) Theorem 5.41 can be regarded as folklore. Indeed, it is an unpublished result with no clear originator, but which is believed to be true among the specialists. More specifically, folk mathematics, or mathematical folklore, is the body of theorems, definitions, proofs, facts, or techniques that circulate among mathematicians by word of mouth, but have not yet appeared in print, either in books or in scholarly journals. As remarked in [23], if an aspiring researcher is so presumptuous as to write out a proper proof and attempt to publish it, they will get shot down in flames. However, if one manages to find a deep application of a folklore result which is otherwise not easily accessible, then it might be worth publishing such a result. A recurring theme in this book are orientation issues in Morse–Bott homology. Maybe the process of constructing coherent orientations in Morse–Bott homology gets clearer by invoking orientations in the polyfold setup as explained in [109, Chapter 6].

Exercise 5.43 Compare the statement of Theorem 5.41 with the statement of [73, Theorem A.14]. What is the crucial difference between the classical approach via Sard-Smale theory and the abstract perturbation provided by polyfold theory? What are the advantages and disadvantages of the two theories in this scenario?

Problems

5.44 (Motivation of Scale Differentiability) For every $k \in \mathbb{N}_0$ consider the modified shift map [78, Equation (2)]

$$\tau\colon \mathbb{R} \times C^k(\mathbb{T}) \to C^k(\mathbb{T}), \qquad \tau(t, u)(s) := u(s + t).$$

Note that τ fails to be differentiable for any $k \in \mathbb{N}_0$.

1. Show that τ is continuous for every $k \in \mathbb{N}_0$.
2. Show that $\tau|_{\mathbb{R}\times C^{k+1}(\mathbb{T})}$ is differentiable for each $k \in \mathbb{N}_0$ with derivative

$$D_{(t,u)}\tau(r, X) = r\tau(s, u') + \tau(t, X).$$

3. For every $k \in \mathbb{N}_0$ and $(t_0, u_0) \in \mathbb{R} \times C^{k+1}(\mathbb{T})$, the derivative $D_{(t_0,u_0)}\tau$ extends to a bounded linear operator

$$D_{(t_0,u_0)}\tau\colon \mathbb{R} \times C^k(\mathbb{T}) \to C^k(\mathbb{T}).$$

4. The map

$$(\mathbb{R} \times C^{k+1}(\mathbb{T})) \times (\mathbb{R} \times C^k(\mathbb{T})) \to C^k(\mathbb{T}), \qquad (t, u, r, X) \mapsto D_{(t,u)}\tau(r, X)$$

   is continuous for each $k \in \mathbb{N}_0$.
5. Conclude that τ is infinitely often scale differentiable.

5.45 (Chain Rule in Scale Calculus) Prove Proposition 5.30 and motivate the definition of scale differentiability in view of the chain rule. *Hint:* See [109, Section 1.4.2].

5.46 ($\mathbb{Z}$-Coefficients) Let σ be a coherent orientation of a compact smooth manifold M as in [170, Definition 3.15]. Such coherent orientations always exist by [170, Proposition 3.16]. Let $h \in C^\infty(M)$ be a Morse function satisfying the Morse–Smale condition with respect to a Riemannian metric and set

$$\mathrm{Crit}_k(h) := \{x \in \mathrm{Crit}(h) : \mathrm{ind}_h(x) = k\} \qquad \forall k \in \mathbb{Z}$$

and

$$\mathrm{CM}_k(h) := \mathrm{Crit}_k(h) \otimes \mathbb{Z}.$$

For every $k \in \mathbb{Z}$ define a map $\partial_k^\sigma : \mathrm{CM}_k(h) \to \mathrm{CM}_{k-1}(h)$ by

$$\partial_k^\sigma x^- := \sum_{x^+ \in \mathrm{Crit}_{k-1}(h)} \sum_{u \in \mathcal{M}(x^-, x^+)} \tau_\sigma(u) x^+,$$

where $\tau_\sigma(u) = \pm 1$ denote the characteristic signs induced by the coherent orientation σ. Geometrically, the characteristic signs $\tau_\sigma(u)$ tell us whether the vector space isomorphism

$$T_x W^u(x^-) \cap \mathrm{grad}_m h(x)^\perp \to T_{x^+} W^u(x^+)$$

induced by the gradient flow of h for sufficiently large time is orientation preserving or orientation reversing.

1. Redo the computations in Exercise 5.1 for the case of a coherent orientation. Can you guess what the correct Morse homology should be?
2. Show that the Morse–Smale–Witten chain complex $(\mathrm{CM}_*(f), \partial_*^\sigma)$ is a chain complex, that is, there holds

$$\partial_{k-1}^\sigma \circ \partial_k^\sigma = 0 \qquad \forall k \in \mathbb{Z}.$$

   Hint: See [170, Theorem 7] or [189].
3. Denote the induced homology by $\mathrm{HM}_*^\sigma(h)$. Show that there is a natural isomorphism

$$\mathrm{HM}_*^\sigma(h) \cong \mathrm{H}_*(M; \mathbb{Z}).$$

   Hint: See [170, Theorem 9] or [29, Chapter 7].

5.47 (Generic Properties) Let X be a topological space. A subset $\mathcal{U} \subseteq X$ is said to be of the ***second category***, if we can write $\mathcal{U} = \cap_{i=1}^\infty U_i$ for open and dense subsets $U_i \subseteq X$. A property is called ***generic***, if it holds for a subset of the second category.

1. Let $h \in C^\infty(M)$ be a Morse function on a compact smooth manifold M and denote by $\mathcal{R}(M)$ the space of all Riemannian metrics on M with respect to the C^∞-topology. As explained in [36] it is an interesting question on its own to study the space $\mathcal{R}(M)$. However, we will not pursue this question in this problem. Show that there exists a subset $\mathcal{U} \subseteq \mathcal{R}(M)$ depending on h such that (h, m_0) is a Morse–Smale pair for all $m_0 \in \mathcal{U}$. *Hint:* See [75, Corollary 7.6].
2. Consider an exact symplectic manifold (M^{2n}, λ), $n \geq 1$. Define the vector space

$$X := \{H \in C^\infty(M) : dH \text{ compactly supported}\}.$$

Show that there exists a subset $\mathscr{U} \subseteq X$ of the second category, such that for every $H \in \mathscr{U}$ the Rabinowitz action functional

$$\mathscr{A}^H : \mathscr{L}M \times \mathbb{R} \to \mathbb{R}, \qquad \mathscr{A}^H(\gamma, \tau) := \int_0^1 \gamma^*\lambda - \tau \int_0^1 H(\gamma(t))dt$$

is a Morse–Bott function and its critical set consists of $H^{-1}(0)$ together with a disjoint union of circles. Thus the Rabinowitz action functional is generically a Morse–Bott function. *Hint:* See [51, Theorem B.1].

3. Let K be an action-constant isolated component of the set of all periodic orbits on a regular energy surface $\Sigma = H^{-1}(0)$. For some sufficiently small $\delta > 0$ we consider the space of perturbations

$$\mathscr{F}(U, \delta) := \{F \in C_c^\infty(M \times \mathbb{T}) : \operatorname{supp}(F) \subseteq U \times \mathbb{T} \text{ and } \|F\|_{C^1} < \delta\}$$

for some isolating neighbourhood U of K. Show that for all $\delta > 0$ there exists a subset $\mathscr{U} \subseteq \mathscr{F}(U, \delta)$ of second category such that the perturbed Rabinowitz action functional

$$\mathscr{A}_F^H : \mathscr{L}M \times \mathbb{R} \to \mathbb{R}, \quad \mathscr{A}_F^H(\gamma, \tau) := \int_0^1 \gamma^*\lambda - \tau \int_0^1 H(\gamma(t))dt - \int_0^1 F_t(\gamma(t))dt$$

is a Morse function for all $F \in \mathscr{U}$. Thus the perturbed Rabinowitz action functional is generically a Morse function. *Hint:* See [146, Proposition 3.2.5].

4. Show that the symplectic action functional

$$\mathscr{A}_F : \mathscr{L}M \to \mathbb{R}, \qquad \mathscr{A}_F(\gamma) := \int_{\mathbb{D}} \bar{\gamma}^*\omega - \int_0^1 F_t(\gamma(t))dt$$

is generically a Morse function for a symplectically aspherical compact symplectic manifold (M, ω) with respect to an ω-compatible almost complex structure J. *Hint:* See [191, Example 3.3.32].

6 Bubbling Analysis

In this second technical chapter, we explain how one adapts Theorem 5.41 to the semi-infinite-dimensional case encountered in Floer homology. The main technical difficulty is to achieve compactness of the moduli space of gradient flow lines of the symplectic action functional, as apart from breaking, a new phenomenon occurs as we explain now. Suppose (M, ω) is a compact connected symplectically aspherical symplectic manifold without boundary as in Exercise 5.2 and let J be an ω-compatible almost complex structure. On the connected component of all contractible free loops $\Lambda M \subseteq \mathscr{L}M$ we define the L^2-metric at $\gamma \in \Lambda M$ by

$$\langle X, Y\rangle_J := \int_0^1 \omega(JX(t), Y(t))dt \qquad \forall X, Y \in \Gamma(\gamma^*TM).$$

With respect to this metric, the gradient $\operatorname{grad}_J \mathscr{A}_H \in \mathfrak{X}(\Lambda M)$ of the symplectic action functional

$$\mathscr{A}_H \colon \Lambda M \to \mathbb{R}, \qquad \mathscr{A}_H(\gamma) = \int_{\mathbb{D}} \bar{\gamma}^*\omega - \int_0^1 H_t(\gamma(t))dt$$

for some $H \in C^\infty(M \times \mathbb{T})$ is by Exercise 5.4 given by

$$\operatorname{grad}_J \mathscr{A}_H|_\gamma(t) = J(\dot{\gamma}(t) - X_{H_t}(\gamma(t))) \qquad \forall t \in \mathbb{T}.$$

Thus a negative gradient flow line can be interpreted as a map $u \in C^\infty(\mathbb{R} \times \mathbb{T}, M)$ that is a solution of the ***Floer equation***

$$\partial_s u(s,t) + J(\partial_t u(s,t) - X_{H_t}(u(s,t))) = 0 \qquad \forall (s,t) \in \mathbb{R} \times \mathbb{T}. \tag{6.1}$$

Y. Bähni, *Lectures on Twisted Rabinowitz-Floer Homology*, Pathways in Mathematics, https://doi.org/10.1007/978-3-032-10673-5_6

Exercise 6.1 Consider the Floer equation (6.1).

1. Derive the Floer equation from the symplectic action functional $\mathscr{A}_H$ by using the defining formula $\langle \operatorname{grad}_J \mathscr{A}_H(\gamma), X\rangle_J = d\mathscr{A}_H|_\gamma(X)$ for all $X \in \Gamma(\gamma^*TM)$.
2. Can you think of three dynamical theories generalised by the Floer equation?

It is quite miraculous why this gives a well-defined gradient flow under such a weak regularity assumption. Indeed, the L^2-metric does not contain any derivatives! The Floer equation (6.1) is a zero-order perturbation of a Cauchy–Riemann equation and thus an elliptic partial differential equation. By regularity theory for this type of equations we achieve smooth solutions as observed by Gromov in [96] shaping the landscape of modern symplectic geometry. What happens if we consider sequences of such solutions? Assume that (u_k) is a sequence of negative gradient flow lines of the symplectic action functional $\mathscr{A}_H$ such that the derivatives Du_k are uniformly bounded. Then by [136, Theorem 4.1.1] there exists a negative gradient flow line u of the symplectic action functional such that

$$u_k \xrightarrow{C^\infty_{\text{loc}}} u, \qquad k \to \infty,$$

up to a subsequence. Thus, if the derivatives of the sequence of negative gradient flow lines are uniformly bounded, the sequence (u_k) converges to a broken negative gradient flow line in the Floer–Gromov sense. In contrast to finite-dimensional Morse–Bott homology, a new phenomenon occurs if the derivatives explode. In fact, if the derivatives explode, by Proposition 6.5 below, there exists a non-constant J-holomorphic sphere $v \in C^\infty(\mathbb{S}^2, M)$ and bubbling occurs in the limit as illustrated in Fig. 6.1.

This phenomenon cannot happen under our assumption that (M, ω) is symplectically aspherical! Indeed, by [136, Lemma 2.2.1] we would have that

$$\int_{\mathbb{S}^2} v^*\omega > 0$$

contradicting $[\omega]|_{\pi_2(M)} = 0$ as in Exercise 5.2. Assume that the symplectic action functional $\mathscr{A}_H$ is a Morse–Bott function. By Problem 5.47 this is generically

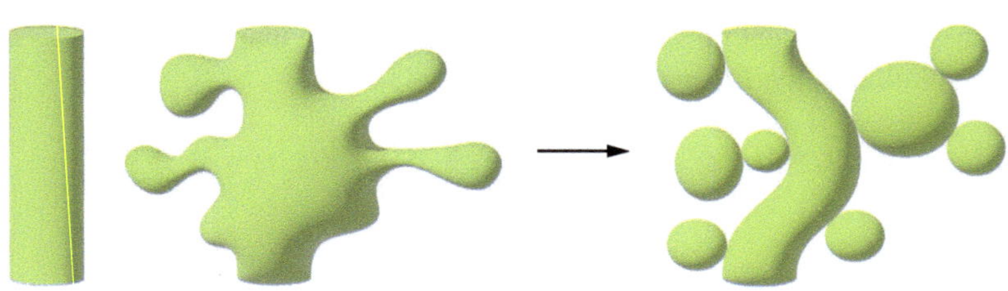

Fig. 6.1 The bubbling phenomenon of a sequence of solutions of the Floer equation

satisfied. In analogy to Morse–Bott homology in the finite-dimensional case as discussed in Chap. 5, we can define the ***Floer Homology of*** H by

$$\mathrm{HF}(H) := \mathrm{HM}(\mathscr{A}_H).$$

More precisely, we choose an additional Morse function $h \in C^\infty(\mathrm{Crit}(\mathscr{A}_H))$ on the critical manifold $\mathrm{Crit}(\mathscr{A}_H) \subseteq M$ via the obvious identification $\gamma \mapsto \gamma(0)$ and define a $\mathbb{Z}_2$-vector space $\mathrm{CF}(H) := \mathrm{Crit}(h) \otimes \mathbb{Z}_2$ together with a boundary operator

$$\partial\colon \mathrm{CF}(\mathscr{A}_H) \to \mathrm{CF}(\mathscr{A}_H), \qquad \partial\gamma^- := \sum_{\gamma^+ \in \mathrm{Crit}(h)} n(\gamma^-, \gamma^+)\gamma^+,$$

where

$$n(\gamma^-, \gamma^+) := \#_2\mathscr{M}^0(\gamma^-, \gamma^+) \in \mathbb{Z}_2$$

denotes the $\mathbb{Z}_2$-count of the zero-dimensional component of the moduli space of all abstractly perturbed unparametrised negative gradient flow lines with cascades. Then the ungraded Floer homology of $H \in C^\infty(M \times \mathbb{T})$ is given by

$$\mathrm{HF}(H) = \frac{\ker \partial}{\operatorname{im} \partial}.$$

Problem 6.16 shows that the definition of Hamiltonian Floer homology does not depend on any auxiliary choices. In fact, Hamiltonian Floer homology is also independent of the choice of Hamiltonian function H by Problem 6.16. Consequently, we have a chain of natural isomorphisms

$$\mathrm{HF}(H) = \mathrm{HM}(\mathscr{A}_H) \cong \mathrm{HM}(\mathscr{A}_0) = \bigoplus_{k \geq 0} \mathrm{HM}_k(h) \cong \bigoplus_{k \geq 0} \mathrm{H}_k(M; \mathbb{Z}_2), \tag{6.2}$$

where $\mathscr{A}_0$ is the symplectic area functional with $\mathrm{Crit}(\mathscr{A}_0) \cong M$. In view of (6.2), one might wonder what the use of the Floer homology $\mathrm{HF}(H)$ of H is. In fact, topologically we do not get any new information about the manifold M. However, from (6.2) we can infer sensible data for the time-dependent Hamiltonian function H itself. Assume for a moment that the symplectic action functional is a Morse function. Then by the Morse inequalities (Exercise 5.18) and (6.2) we conclude

$$\#\mathrm{Crit}(\mathscr{A}_H) = \dim_{\mathbb{Z}_2} \mathrm{CF}(H) \geq \dim_{\mathbb{Z}_2} \mathrm{HF}(H) = \sum_{k=0}^{\dim M} \dim_{\mathbb{Z}_2} \mathrm{H}_k(M; \mathbb{Z}_2).$$

So we get a topological lower bound for the number of nondegenerate periodic orbits of the time-dependent Hamiltonian function H only depending on the compact manifold M (see Problem 6.17). This is a particular instance of the famous

Hamiltonian Arnol'd Conjecture [5, Conjecture 4.1] about the number of fixed points of a nondegenerate Hamiltonian symplectomorphism being at least the sum of the Betti numbers of M. This homological Arnol'd conjecture was the driving force behind the construction of Floer homology. A historical overview and a proof formulated in the language of polyfold theory of the homological Arnol'd conjecture with rational coefficients for general compact symplectic manifolds without further assumptions can be found in the paper [71]. However, rigorously defining Morse–Bott homology for the symplectic action functional $\mathscr{A}_H$ faces us with many analytical difficulties. We start with the easiest one: A possible $\mathbb{Z}$-grading of the Floer homology.

Exercise 6.2 Show that both the Morse index and the Morse coindex of the symplectic action functional $\mathscr{A}_H$ are infinite at critical points. *Hint:* See [5, pp. 170–171].

Let $\gamma \in \operatorname{Crit}(\mathscr{A}_H)$ be a critical point of the symplectic action functional and choose a filling disc $\bar{\gamma} \in C^\infty(\mathbb{D}, M^{2n})$ of γ. Fix a symplectic trivialisation [137, Proposition 2.6.7] of the pullback tangent bundle $\Phi\colon \mathbb{D} \times \mathbb{R}^{2n} \to \bar{\gamma}^*TM$. Then we can associate to the pair $(\gamma, \bar{\gamma})$ an integer

$$\mu(\gamma, \bar{\gamma}) := \mu_{\mathrm{CZ}}(\Psi) \in \mathbb{Z},$$

where μ_{CZ} denotes the Conley–Zehnder index [77, Definition 10.4.1] of the path of symplectic matrices Ψ defined by

$$\Psi\colon [0,1] \to \mathrm{Sp}(2n), \qquad \Psi(t) := \Phi^{-1}_{e^{2\pi i t}} \circ D\phi_t^{X_{H_t}} \circ \Phi_1.$$

If we choose another filling disc $\bar{\gamma}' \in C^\infty(\mathbb{D}, M)$ of γ, then we have the formula

$$\mu(\gamma, \bar{\gamma}) - \mu(\gamma, \bar{\gamma}') = 2c_1([\bar{\gamma}' \# \bar{\gamma}^-]),$$

where $[\bar{\gamma}' \# \overline{\gamma}^-] \in \pi_2(M)$ and c_1 denotes the first Chern class [137, p. 85]. For more topological properties of the Chern classes, see [105, Section 3.1]. So, there is always a well-defined $\mathbb{Z}_2$-grading of $\mathrm{CF}(H)$. If $c_1(TM)|_{\pi_2(M)} = 0$, then there is also a well-defined $\mathbb{Z}$-grading by [51, p. 297]. More precisely, if $c_1(TM)|_{\pi_2(M)} = 0$, then the Floer chain group $\mathrm{CF}(H)$ is graded by the signature index

$$\operatorname{ind}(\gamma) := \mu_{\mathrm{CZ}}(\gamma) - \frac{1}{2} \operatorname{sgn} \operatorname{Hess}_h(\gamma) \qquad \forall \gamma \in \operatorname{Crit}(h).$$

The Conley–Zehnder index can be thought of as a kind of winding number of the linearised Hamiltonian flow along the corresponding orbit by [77, Theorem 10.6.1] and was recently used for promoting the interaction between symplectic geometry and practical problems like space mission design via studying bifurcations in Hill's lunar problem as in [26] and [25]. With respect to the above grading, the chain of canonical isomorphisms (6.2) gives by Problem 6.18 rise to a natural isomorphism

$$\mathrm{HF}_*(H) \cong \mathrm{H}_{*+n}(M; \mathbb{Z}_2). \tag{6.3}$$

Another way of proving (6.3) is provided in Problem 7.63 via the notion of C^2-small Hamiltonian functions or as in [136, Theorem 12.1.4] via the Piunikhin–Salamon–Schwarz isomorphism

$$\Phi_{\mathrm{PSS}}\colon \mathrm{HM}_*(M) \to \mathrm{HF}_{*-n}(M, \omega)$$

relating Morse flow lines with Floer cylinders via spiked spheres. The dependence of the Floer homology of M on a Hamiltonian symplectomorphism can be studied by using the Seidel representation $\pi_1(\mathrm{Ham}(M, \omega)) \to \mathrm{H}^*(M)$ as in [136, Example 8.6.8]. Using the above grading we can finally state the semi-infinite dimensional analogue of Theorem 5.41 for the symplectic action functional $\mathscr{A}_H$.

Theorem 6.3 (M-Polyfold Setup for Hamiltonian Floer Homology) *Let (M, ω) be a compact symplectically aspherical symplectic manifold without boundary such that $c_1(TM)|_{\pi_2(M)} = 0$ and let $H \in C^\infty(M \times \mathbb{T})$ be a time-dependent Hamiltonian function such that the symplectic action functional $\mathscr{A}_H$ is a Morse–Bott function. Choose an ω-compatible almost complex structure J on M and an additional Morse function $h \in C^\infty(\mathrm{Crit}(\mathscr{A}_H))$ together with a Riemannian metric m on $\mathrm{Crit}(\mathscr{A}_H)$ such that (h, m) is a Morse–Smale pair. For all critical points $\gamma^\pm \in \mathrm{Crit}(h)$, the moduli space*

$$\mathscr{M}(\gamma^-, \gamma^+) := \{u \in X(\gamma^-, \gamma^+) : (\partial + \sigma)(u) = 0\}$$

is a smooth compact manifold with corners of dimension

$$\dim \mathscr{M}(\gamma^-, \gamma^+) = \mathrm{ind}(\gamma^-) - \mathrm{ind}(\gamma^+) - 1,$$

for $X(\gamma^-, \gamma^+)$ as in (5.9), where

$$\mathrm{ind}(\gamma) = \mu_{\mathrm{CZ}}(\gamma) - \frac{1}{2}\operatorname{sgn}\mathrm{Hess}_h(\gamma) \qquad \forall \gamma \in \mathrm{Crit}(h).$$

Moreover, there is a canonical diffeomorphism

$$\partial\mathscr{M}(\gamma^-, \gamma^+) = \mathscr{M}(\gamma^-, \gamma^+) \cap \partial X \cong \coprod_{\gamma \in \mathrm{Crit}(h)} \mathscr{M}(\gamma^-, \gamma) \times \mathscr{M}(\gamma, \gamma^+).$$

Remark 6.4 (Numerical Implementation of Conley–Zehnder Indices) The Conley–Zehnder index assigns a winding number to every Hamiltonian orbit [77, Section 10.6] by linearising the orbit along its flow and invoking the topology of the symplectic group $\mathrm{Sp}(2n)$ as discussed in Sect. A.1. In practice, it helps to understand which families of orbits connect to each other as the Conley–Zehnder index remains constant if no bifurcation occurs [143]. The Conley–Zehnder index is usually very hard to compute explicitly, but in the case of the restricted three-body problem, there is a numerical implementation based on [142, Appendix A] that could be used for classifying newly generated orbits via generative artificial intelligence [86]. However, finding a symplectic trivialisation for arbitrary spatial orbits is difficult.

The proof of Theorem 6.3 is based on the following result, which implies compactness of the moduli space of unparametrised gradient flow lines of the symplectic action functional, and the details are part of Problem 6.19.

Proposition 6.5 (Bubbling, [24, Proposition 6.6.2]) *Let (M, ω) be a compact symplectically aspherical symplectic manifold and let (u_k) be a sequence of gradient flow lines of the symplectic action functional $\mathscr{A}_H$ for some $H \in C^\infty(M \times \mathbb{T})$ with uniformly bounded energy*

$$E_J(u_k) := \int_{-\infty}^{+\infty} \|\partial_s u_k\|_J^2 ds = \int_{-\infty}^{+\infty} \int_0^1 \omega(J\partial_s u_k, \partial_s u_k) dt ds$$

for some, and hence every, ω-compatible almost complex structure J. Then the sequence of derivatives (Du_k) is uniformly bounded.

Remark 6.6 If the sequence of derivatives is unbounded, we will see that a holomorphic sphere bubbles off as in Fig. 6.2. For a simple example where this phenomenon occurs, consider [24, Section 6.6.a]. More sophisticated examples can be found in [136, Section 4.2]. Studying Floer homology where bubbling occurs is usually quite difficult. Our proof of Proposition 6.5 differs from the existing literature in that it makes use of the Morse–Bott condition and is due to Urs Frauenfelder.

The main idea of the proof is to assume that the derivatives (Du_k) explode and then construct a non-constant J-holomorphic sphere. Indeed, assume that there exists a sequence (s_k, t_k) in $\mathbb{R} \times \mathbb{T}$ such that

$$\lim_{k\to\infty} \|\partial_s u_k(s_k, t_k)\|_J \to +\infty.$$

We rescale the sequence (u_k) by "zooming in with a magnifying glass" as depicted in Fig. 6.2. In order to analytically achieve this, set

$$m_k := \|\partial_s u_k(s_k, t_k)\| \qquad \text{and} \qquad v_k(\sigma, \tau) := u_k\left(\frac{\sigma}{m_k} + s_k, \frac{\tau}{m_k} + t_k\right)$$

Fig. 6.2 Looking at the sequence (u_k) of negative gradient flow lines of the symplectic action functional with a magnifying glass via rescaling

for all $(\sigma, \tau) \in \mathbb{C}$. Then we compute

$$m_k \partial_\sigma v_k(\sigma, \tau) = \partial_s u_k \left(\frac{\sigma}{m_k} + s_k, \frac{\tau}{m_k} + t_k \right),$$

$$m_k \partial_\tau v_k(\sigma, \tau) = \partial_t u_k \left(\frac{\sigma}{m_k} + s_k, \frac{\tau}{m_k} + t_k \right).$$

In particular $\|\partial_\sigma v_k(0,0)\| = 1$ for all $k \in \mathbb{N}$. Moreover, every v_k solves

$$\partial_\sigma v_k(\sigma, \tau) + J \partial_\tau v_k(\sigma, \tau) = \frac{1}{m_k} J X_{H_{\frac{\tau}{m_k} + t_k}}(v_k(\sigma, \tau)) \qquad \forall (\sigma, \tau) \in \mathbb{C}$$

as u_k satisfies the Floer equation (6.1). If there exists $v_\infty \in C^\infty(\mathbb{C}, M)$ such that

$$v_k \xrightarrow{C^\infty_{\mathrm{loc}}} v_\infty, \qquad k \to \infty,$$

modulo subsequences, then v_∞ satisfies

$$\partial_\sigma v_\infty(\sigma, \tau) + J \partial_\tau v_\infty(\sigma, \tau) = 0 \qquad \forall (\sigma, \tau) \in \mathbb{C}.$$

Consequently, v_∞ is a J-holomorphic plane. Using the assumption that the energy of the sequence (u_k) is uniformly bounded, v_∞ can be extended to a J-holomorphic sphere $v \in C^\infty(\mathbb{S}^2, M)$ such that $v|_\mathbb{C} = v_\infty$ by identifying $\mathbb{S}^2 \cong \mathbb{C} \cup \{\infty\}$.

6.1 Elliptic Bootstrapping

We first discuss the reparametrisation sequence (s_k, t_k) in $\mathbb{R} \times \mathbb{T}$ for rescaling the sequence (u_k). One has to carefully choose this reparametrisation in order to get a uniformly bounded gradient of the reparametrised sequence. A prototypical result that achieves this is provided by a classical result.

Lemma 6.7 ([1, Lemma 4.4.4], [24, Lemma 6.6.3]) *Let (X, d) be a complete metric space and let*

$$f: X \to [0, +\infty)$$

be a continuous function. Given $\varepsilon > 0$ and $x \in X$, there exist $0 < \delta \leq \varepsilon$ and $y \in X$ such that

$$d(x, y) \leq 2\delta, \qquad \delta f(y) \geq \varepsilon f(x), \qquad \text{and} \qquad \sup_{z \in \bar{B}_\delta(y)} f(z) \leq 2f(y).$$

Remark 6.8 The converse of Lemma 6.7 is also true; that is, it provides a characterisation of complete metric spaces.

Using Lemma 6.7 we are able to construct a continuous J-holomorphic plane via Ascoli's theorem.

Lemma 6.9 (Rescaling) *Let (u_k) be a sequence of negative gradient flow lines of the symplectic action functional $\mathscr{A}_H$ for some $H \in C^\infty(M \times \mathbb{T})$ and J some ω-compatible almost complex structure. Then there exists a sequence (R_k) in $(0, +\infty)$ with $R_k \to +\infty$ and a sequence $v_k \in C^\infty(B_{R_k}(0), M)$ such that*

$$\lim_{k \to \infty} \|\partial_\sigma v_k + J \partial_\tau v_k\|_\infty = 0 \qquad \text{and} \qquad 1 \leq \|\partial_\sigma v_k\|_\infty \leq 2.$$

Proof Abbreviate $z_k := (s_k, t_k) \in \mathbb{R} \times \mathbb{T}$. Apply Hofer's Lemma 6.7 with

$$f: \mathbb{R} \times \mathbb{T} \to [0, +\infty), \qquad f(s, t) := \|\partial_s u_k(s, t)\|,$$

as well as

$$\varepsilon_k := \frac{1}{\sqrt{m_k}}, \qquad \text{and} \qquad x_k := z_k.$$

Thus there exists a nearby sequence $z_k' = (s_k', t_k') \in \mathbb{R} \times \mathbb{T}$ and $0 < \delta_k \leq \varepsilon_k$ with

$$\delta_k \|\partial_s u_k(z_k')\| \geq \varepsilon_k \|\partial_s u_k(z_k)\| = \sqrt{m_k}$$

and

$$\sup_{z \in B_{\delta_k}(z'_k)} \|\partial_s u_k(z)\| \leq 2\|\partial_s u_k(z'_k)\|.$$

Rescale $m'_k := \|\partial_s u_k(z'_k)\|$, set $R_k := \delta_k m'_k$ and

$$v_k(\sigma, \tau) := u_k\left(\frac{\sigma}{m'_k} + s'_k, \frac{\tau}{m'_k} + t'_k\right) \qquad \forall (\sigma, \tau) \in B_{R_k}(0).$$

As u_k is a solution to the Floer equation (6.1), v_k satisfies the equation

$$\partial_\sigma v_k(\sigma, \tau) + J\partial_\tau v_k(\sigma, \tau) = \frac{1}{m'_k} J X_{H_{\frac{\tau}{m'_k}+t'_k}}(v_k(\sigma, \tau)) \tag{6.4}$$

for all $(\sigma, \tau) \in B_{R_k}(0)$. As $m'_k \geq m_k \to +\infty$, we conclude

$$\lim_{k \to \infty} \|\partial_\sigma v_k + J\partial_\tau v_k\|_\infty = 0.$$

Finally, we compute

$$\begin{aligned}
\|\partial_\sigma v_k\|_\infty &= \sup_{z \in B_{R_k}(0)} \|\partial_\sigma v_k(z)\| \\
&= \frac{1}{m'_k} \sup_{z \in B_{\delta_k}(z'_k)} \|\partial_s u_k(z)\| \\
&\leq \frac{2}{m'_k} \|\partial_s u_k(z'_k)\| \\
&= 2,
\end{aligned}$$

and

$$m'_k \|\partial_\sigma v_k(0, 0)\| = \|\partial_s u_k(z'_k)\| = m'_k,$$

implying

$$\|\partial_\sigma v_k(0, 0)\| = 1.$$

This concludes the proof of the lemma. □

As in Theorem 4.21 we can use Ascoli's theorem to show that the sequence constructed in the Rescaling Lemma 6.9 converges to a continuous holomorphic plane in the C^0_{loc}-topology of local uniform convergence.

Lemma 6.10 *Let (v_k) be the sequence constructed in the Rescaling Lemma 6.9. Then there exists $v_\infty \in C^0(\mathbb{C}, M)$ such that*

$$v_k \xrightarrow{C^0_{\text{loc}}} v_\infty, \qquad k \to \infty,$$

up to a subsequence.

Proof By the rescaling Lemma 6.9 and the compactness of M there exists $C > 0$ such that

$$\|\partial_\tau v_k\|_\infty \leq C \qquad \forall k \in \mathbb{N}.$$

Fix $R > 0$ and choose $K_R \in \mathbb{N}$ such that $R_k \geq R$ for all $k \geq K_R$ and consider restrictions $v_k|_{B_R(0)}$ for all $k \geq K_R$. As the derivatives of v_k are uniformly bounded, we conclude that $v_k|_{B_R(0)}$ belongs to the Sobolev space

$$W^{1,\infty}(B_R(0)) := \{f \in W^{1,\infty}(B_R(0), \mathbb{R}^{4n+1}) : f(B_R(0)) \subseteq M\}$$

for all $k \geq K_k$ where we consider $M^{2n} \hookrightarrow \mathbb{R}^{4n+1}$ via the Whitney embedding Theorem. Thus, by Morrey's inequality [43, Corollary 9.14], every $v_k|_{B_R(0)}$ is of the Hölder class $C^{0,1}(B_R(0))$, and hence $v_k|_{B_R(0)}$ is equicontinuous. As M is compact, Ascoli's Theorem implies the existence of $v^R \in C^0(B_R(0), M)$ with

$$v_k|_{B_R(0)} \xrightarrow{C^0} v^R, \qquad k \to \infty,$$

up to a subsequence. Choose a subsequence (k_j^1) with $k_j^1 \geq K_1$ for all $j \in \mathbb{N}$ and such that there exists $v^1 \in C^0(B_1(0), M)$ with

$$v_{k_j^1}|_{B_1(0)} \xrightarrow{C^0} v^1, \qquad j \to \infty.$$

Inductively, choose a subsequence $(k_j^{\mu+1})$ of (k_j^μ) for all $\mu \in \mathbb{N}$ with $k_j^{\mu+1} \geq K_{\mu+1}$ for all $j \in \mathbb{N}$ and such that there exists $v^{\mu+1} \in C^0(B_{\mu+1}(0), M)$ with

$$v_{k_j^{\mu+1}}|_{B_{\mu+1}(0)} \xrightarrow{C^0} v^{\mu+1}, \qquad j \to \infty.$$

Finally, taking the diagonal subsequence yields

$$v_{k_j^j} \xrightarrow{C^0_{\text{loc}}} v_\infty \in C^0(\mathbb{C}, M), \qquad j \to \infty,$$

implying the existence of a continuous holomorphic plane. □

Next, we gradually improve the regularity of the map $v_\infty \in C^0(\mathbb{C}, M)$ constructed in Lemma 6.10 by bootstrapping. The idea is the same as in Theorem 4.21 to iteratively improve the regularity of v_∞ using the Floer equation (6.1). However, in contrast to the bootstrapping performed in the proof of Theorem 4.21 for a Hamiltonian equation—an ordinary differential equation—we are now dealing with a perturbed elliptic partial differential equation.

Lemma 6.11 (Elliptic Bootstrapping) *Denote by $v_\infty \in C^0(\mathbb{C}, M)$ the map constructed in Lemma 6.10. Then $v_\infty \in C^\infty(\mathbb{C}, M)$ is a nonconstant J-holomorphic plane.*

Proof Fix sequences $(z_j) \subseteq \mathbb{C}$ and $r_j \subseteq (0, +\infty)$ such that the following conditions hold:

- $v_\infty|_{B_{4r_j}(z_j)}$ is contained in a chart U_j of M for all $j \in \mathbb{N}$.
- $\bigcup_{j \in \mathbb{N}} B_{r_j}(z_j) = \mathbb{C}$.
- For all $j \in \mathbb{N}$ there exists K_j such that $v_k|_{B_{2r_j}(z_j)} \subseteq U_j$ for all $k \geq K_j$.

Fix smooth bump functions $\beta_j \in C^\infty(\mathbb{C}, [0, 1])$ for $\overline{B}_{r_j}(z_j)$ supported in $B_{2r_j}(z_j)$ and define

$$\overline{v}_k^j := \beta_j v_k \in C_c^\infty(\mathbb{C}, \mathbb{R}^{2n})$$

for all $k, j \in \mathbb{N}$. We compute

$$\Delta \overline{v}_k^j = (\Delta \beta_j) v_k + 2\partial_\sigma \beta_j \partial_\sigma v_k + 2\partial_\tau \beta_j \partial_\tau v_k + \beta_j \Delta v_k. \tag{6.5}$$

To compute Δv_k, we differentiate (6.4) in charts. Applying ∂_σ we get that

$$\partial_\sigma^2 v_k(\sigma, \tau) + DJ \partial_\sigma v_k(\sigma, \tau) \partial_\tau v_k(\sigma, \tau) + J \partial_\sigma \partial_\tau v_k(\sigma, \tau)$$

is equal to

$$\frac{1}{m_k'} D \operatorname{grad}_J H_{\frac{\tau}{m_k'} + t_k'}(v_k(\sigma, \tau)) \partial_\sigma v_k(\sigma, \tau),$$

where $\operatorname{grad}_J H_t$ denotes the gradient of H_t with respect to the Riemannian metric induced by J, that is, the Riemannian metric $\omega(J\cdot, \cdot)$. Applying ∂_τ to (6.4) we get that

$$\partial_\tau \partial_\sigma v_k(\sigma, \tau) + J \partial_\tau^2 v_k(\sigma, \tau) + DJ(\partial_\tau v_k(\sigma, \tau))^2$$

is equal to

$$\frac{1}{m_k'} D\,\mathrm{grad}_J\, H_{\frac{\tau}{m_k'}+t_k'}(v_k(\sigma,\tau))\partial_\tau v_k(\sigma,\tau) + \frac{1}{m_k'}\partial_\tau\,\mathrm{grad}_J\, H_{\frac{\tau}{m_k'}+t_k'}(v_k(\sigma,\tau)).$$

Hence

$$\Delta v_k = P(\partial_\sigma v_k, \partial_\tau v_k),$$

where P is a polynomial of degree 2 with C^∞-coefficients and so

$$\Delta \overline{v}_k^j = P(\partial_\sigma \overline{v}_k^j, \partial \overline{v}_k^j)$$

by (6.5). As the derivatives $\partial_\sigma v_k$ and $\partial_\tau v_k$ are uniformly bounded, we conclude that $\Delta\overline{v}_k^j$ is also uniformly bounded. Thus $\|\Delta\overline{v}_k^j\|_{L^p}$ is uniformly bounded for all exponents $p \in [1, +\infty]$ as $\Delta\overline{v}_k^j$ is compactly supported by construction. Therefore, the Calderon–Zygmund inequality [136, Corollary B.2.7] implies that $\|\overline{v}_k^j\|_{W^{2,p}}$ is uniformly bounded for all $1 < p < +\infty$. In particular,

$$\|\partial_\sigma \overline{v}_k^j\|_{W^{1,p}} \qquad \text{and} \qquad \|\partial_\tau \overline{v}_k^j\|_{W^{1,p}}$$

are uniformly bounded. Thus again by Morrey's inequality, $\partial_\sigma\overline{v}_k^j$ and $\partial_\tau\overline{v}_k^j$ belong to the Hölder class $C^{0,\alpha}$ for all $0 < \alpha < 1$. By Ascoli's Theorem, $\partial_\sigma\overline{v}_k^j$ and $\partial_\tau\overline{v}_k^j$ admit convergent subsequences and

$$\overline{v}_\infty^j := \beta_j v_\infty \in C_c^1(\mathbb{C}, \mathbb{R}^{2n})$$

satisfies

$$v_k|_{B_{r_j}(z_j)} \xrightarrow{C^1} \overline{v}_\infty^j$$

for all $j \in \mathbb{N}$. Hence we have showed that $v_\infty \in C^1(\mathbb{C}, M)$ and

$$v_k \xrightarrow{C^1_{\mathrm{loc}}} v_\infty$$

up to subsequences. This procedure can be generalised to higher derivatives and is referred to as *elliptic bootstrapping*. Note that taking higher order derivatives makes only sense locally in a chart if we do not refer to a particular connection. Thus we get $v_\infty \in C^\infty(\mathbb{C}, M)$ and

$$v_k \xrightarrow{C^\infty_{\mathrm{loc}}} v_\infty, \qquad k \to \infty,$$

up to subsequences. Finally, v_∞ is nonconstant as

$$\|\partial_\sigma v_\infty\|_\infty = \lim_{k\to\infty} \|\partial_\sigma v_k\|_\infty \geq 1$$

and satisfies

$$\partial_\sigma v_\infty + J\partial_\tau v_\infty = 0$$

by the Rescaling Lemma 6.9. □

6.2 Removal of Singularities

Many results of classical complex analysis carry over to J-holomorphic curves. An example is the removal of singularities [136, Theorem 4.1.2], implying that a J-holomorphic curve from a punctured disc can be smoothly extended to the whole disc provided it has finite energy.

Exercise 6.12 Which properties do complex and J-holomorphic functions share?

Consider the conformal diffeomorphism

$$\varphi\colon \mathbb{C}\setminus\{0\} \to \mathbb{R}\times\mathbb{T}, \qquad \varphi(re^{2\pi i\theta}) := \mathrm{Log}(re^{2\pi i\theta}),$$

and define

$$w\colon \mathbb{R}\times\mathbb{T} \to M, \qquad w := v_\infty \circ \varphi^{-1},$$

where $v_\infty \in C^\infty(\mathbb{C}, M)$ denotes the nonconstant J-holomorphic plane constructed in Lemma 6.10. Then

$$\lim_{s\to-\infty} w(s,\cdot) = v_\infty(0)$$

and w is of finite energy. Indeed, we compute

$$E_J(w) = \int_{\mathbb{R}\times\mathbb{T}} w^*\omega = \int_{-\infty}^{+\infty}\int_0^1 \omega(J\partial_s w(s,t), \partial_s w(s,t))dtds \leq \sup_{k\in\mathbb{N}} E_J(u_k)$$

by [136, Lemma 2.2.1]. Moreover, w is a negative gradient flow line of the symplectic area functional

$$\mathscr{A}_0\colon \Lambda M \to \mathbb{R}, \qquad \mathscr{A}_0(\gamma) = \int_{\mathbb{D}} \bar{\gamma}^*\omega.$$

Lemma 6.13 *For every sequence $(r_k) \subseteq (0, +\infty)$ with $r_k \to +\infty$ as $k \to \infty$ define a sequence (w_k) by*

$$w_k(s,t) := w(s + r_k, t) \qquad \forall (s,t) \in \mathbb{R} \times \mathbb{T}.$$

Then there exists a point $w_\infty \in M$ such that

$$w_k \xrightarrow{C^\infty_{\text{loc}}} w_\infty, \qquad k \to \infty,$$

up to a subsequence.

Proof By [136, pp. 89–90] there exists a constant $a \geq 0$ such that $\Delta e \geq -ae^2$ for

$$e \colon \mathbb{R} \times \mathbb{T} \to [0, +\infty), \qquad e(s,t) := \|\partial_s w(s,t)\|^2$$

being the energy density. As $E_J(w) < +\infty$, there exists $R > 0$ such that

$$\int_R^{+\infty} \int_0^1 e(s,t)dtds \leq \min\left\{\frac{\pi}{8a}, 1\right\}$$

and

$$\int_{-\infty}^{-R} \int_0^1 e(s,t)dtds \leq \min\left\{\frac{\pi}{8a}, 1\right\}.$$

Let $z = (s,t) \in \mathbb{R} \times \mathbb{T}$ be such that $|s| \geq R + 1$. Then

$$\int_{B_1(z)} e < \frac{\pi}{8a}.$$

By [136, Lemma 4.3.3] we conclude

$$e(z) \leq \frac{8}{\pi} \int_{B_1(z)} e \leq \frac{8}{\pi}.$$

Consequently, e is uniformly bounded, as $e|_{[-R-1,R+1]\times\mathbb{T}}$ is uniformly bounded by continuity of e. Hence $\|\partial_s w\|$ is uniformly bounded by the definition of e and $\|\partial_t w\|$ is uniformly bounded as w is a J-holomorphic curve. By an elliptic bootstrapping argument as in Lemma 6.11, we conclude

$$w_k \xrightarrow{C^\infty_{\text{loc}}} w_\infty, \qquad k \to \infty,$$

up to a subsequence, where w_∞ is a negative gradient flow line of the symplectic area functional $\mathscr{A}_0$. We claim that w_∞ is constant. In fact, assume that w_∞ is not constant. Then there exists $s < s'$ such that

$$\varepsilon := \mathscr{A}_0(w_\infty(s)) - \mathscr{A}_0(w_\infty(s')) > 0.$$

Moreover, there exists $K \in \mathbb{N}$ such that

$$\mathscr{A}_0(w_k(s)) - \mathscr{A}_0(w_k(s')) = \mathscr{A}_0(w(s + r_k)) - \mathscr{A}_0(w(s' + r_k)) \geq \frac{\varepsilon}{2}$$

for all $k \geq K$. Define a subsequence (r_{k_j}) of (r_k) recursively by

$$k_0 := K \qquad \text{and} \qquad k_j := \min\left\{l \in \mathbb{N} : s' + r_{k_{j-1}} \leq s + r_l\right\}.$$

This works as $r_k \to +\infty$ for $k \to \infty$. Fix $l \in \mathbb{N}$. Then we compute

$$\begin{aligned} E(w) &= \sup_{s\in\mathbb{R}} \mathscr{A}_0(w(s)) - \inf_{s\in\mathbb{R}} \mathscr{A}_0(w(s)) \\ &\geq \mathscr{A}_0(w(s + r_{k_0})) - \mathscr{A}_0(w(s + r_{k_l})) \\ &= \sum_{\nu=1}^{l} \big(\mathscr{A}_0(w(s + r_{k_{\nu-1}})) - \mathscr{A}_0(w(s + r_{k_\nu}))\big) \\ &\geq \sum_{\nu=1}^{l} \big(\mathscr{A}_0(w(s + r_{k_{\nu-1}})) - \mathscr{A}_0(w(s' + r_{k_{\nu-1}}))\big) \\ &\geq \frac{\varepsilon l}{2}. \end{aligned}$$

As $l \in \mathbb{N}$ was arbitrary, we conclude that $E(w) = +\infty$. □

Proof of Proposition 6.5 By the compactness of M, there exists a finite open cover $U_1, \ldots, U_m$ of M such that $\bar{U}_j$ is contained in a Darboux chart. Define

$$\mathscr{U} := \{\gamma \in \Lambda M : \gamma(\mathbb{T}) \subseteq U_j \text{ for some } j = 1, \ldots, m\}.$$

Step 1: There exists $s_0 \in \mathbb{R}$ such that $w(s) \in \mathscr{U}$ for all $s \geq s_0$. Assume that there exists a sequence $(r_k) \subseteq (0, +\infty)$ with $r_k \to +\infty$ as $k \to \infty$ and $w(r_k) \notin \mathscr{U}$ for all $k \in \mathbb{N}$. By Lemma 6.13, there exists $w_\infty \in M$ such that

$$w_k \xrightarrow{C^\infty_{\text{loc}}} w_\infty, \qquad k \to \infty,$$

up to a subsequence. But $w_\infty \in U_j$ for some j and therefore there exists $K \in \mathbb{N}$ such that $w(r_k, t) \in U_j$ for all $t \in \mathbb{T}$ and $k \geq K$. Consequently, $w(r_k) \in \mathscr{U}$ for all $k \geq K$.

Step 2: There exist constants $C > 0$ and $\kappa > 0$ such that

$$d_{L^2}(w(s), w_\infty) \le Ce^{-\kappa s} \qquad \forall s \ge s_0,$$

where w_∞ is the limit of $w(k) = w_k(0)$ as $k \to \infty$ up to a subsequence (k_j). This is precisely where we use the fact that the area functional $\mathscr{A}_0$ is a Morse–Bott function and thus its gradient flow lines decay exponentially as in the finite-dimensional case by Proposition 5.33. By Exercises 5.36 and 5.37, we have the action-energy inequality

$$|\mathscr{A}_0(\gamma)| \le C_0 \| \operatorname{grad}_J \mathscr{A}_0(\gamma)\|_J^2 \qquad \forall \gamma \in \mathscr{U}$$

for some constant $C_0 > 0$ as the symplectic area functional $\mathscr{A}_0$ is Morse–Bott. See Remark 5.6. Let $s \ge s_0$. Choose j_0 such that $k_{j_0} > s$. We estimate

$$\mathscr{A}_0(w(s)) > \mathscr{A}_0(w(k_{j_0})) \ge \lim_{j\to\infty} \mathscr{A}_0(w(k_j)) = \lim_{j\to\infty} \mathscr{A}_0(w_{k_j}(0)) = \mathscr{A}_0(w_\infty) = 0.$$

Let $s_2 > s_1 > s_0$. Using Lemmas 5.38 and 5.39 we compute

$$\begin{aligned} d(w(s_1), w(s_2)) &\le \frac{2}{\sqrt{C_0}} \left(\sqrt{\mathscr{A}_0(w(s_1))} - \sqrt{\mathscr{A}_0(w(s_2))} \right) \\ &\le \frac{2}{\sqrt{C_0}} \sqrt{\mathscr{A}_0(w(s_1))} \\ &\le \frac{2}{\sqrt{C_0}} \sqrt{\mathscr{A}_0(w(s_0))} e^{\frac{1}{2C_0}(s_0 - s_1)}. \end{aligned}$$

Choose j_1 such that $k_{j_1} \ge s_1$. Then for all $j \ge j_1$, we have that

$$d_{L^2}(w(s_1), w(k_j)) \le Ce^{-\kappa s_1},$$

where

$$C := \frac{2}{\sqrt{C_0}} \sqrt{\mathscr{A}_0(w(s_0))} e^{\frac{1}{2C_0} s_0} \qquad \text{and} \qquad \kappa := \frac{1}{2C_0}.$$

Thus

$$d_{L^2}(w(s_1), w_\infty) = \lim_{j\to\infty} d_{L^2}(w(s_1), w(k_j)) \le Ce^{-\kappa s_1}.$$

Step 3: There exists a unique point $w_\infty \in M$ such that

$$w(s) \xrightarrow{C^\infty} w_\infty, \qquad s \to +\infty.$$

Uniqueness follows immediately from the exponential decay established in Step 2. In fact, a priori, w_∞ depends on the choice of subsequence (k_j). Let $w'_\infty \in M$ be the limit of a different subsequence. Then we compute

$$d_{L^2}(w_\infty, w'_\infty) \leq d_{L^2}(w(s), w_\infty) + d_{L^2}(w(s), w'_\infty) \leq Ce^{-\kappa s} + C'e^{-\kappa' s} \to 0$$

as $s \to +\infty$. That the limit can also be taken with respect to the C^∞-topology follows from [24, Proposition 6.5.3]. □

An immediate corollary to the Proposition 6.5 is a compactness result for sequences of negative gradient flow lines of the Rabinowitz action functional in Definition 4.23. This result will be used throughout the last part of the book.

Corollary 6.14 *Let (M, ω) be a symplectically aspherical symplectic manifold and let (u_k, τ_k) be a sequence of solutions $(u_k, \tau_k) \in C^\infty(\mathbb{R} \times \mathbb{T}, M) \times C^\infty(\mathbb{R}, \mathbb{R})$ of*

$$\begin{cases} \partial_s u_k(s,t) + J(\partial_t u_k(s,t) - \tau_k(s) X_H(u_k(s,t))) = 0, \\ \partial_s \tau_k(s) = \int_0^1 H(u_k(s,t))dt, \end{cases}$$

for all $s \in \mathbb{R}$ and $k \in \mathbb{N}$ for some $H \in C^\infty(M)$, with uniformly bounded energy

$$E_J(u_k) := \int_{-\infty}^{+\infty} \int_0^1 \omega(J\partial_s u_k(s,t), \partial_s u_k(s,t))dtds + \int_{-\infty}^{+\infty} (\partial_s \tau_k(s))^2 ds.$$

If there exists a compact subset $K \subseteq M \times \mathbb{R}$ such that

$$\operatorname{im}(u_k, \tau_k) \subseteq K \qquad \forall k \in \mathbb{N},$$

then the derivatives of (u_k) are uniformly bounded.

Proof Crucial is the observation that under the above assumptions, the limit of (6.4) is still a J-holomorphic curve. □

We conclude Part II with an exercise comparing the construction of Hamiltonian Floer homology in contrast to the finite-dimensional setting of Morse–Bott homology encountered in the last chapter.

Exercise 6.15 (Two Viewpoints on Hamiltonian Floer Homology) Let (M^{2n}, ω) be a compact symplectic manifold and let $H \in C^\infty(M \times \mathbb{T})$ be a time-dependent Hamiltonian function such that the symplectic action functional $\mathscr{A}_H$ is a Morse–Bott function. Choose some additional Morse function $h \in C^\infty(\operatorname{Crit}(\mathscr{A}_H))$ together with a Riemannian metric m on the critical manifold $\operatorname{Crit}(\mathscr{A}_H)$ such that (h, m) is a Morse–Smale pair. Below we describe two models of the construction of a Morse–

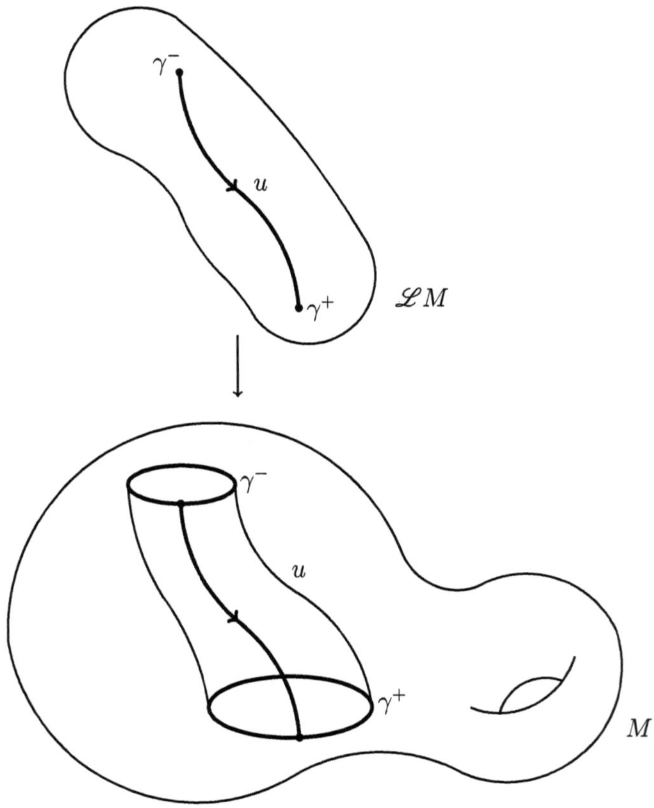

Fig. 6.3 Two viewpoints of a negative gradient flow line u of the symplectic action functional $\mathscr{A}_H$ connecting two critical points $\gamma^{\pm} \in \mathrm{Crit}(\mathscr{A}_H)$

Bott homology for $\mathscr{A}_H$. Compare these two models and explain in detail what is correct in each model and what is a simplification. Are there also any misleading concepts? See also Fig. 6.3 for an illustration.

1. As the symplectic action functional $\mathscr{A}_H$ is a Morse–Bott function, we can simply define

$$\mathrm{HF}_*(H) := \mathrm{HM}_*(\mathscr{A}_H).$$

   Here negative gradient flow lines are maps $u \in C^\infty(\mathbb{R}, \Lambda M)$ satisfying

$$\dot{u}(s) = -\operatorname{grad}_J \mathscr{A}_H(u(s)) \qquad \forall s \in \mathbb{R}, \tag{6.6}$$

   where J is an ω-compatible almost complex structure and $\operatorname{grad}_J \mathscr{A}_H \in \mathfrak{X}(\Lambda M)$ denotes the gradient of $\mathscr{A}_H$. If the gradient flow line has the asymptotics

$$\lim_{s \to \pm\infty} u(s) = \gamma^{\pm} \in \operatorname{Crit}(\mathscr{A}_H) \subseteq \Lambda M,$$

with respect to the C^0-topology, that is, convergence is uniform in $t \in \mathbb{T}$, the dimension of the moduli space is given by

$$\operatorname{ind}(\gamma^-) - \operatorname{ind}(\gamma^+) - 1,$$

where

$$\operatorname{ind}(\gamma) := \mu_{\mathrm{CZ}}(\gamma) - \frac{1}{2} \operatorname{sgn} \operatorname{Hess}_h(\gamma) \qquad \forall \gamma \in \operatorname{Crit}(h).$$

Compactness, transversality and gluing follow from the abstract perturbation techniques provided by polyfold theory. Again, compactness is understood up to broken gradient flow lines. As Floer homology does not depend on the Hamiltonian function H, we conclude

$$\mathrm{HF}_*(H) \cong \mathrm{H}_{*+n}(M; \mathbb{Z}_2)$$

by the observation $\operatorname{Crit}(\mathscr{A}_0) \cong M$ under the canonical identification $\gamma \mapsto \gamma(0)$. Hence the Floer homology of a time-dependent Hamiltonian function does not give an interesting global invariant of (M, ω).

2. We assume that the symplectic manifold is symplectically aspherical and satisfies the condition $c_1(TM)|_{\pi_2(M)} = 0$. Then we can define

$$\mathrm{HF}_*(H) := \mathrm{HM}_*(\mathscr{A}_H).$$

One needs to be careful, as ΛM is an infinite-dimensional Fréchet manifold and thus behaves very badly. A negative gradient flow line is best interpreted as a Floer cylinder $u \in C^\infty(\mathbb{R} \times \mathbb{T}, M)$ satisfying the Floer equation

$$\partial_s u(s,t) + J(\partial_t u(s,t) - X_{H_t}(u(s,t))) = 0 \qquad \forall (s,t) \in \mathbb{R} \times \mathbb{T},$$

where J is an ω-compatible almost complex structure. This interpretation allows us to use elliptic bootstrapping for ensuring compactness modulo breaking of the relevant moduli spaces in absence of bubbling. If the gradient flow line has the asymptotics

$$\lim_{s \to \pm\infty} u(s) = \gamma^{\pm} \in \operatorname{Crit}(\mathscr{A}_H) \subseteq \Lambda M,$$

the dimension of the moduli space is given by

$$\operatorname{ind}(\gamma^-) - \operatorname{ind}(\gamma^+) - 1,$$

where

$$\mathrm{ind}(\gamma) := \mu_{\mathrm{CZ}}(\gamma) - \frac{1}{2}\,\mathrm{sgn}\,\mathrm{Hess}_h(\gamma) \qquad \forall \gamma \in \mathrm{Crit}(h).$$

By invariance and the observation that $\mathrm{Crit}(\mathscr{A}_0) \cong M$ by the canonical identification $\gamma \mapsto \gamma(0)$ we conclude

$$\mathrm{HF}_*(H) \cong \mathrm{H}_{*+n}(M; \mathbb{Z}_2).$$

Hence the Floer homology of a time-dependent Hamiltonian function does not give an interesting global invariant of (M, ω).

Problems

6.16 (Invariance of Hamiltonian Floer Homology) Let (M, ω) be a compact symplectically aspherical symplectic manifold.

1. Let $H \in C^\infty(M \times \mathbb{T})$ be such that the symplectic action functional $\mathscr{A}_H$ is a Morse–Bott function and pick two Morse functions $h^\pm \in \mathrm{Crit}(\mathscr{A}_H)$. Assume that $J^\pm$ are two ω-compatible almost complex structures giving us transversality of the relevant moduli spaces of negative gradient flow lines of the symplectic action functional $\mathscr{A}_H$. Choose a generic homotopy from J^- to J^+ of ω-compatible almost complex structures and consider solutions to the s-dependent Floer equation

$$\partial_s u(s,t) + J_s(\partial_t u(s,t) - X_{H_t}(u(s,t))) = 0 \qquad \forall (s,t) \in \mathbb{R} \times \mathbb{T}. \tag{6.7}$$

For two critical points $\gamma^\pm \in \mathrm{Crit}(h^\pm)$ denote the moduli space of all solutions to the s-dependent Floer equation (6.7) by $\mathscr{M}(\gamma^-, \gamma^+)$. Define a linear map

$$\Phi\colon \mathrm{CF}(H) \to \mathrm{CF}(H), \qquad \Phi(\gamma^-) := \sum_{\gamma^+ \in \mathrm{Crit}(h^+)} \#_2 \mathscr{M}^0(\gamma^-, \gamma^+)\gamma^+.$$

(a) Show that the map Φ induces an isomorphism.
(b) Assume that $c_1(TM)|_{\pi_2(M)} = 0$. Then one can consider the restriction

$$\Phi_k\colon \mathrm{CF}_k(H) \to \mathrm{CF}_k(H), \qquad \Phi(\gamma^-) := \sum_{\substack{\gamma^+ \in \mathrm{Crit}(h^+)\\ \mathrm{ind}(\gamma^+) = k}} \#_2 \mathscr{M}^0(\gamma^-, \gamma^+)\gamma^+,$$

for all $k \in \mathbb{Z}$. Show that then ϕ induces a chain complex isomorphism from the Floer complex of (H, J^-) to (H, J^+).

2. Let $H^\pm \in C^\infty(M \times \mathbb{T})$ be such that the symplectic action functionals $\mathscr{A}_{H^\pm}$ are Morse–Bott functions and pick two Morse functions $h^\pm \in \mathrm{Crit}(\mathscr{A}_H)$. Choose

a generic homotopy from H^- to H^+ and consider solutions to the s-dependent Floer equation

$$\partial_s u(s,t) + J(\partial_t u(s,t) - X_{H_{s,t}}(u(s,t))) = 0 \qquad \forall (s,t) \in \mathbb{R} \times \mathbb{T}. \tag{6.8}$$

For two critical points $\gamma^\pm \in \mathrm{Crit}(h^\pm)$ denote the moduli space of all solutions to the s-dependent Floer equation (6.8) by $\mathscr{M}(\gamma^-, \gamma^+)$. Define a linear map

$$\Phi_{H^-,H^+} \colon \mathrm{CF}(H^-) \to \mathrm{CF}(H^+), \qquad \Phi(\gamma^-) := \sum_{\gamma^+ \in \mathrm{Crit}(h^+)} \#_2 \mathscr{M}^0(\gamma^-, \gamma^+)\gamma^+.$$

(a) By considering "homotopies of homotopies", show that the chain map Φ_{H^+,H^-} is a homotopy inverse to Φ_{H^-,H^+}. The induced isomorphism of Φ_{H^-,H^+} at the level of homology is called a ***continuation homomorphism***.

(b) Assume that $c_1(TM)|_{\pi_2(M)} = 0$. Then one can consider the restriction

$$\Phi_k \colon \mathrm{CF}_k(H^-) \to \mathrm{CF}_k(H^+), \qquad \Phi(\gamma^-) := \sum_{\substack{\gamma^+ \in \mathrm{Crit}(h^+) \\ \mathrm{ind}(\gamma^+) = k}} \#_2 \mathscr{M}^0(\gamma^-, \gamma^+)\gamma^+,$$

for all $k \in \mathbb{Z}$. Show that then ϕ_{H^-,H^+} induces an isomorphism

$$\mathrm{HF}_k(H^-) \cong \mathrm{HF}_k(H^+) \qquad \forall k \in \mathbb{Z}.$$

6.17 (Nondegenerate Periodic Orbits) Let (M, ω) be a symplectically aspherical symplectic manifold and let $H \in C^\infty(M \times \mathbb{T})$ be a time-dependent Hamiltonian function. A critical point $\gamma \in \mathrm{Crit}(\mathscr{A}_H)$ is said to be ***nondegenerate***, if

$$\ker\left(D\phi_1^H(\gamma(0)) - \mathrm{id}_{T_{\gamma(0)}M}\right) = \{0\}.$$

We say that H is nondegenerate, if every critical point of $\mathscr{A}_H$ is nondegenerate.

1. Assume that $H \in C^\infty(M)$ is a time-independent Hamiltonian function. Can a critical point $\gamma \in \mathrm{Crit}(\mathscr{A}_H)$ be nondegenerate? Explain your answer!
2. Assume that $H \in C^\infty(M)$ is a time-independent Hamiltonian function. Show that if a critical point $x \in \mathrm{Crit}(H)$ is nondegenerate as a constant periodic orbit, then it is nondegenerate as a critical point of H, meaning that the Hessian of H is invertible at x. Conversely, show that if the Hamiltonian function H is sufficiently C^2-small, then every nondegenerate critical point of H is also nondegenerate as a constant periodic orbit. *Hint:* See [24, Proposition 5.4.5] and [24, Remark 5.4.7].
3. Let $H \in C^\infty(M \times \mathbb{T})$ be a nondegenerate Hamiltonian function. Show that the symplectic action functional $\mathscr{A}_H$ is a Morse function.

6.18 (Fredholm Index) Let (M^{2n}, ω) be a compact symplectically aspherical symplectic manifold such that $c_1(TM)|_{\pi_2(M)} = 0$ holds. Let $H \in C^\infty(M \times \mathbb{T})$ be such that the symplectic action functional $\mathscr{A}_H$ is a Morse function. Suppose that $u \in C^\infty(\mathbb{R}\times\mathbb{T}, M)$ is a solution to the Floer equation (6.1) satisfying the asymptotic conditions

$$\lim_{s\to\pm\infty} u(s,\cdot) = \gamma^\pm(\cdot) \in \mathrm{Crit}(\mathscr{A}_H).$$

By choosing a unitary trivialisation of u^*TM and linearising the Floer equation one gets a bounded linear operator of the form

$$W^{1,p}(\mathbb{R}\times\mathbb{T}, \mathbb{R}^{2n}) \to L^p(\mathbb{R}\times\mathbb{T}, \mathbb{R}^{2n}), \quad v \mapsto \partial_s v(s,t) + J_0\partial_t v(s,t) - A(s,t)v(s,t),$$

for all $p \in (2, \infty)$, where

$$J_0 := \begin{pmatrix} 0 & I_n \\ -I_n & 0 \end{pmatrix} \in \mathbb{R}^{2n\times 2n}$$

for the identity matrix $I_n \in \mathbb{R}^{n\times n}$, and

$$A\colon [-\infty, +\infty] \times \mathbb{T} \to L(\mathbb{R}^{2n}, \mathbb{R}^{2n})$$

is continuous such that

$$\dot{w}(t) = J_0 A(-\infty, t) w(t) \qquad \text{and} \qquad \dot{w}(t) = J_0 A(+\infty, t) w(t)$$

are the linearisations of Hamilton's equations along $\gamma^\pm$. Show that the Fredholm index of the linearised Floer equation is $\mu_{\mathrm{CZ}}(\gamma^-) - \mu_{\mathrm{CZ}}(\gamma^+)$. How must one proceed if $\mathscr{A}_H$ is a Morse–Bott function? *Hint:* See [51, Section 4]. The restriction $p > 2$ is needed to find continuous representatives in $W^{1,p}(\mathbb{R} \times \mathbb{T}, \mathbb{R}^{2n})$ since in this case we have that $\dim(\mathbb{R} \times \mathbb{T}) = 2$.

6.19 (M-Polyfold Setup for Morse–Bott Homology) Prove Theorems 5.41 and 6.3 in full detail. Can you suggest a formulation if we want the analogue statement for coefficients in $\mathbb{Z}$?

Part III
Twisted Rabinowitz–Floer Homology

Rabinowitz–Floer homology is the Morse–Bott homology in the sense of Floer associated with the Rabinowitz action functional introduced by Kai Cieliebak and Urs Frauenfelder in [51] and [12]. For an overview see the survey article [16]. In this last part, we consider a generalisation of this theory to a Rabinowitz–Floer homology of a Liouville automorphism as in [27]. The first application is an existence result of noncontractible periodic Reeb orbits on quotients of symmetric star-shaped hypersurfaces. In particular, our result applies to lens spaces. The second application is a generalisation of Schlenk's theorem [56, Theorem 4.9] about the existence of contractible periodic Reeb orbits on stable, displaceable hypersurfaces in symplectically aspherical, geometrically bounded, symplectic manifolds, to a forcing result for contractible twisted periodic Reeb orbits as in [28]. We make use of holomorphic curve techniques for a suitable generalisation of the Rabinowitz action functional in the stable case in order to prove the forcing result. As in Schlenk's theorem, we derive a lower bound for the displacement energy of the displaceable hypersurface in terms of the action value of such periodic orbits. In the end, we point out how this estimate can be used for deriving bounds for the systolic ratio of convex symmetric hypersurfaces based on ideas established in [119] and the strong Viterbo conjecture reviewed in [100].

Definition of Twisted Rabinowitz–Floer Homology

7

In the last part of this book, we introduce an analogue of the twisted Floer homology [185] in the Rabinowitz–Floer setting. Following [51] and [12], we construct a Morse–Bott homology for a suitable generalisation of the Rabinowitz action functional. The main two goals you should have in mind are detecting noncontractible periodic Reeb orbits on lens spaces and showing the existence of symmetric periodic Reeb orbits on completions of Liouville domains. This setup is similar to the one in symplectic homology as described in Problem 7.61, but requires some additional analytical considerations by bounding Lagrange multipliers. As Liouville domains will be required in this part, we first introduce this important concept (Fig. 7.1).

Definition 7.1 (Liouville Domain) A ***Liouville domain*** is defined as a compact connected exact symplectic manifold (W, λ) with a connected boundary such that the Liouville vector field $X = X_\lambda \in \mathfrak{X}(W)$ implicitly defined by $i_X d\lambda = \lambda$ points outward along the boundary ∂W.

Exercise 7.2 Let (W, λ) be a Liouville domain. Show that $\partial W \neq \varnothing$ and that the boundary $(\partial W, \lambda|_{\partial W})$ is a contact manifold.

Example 7.3 (Star-Shaped Hypersurface) Let $\Sigma = H_\Sigma^{-1}(0) \subseteq \mathbb{R}^{2n}$ be a star-shaped hypersurface. Then the sublevel set $W_\Sigma := H_\Sigma^{-1}((-\infty, 0])$ is a properly embedded submanifold of dimension $2n$ with boundary $\partial W = \Sigma$ by [129, Proposition 5.47 (a)]. Moreover, W_Σ is compact and connected with a connected boundary $\partial W_\Sigma = \Sigma$. The corresponding Liouville vector field from Example 2.8 is outward-pointing along Σ since

$$dH_\Sigma(X)|_\Sigma = -i_{X_{H_\Sigma}}(X)|_\Sigma = d\lambda_0(X, X_{H_\Sigma})|_\Sigma = \lambda_0(X_{H_\Sigma})|_\Sigma = 1.$$

Thus $(W_\Sigma, \lambda_0|_{W_\Sigma})$ is a Liouville domain.

Y. Bähni, *Lectures on Twisted Rabinowitz-Floer Homology*, Pathways in Mathematics, https://doi.org/10.1007/978-3-032-10673-5_7

Fig. 7.1 A Liouville domain

Definition 7.4 (Liouville Automorphism) Let (W, λ) be a Liouville domain. A diffeomorphism $\varphi \in \mathrm{Diff}(W)$ of finite order is said to be a ***Liouville automorphism***, if $\varphi^*\lambda - \lambda$ is exact and compactly supported in the interior $\mathrm{Int}\, W$. The set of all Liouville automorphisms of the Liouville domain (W, λ) is denoted by $\mathrm{Aut}(W, \lambda)$.

Remark 7.5 By diffeomorphism invariance of the boundary [129, Theorem 2.18], any Liouville automorphism $\varphi \in \mathrm{Aut}(W, \lambda)$ restricts itself to a diffeomorphism of the boundary $\varphi|_{\partial W} \in \mathrm{Diff}(\partial W)$.

Example 7.6 (Star-Shaped Hypersurface) For $m \geq 1$ and $k_1, \dots, k_n \in \mathbb{Z}$ coprime to m, consider the rotation

$$\varphi\colon \mathbb{C}^n \to \mathbb{C}^n, \qquad \varphi(z_1, \dots, z_n) := \left(e^{2\pi i k_1/m} z_1, \dots, e^{2\pi i k_n/m} z_n\right).$$

Let $W_\Sigma := H_\Sigma^{-1}((-\infty, 0]) \subseteq \mathbb{R}^{2n}$ be the Liouville domain from Example 7.3 associated with a star-shaped hypersurface $\Sigma = \Sigma_\rho$ such that $\rho \circ \varphi|_{\mathbb{S}^{2n-1}} = \rho$. Then $\varphi|_{W_\Sigma}$ is a Liouville automorphism of $(W_\Sigma, \lambda_0|_{W_\Sigma})$.

Exercise 7.7 Let $\varphi \in \mathrm{Aut}(W, \lambda)$ be a Liouville automorphism.

1. Show that there exists a unique function $f_\varphi \in C_c^\infty(\mathrm{Int}\, W)$ such that

$$\varphi^*\lambda - \lambda = df_\varphi.$$

2. Show that the set $\mathrm{Aut}(W, \lambda)$ of Liouville automorphisms is in general not a group.

Exercise 7.8 Let (W, λ) be a Liouville domain. Show that there exists an exact symplectic embedding

$$\psi\colon \left((-\infty, 0] \times \partial W, e^r \lambda|_{\partial W}\right) \hookrightarrow (W, \lambda), \qquad \psi(r, t) := \phi_r^X(x),$$

where $X = X_\lambda \in \mathfrak{X}(W)$ denotes the Liouville vector field.

Fig. 7.2 The completion of a Liouville domain

Definition 7.9 (Completion of a Liouville Domain) The ***completion of a Liouville domain*** (W, λ) is defined to be the exact symplectic manifold $(\widehat{W}, \lambda)$ given by

$$\widehat{W} := W \cup_{\partial W} [0, +\infty) \times \partial W \qquad \text{and} \qquad \lambda|_{[0,+\infty)\times\partial W} := e^r \lambda|_{\partial W}.$$

Exercise 7.10 Explain why the completion $(\widehat{W}, \lambda)$ is well-defined (Fig. 7.2).

Example 7.11 (Star-Shaped Hypersurface) Let $(W_\Sigma, \lambda_0|_{W_\Sigma})$ be the Liouville domain from Example 7.3. The completion is symplectomorphic to $(\mathbb{R}^{2n}, \lambda_0)$ by considering the explicit symplectomorphism

$$(0, +\infty) \times \Sigma \to \mathbb{R}^{2n} \setminus W_\Sigma, \qquad (r, z) \mapsto e^{r/2} z.$$

Remark 7.12 An example of a Liouville domain that we have not discussed so far is provided in Problem 7.60. This example is particularly interesting from the point of view of algebraic geometry.

Remark 7.13 Let $\varphi \in \operatorname{Aut}(W, \lambda)$ be a Liouville automorphism and let $(\widehat{W}, \lambda)$ be the completion of the Liouville domain (W, λ). Then φ naturally extends to $[0, +\infty) \times \partial W$ by setting $\varphi|_{[0,+\infty)\times\partial W} := \varphi(x)$, as by Remark 7.5 we have $\varphi|_{\partial W} \in \operatorname{Diff}(\partial W)$.

Let $\varphi \in \operatorname{Aut}(W, \lambda)$ be a Liouville automorphism and let $(\widehat{W}, \lambda)$ be the completion of the Liouville domain (W, λ). As in Exercise 4.43, denote by

$$\mathscr{A}_\varphi^0 \colon \mathscr{L}_\varphi \widehat{W} \to \mathbb{R}, \qquad \mathscr{A}_\varphi^0(\gamma) := \int_0^1 \gamma^* \lambda,$$

the twisted symplectic area functional. This functional can be used to detect fixed points in $\widehat{W}$, and if we add a Lagrange multiplier to it similar to the Rabinowitz action functional (Definition 4.23) we can detect twisted periodic orbits on the boundary ∂W. In order for this to work, we need to represent the boundary ∂W of the Liouville domain W as a regular energy hypersurface in the completion $\widehat{W}$.

Definition 7.14 (Twisted Hamiltonian Function) Let $(\widehat{W}, \lambda)$ be the completion of a Liouville domain (W, λ) and let $\varphi \in \mathrm{Aut}(W, \lambda)$ be a Liouville automorphism. A function $H \in C^\infty(\widehat{W} \times \mathbb{R})$ is a ***twisted Hamiltonian function***, if $\varphi^* H_{t+1} = H_t$ holds for all $t \in \mathbb{R}$. We denote the space of all twisted Hamiltonian functions by $C^\infty_\varphi(\widehat{W} \times \mathbb{R})$ and the subspace of all autonomous twisted Hamiltonian functions by $C^\infty_\varphi(\widehat{W})$.

Definition 7.15 (Twisted Defining Hamiltonian Function) Let (W, λ) be a Liouville domain and let $\varphi \in \mathrm{Aut}(W, \lambda)$ be a Liouville automorphism. A ***twisted defining Hamiltonian function for*** $\boldsymbol{\partial W}$ is a twisted Hamiltonian function $H \in C^\infty_\varphi(\widehat{W})$ on the completion $(\widehat{W}, \lambda)$ of (W, λ), satisfying the following conditions:

1. $H^{-1}(0) = \partial W$ and $\partial W \cap \mathrm{Crit}(H) = \varnothing$.
2. dH is compactly supported.
3. $X_H|_{\partial W} = R$ is the Reeb vector field of the contact manifold $(\partial W, \lambda|_{\partial W})$.

We denote by $\mathscr{F}_\varphi(\partial W)$ the set of twisted defining Hamiltonian functions for ∂W.

Remark 7.16 A necessary condition for $\mathscr{F}_\varphi(\partial W) \neq \varnothing$ is that $\varphi^* R = R$ since

$$\varphi^* R = \varphi^* X_H|_{\partial W} = X_{\varphi^* H}|_{\partial W} = X_H|_{\partial W} = R$$

by Lemma 2.17 and the definition of $C^\infty_\varphi(\widehat{W})$. This is not true in general if φ is not a strict contactomorphism on ∂W, that is, if $\varphi^*\lambda - \lambda \neq 0$ on the boundary ∂W.

Example 7.17 (Star-Shaped Hypersurface) Let $(W_\Sigma, \lambda_0|_{W_\Sigma})$ be the Liouville domain associated with a star-shaped hypersurface $\Sigma \subseteq \mathbb{R}^{2n}$ described in Example 7.7. Then the Hamiltonian function H_Σ defined in (1.2) belongs to $\mathscr{F}_\varphi(\Sigma)$.

Lemma 7.18 *Let (W, λ) be a Liouville domain and let $\varphi \in \mathrm{Aut}(W, \lambda)$ be a Liouville automorphism. Then $\mathscr{F}_\varphi(\partial W) \neq \varnothing$.*

Proof The construction of the twisted defining Hamiltonian H for ∂W is inspired by the proof of [56, Proposition 4.1]. Fix $\delta > 0$ such that the exact symplectic embedding

$$\psi \colon \left((-\delta, 0] \times \Sigma, e^r \lambda|_{\partial W}\right) \hookrightarrow (W, \lambda)$$

defined by

$$\psi(r, x) := \phi^X_r(x)$$

satisfies

$$U_\delta := \psi((-\delta, 0] \times \partial W) \cap \operatorname{supp} f_\varphi = \varnothing. \tag{7.1}$$

Indeed, that ψ is an exact symplectic embedding follows from the computation

$$\begin{aligned}
\frac{d}{dr}\psi_r^*\lambda &= \frac{d}{dr}\left(\phi_r^X\right)^*\lambda \\
&= (\phi_r^X)^* L_X\lambda \\
&= \left(\phi_r^X\right)^*(di_X\lambda + i_X d\lambda) \\
&= \left(\phi_r^X\right)^*(di_X i_X d\lambda + \lambda) \\
&= \left(\phi_r^X\right)^*\lambda \\
&= \psi_r^*\lambda
\end{aligned}$$

similar to Exercise 7.8, implying $\psi_r^*\lambda = e^r\lambda|_\Sigma$ for all $r \in (-\delta, 0]$ by $\psi_0 = \iota_{\partial W}$, where $\iota_{\partial W} \colon \partial W \hookrightarrow W$ denotes the inclusion. Note that $\psi_r^* X = \partial_r$ for the Liouville vector field $X \in \mathfrak{X}(W)$. We claim

$$\varphi(\psi(r, x)) = \psi(r, \varphi(x)) \qquad \forall (r, x) \in (-\delta, 0] \times \partial W, \tag{7.2}$$

that is, φ and ψ commute. Note that (7.2) makes sense because $\varphi(\partial W) = \partial W$ because $\varphi|_{\partial W} \in \operatorname{Diff}(\partial W)$ by assumption. Indeed, (7.2) follows from the uniqueness of integral curves and the computation

$$\begin{aligned}
\frac{d}{dr}\varphi(\psi(r, x)) &= \frac{d}{dr}\varphi(\phi_r^X(x)) \\
&= D\varphi(X(\phi_r^X(x))) \\
&= (D\varphi \circ X|_{U_\delta} \circ \varphi^{-1} \circ \varphi)(\phi_r^X(x)) \\
&= (\varphi_* X|_{\varphi(U_\delta)} \circ \varphi)(\phi_r^X(x)) \\
&= (X|_{\varphi(U_\delta)} \circ \varphi)(\phi_r^X(x)) \\
&= X(\varphi(\psi(r, x)))
\end{aligned}$$

where we used the φ-invariance of the Liouville vector field on U_δ, that is,

$$\varphi_* X|_{\varphi(U_\delta)} = X|_{\varphi(U_\delta)},$$

which in turn follows from the computation

$$\begin{aligned} i_{\varphi_* X} d\lambda &= d\lambda(\varphi_* X, \cdot) \\ &= d\lambda(D\varphi \circ X \circ \varphi^{-1}, \cdot) \\ &= d\lambda\left(D\varphi \circ X \circ \varphi^{-1}, D\varphi \circ D\varphi^{-1} \cdot\right) \\ &= \varphi^* d\lambda(X \circ \varphi^{-1}, D\varphi^{-1} \cdot) \\ &= d\lambda(X \circ \varphi^{-1}, D\varphi^{-1} \cdot) \\ &= \varphi_*(i_X d\lambda) \\ &= \varphi_* \lambda \\ &= \lambda - d(f_\varphi \circ \varphi^{-1}) \end{aligned}$$

and assumption (7.1). Next we construct the defining Hamiltonian $H \in C^\infty(\widehat{W})$. Let $h \in C^\infty(\mathbb{R})$ be a sufficiently small mollification of the piecewise linear function depicted in Fig. 7.3. More formally, h is given by

$$h(r) := \begin{cases} r & r \in [-\frac{\delta}{2}, \frac{\delta}{2}], \\ \frac{\delta}{2} & r \in [\frac{\delta}{2}, +\infty), \\ -\frac{\delta}{2} & r \in (-\infty, -\frac{\delta}{2}]. \end{cases}$$

Define $H \in C^\infty(\widehat{W})$ by

$$H(p) := \begin{cases} h(r) & p = \psi(r, x) \in U_\delta, \\ h(r) & p = (r, x) \in [0, +\infty) \times \partial W, \\ -\frac{\delta}{2} & p \in W \setminus U_\delta. \end{cases} \tag{7.3}$$

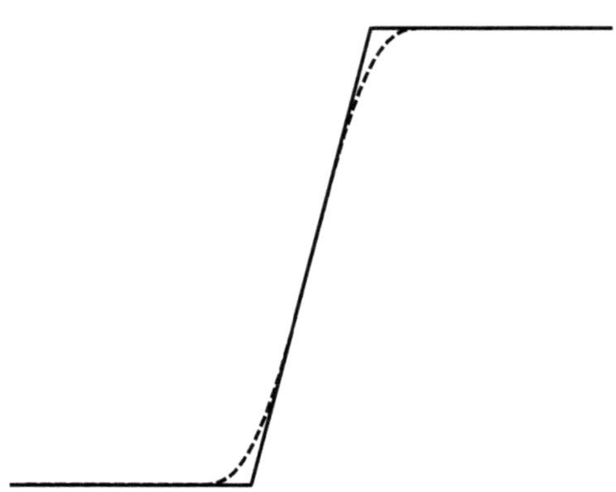

Fig. 7.3 Mollification of the piecewise linear function h

Then $\partial W \cap \mathrm{Crit}(H) = \varnothing$ and dH is compactly supported by construction. Moreover, H is φ-invariant by (7.2). Finally, $X_H|_{\partial W} = R$ follows from $X_H = h'(r)e^{-r}R$. Indeed, on U_δ we compute

$$\begin{aligned} i_{h'(r)e^{-r}R}\psi_r^* d\lambda &= i_{h'(r)e^{-r}R} d(e^r \lambda|_{\partial W}) \\ &= i_{h'(r)e^{-r}R}(e^r dr \wedge \lambda|_{\partial W} + e^r d\lambda|_{\partial W}) \\ &= -h'(r)dr \\ &= -dH. \end{aligned}$$

Hence, it follows that $H \in \mathscr{F}_\varphi(\partial W)$ is a twisted defining Hamiltonian function □

Given that for any completion of a Liouville domain there exists a twisted defining Hamiltonian function for the boundary of the Liouville domain by Lemma 7.18, we can define a suitable version of the Rabinowitz action functional.

Definition 7.19 (Twisted Rabinowitz Action Functional) Let $(\widehat{W}, \lambda)$ be the completion of a Liouville domain (W, λ) and let $\varphi \in \mathrm{Aut}(W, \lambda)$ be a Liouville automorphism. For a twisted defining Hamiltonian function $H \in \mathscr{F}_\varphi(\partial W)$, define the ***twisted Rabinowitz action functional***

$$\mathscr{A}_\varphi^H \colon \mathscr{L}_\varphi \widehat{W} \times \mathbb{R} \to \mathbb{R}, \qquad \mathscr{A}_\varphi^H(\gamma, \tau) := \int_0^1 \gamma^*\lambda - \tau \int_0^1 H(\gamma(t))dt - f_\varphi(\gamma(0)),$$

where $f_\varphi \in C_c^\infty(\mathrm{Int}\, W)$ is the unique function satisfying $\varphi^*\lambda - \lambda = df_\varphi$.

Definition 7.20 (Twisted Homotopy of Liouville Domains) Let $(\widehat{W}, \lambda)$ be the completion of a Liouville domain (W, λ) and let $\varphi \in \mathrm{Aut}(W, \lambda)$ be a Liouville automorphism. A ***twisted homotopy of Liouville domains in*** $\widehat{W}$ is a time-dependent Hamiltonian function $H \in C^\infty(\widehat{W} \times [0, 1])$ such that

1. $W_\sigma := H_\sigma^{-1}((-\infty, 0])$ is a Liouville domain with symplectic form $d\lambda|_{W_\sigma}$ and boundary $\partial W_\sigma := H_\sigma^{-1}(0)$ for all $\sigma \in [0, 1]$,
2. $H_\sigma \in \mathscr{F}_\varphi(\partial W_\sigma)$ for all $\sigma \in [0, 1]$,
3. $\partial W_\sigma \cap \mathrm{supp}\, f_\varphi = \varnothing$ for all $\sigma \in [0, 1]$.

Example 7.21 (Star-Shaped Hypersurfaces) The homotopy of star-shaped hypersurfaces in Example 4.28 shown in Fig. 7.4 is an example of a twisted homotopy of Liouville domains if $\rho_\sigma \circ \varphi|_{\mathbb{S}^{2n-1}} = \rho_\sigma$ holds for all $\sigma \in [0, 1]$.

Similarly to Floer theory, we can construct a Morse–Bott theory associated with the twisted Rabinowitz action functional $\mathscr{A}_\varphi^H$. The following result gives a roadmap to twisted Rabinowitz–Floer homology and was proven in [27, Theorem 1.1].

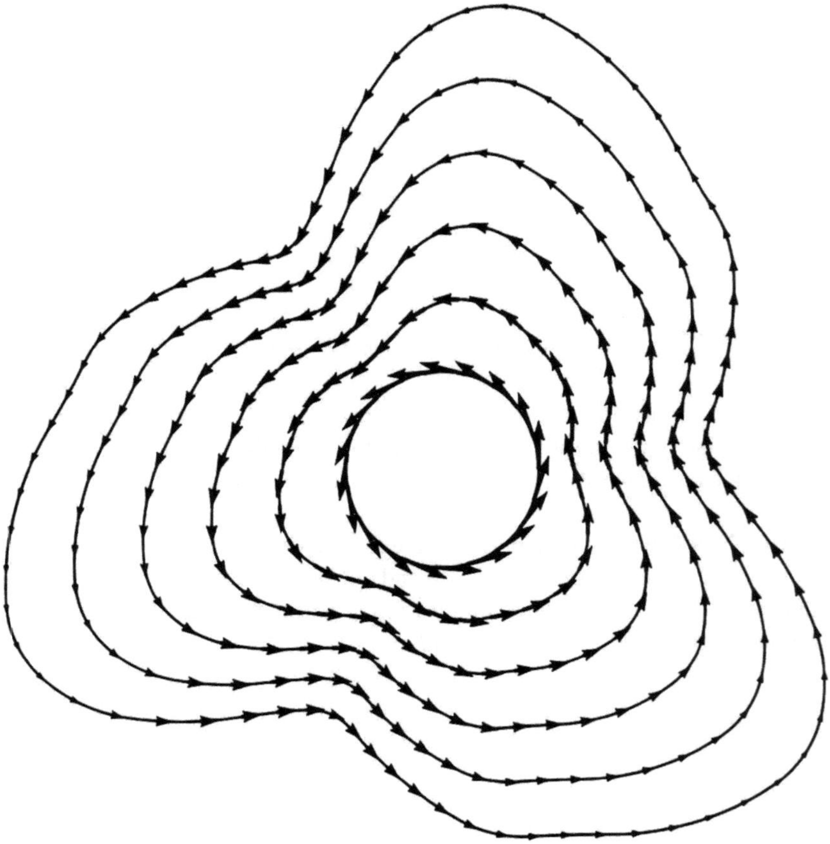

Fig. 7.4 A homotopy of symmetric star-shaped hypersurfaces in $\mathbb{R}^2$

Theorem 7.22 (Twisted Rabinowitz–Floer Homology) *Let $(\widehat{W}, \lambda)$ be the completion of a Liouville domain (W, λ) and let $\varphi \in \mathrm{Aut}(W, \lambda)$ be a Liouville automorphism.*

1. *The semi-infinite dimensional Morse–Bott homology*

$$\mathrm{RFH}^{\varphi}(\partial W, \widehat{W}) := \mathrm{HM}(\mathscr{A}_{\varphi}^{H})$$

in the sense of Floer of the twisted Rabinowitz action functional $\mathscr{A}_{\varphi}^{H}$ exists and is well-defined. Moreover, twisted Rabinowitz–Floer homology is invariant under twisted homotopies of Liouville domains.

2. *If ∂W does not admit any nonconstant twisted periodic Reeb orbits, then*

$$\mathrm{RFH}_{*}^{\varphi}(\partial W, \widehat{W}) \cong \mathrm{H}_{*}(\mathrm{Fix}(\varphi|_{\partial W}); \mathbb{Z}_2).$$

3. *If ∂W is displaceable in $(\widehat{W}, \lambda)$, then $\mathrm{RFH}^{\varphi}(\partial W, \widehat{W}) \cong 0$.*

Remark 7.23 Standard Rabinowitz–Floer homology as introduced in [51] is the special case

$$\mathrm{RFH}(\partial W, \widehat{W}) := \mathrm{RFH}^{\mathrm{id}_W}(\partial W, \widehat{W}).$$

Exercise 7.24 Let $W_\Sigma \subseteq \mathbb{R}^{2n}$ be a star-shaped domain and let $\varphi \in \mathrm{Aut}(W_\Sigma, \lambda_0)$ be a Liouville automorphism.

1. Using Theorem 7.22, prove that if $\varphi = \mathrm{id}_{W_\Sigma}$, then Σ admits a periodic Reeb orbit.
2. Does there always exist a φ-symmetric periodic Reeb orbit on Σ if $\varphi \neq \mathrm{id}_{W_\Sigma}$? Find either a proof under a suitable condition or find counterexamples.

7.1 Towards a Morse–Bott Homology of the Twisted Rabinowitz Action Functional

In order to define a Morse–Bott homology associated with the twisted Rabinowitz action functional $\mathscr{A}^H_\varphi$, we need to study the critical points $\mathrm{Crit}(\mathscr{A}^H_\varphi)$ and the Hessian $\mathrm{Hess}\,\mathscr{A}^H_\varphi$. It is a surprising fact that the kernel of the Hessian of the twisted Rabinowitz action functional at a critical point is always finite-dimensional!

Definition 7.25 (Differential of the Twisted Rabinowitz Action Functional) Let $(\widehat{W}, \lambda)$ be the completion of a Liouville domain (W, λ) and let $\varphi \in \mathrm{Aut}(W, \lambda)$ be a Liouville automorphism. For a twisted defining Hamiltonian function $H \in \mathscr{F}_\varphi(\partial W)$, define the ***differential of the perturbed twisted Rabinowitz action functional***

$$d\mathscr{A}^H_\varphi|_{(\gamma,\tau)} \colon T_\gamma \mathscr{L}_\varphi \widehat{W} \times \mathbb{R} \to \mathbb{R}$$

for all $(\gamma, \tau) \in \mathscr{L}_\varphi \widehat{W} \times \mathbb{R}$ by

$$d\mathscr{A}^H_\varphi|_{(\gamma,\tau)}(X, \eta) := \left.\frac{d}{d\varepsilon}\right|_{\varepsilon=0} \mathscr{A}^H_\varphi(\gamma_\varepsilon, \tau + \varepsilon\eta),$$

where γ_ε is a twisted variation of γ such that $\delta\gamma = X$.

Proposition 7.26 (Differential of the Twisted Rabinowitz Action Functional) *Let $(\widehat{W}, \lambda)$ be the completion of a Liouville domain (W, λ) and let $\varphi \in \mathrm{Aut}(W, \lambda)$ be a Liouville automorphism. For $H \in \mathscr{F}_\varphi(\partial W)$, we compute*

$$d\mathscr{A}^H_\varphi|_{(\gamma,\tau)}(X, \eta) = \int_0^1 d\lambda(X(t), \dot{\gamma}(t) - \tau X_H(\gamma(t)) - \eta \int_0^1 H(\gamma(t)) \tag{7.4}$$

for all $(\gamma,\tau) \in \mathscr{L}_\varphi \widehat{W} \times \mathbb{R}$ *and* $(X,\eta) \in T_\gamma \mathscr{L}_\varphi \widehat{W} \times \mathbb{R}$. *Moreover,* $(\gamma,\tau) \in \operatorname{Crit} \mathscr{A}_\varphi^H$ *if and only if*

$$\gamma(\mathbb{R}) \subseteq \partial W \qquad \text{and} \qquad \dot{\gamma}(t) = \tau X_H(\gamma(t)) \tag{7.5}$$

for all $t \in \mathbb{R}$.

Proof In order to show (7.4), we compute first the differential of the twisted symplectic area functional

$$\mathscr{A}_\varphi^0 \colon \mathscr{L}_\varphi \widehat{W} \to \mathbb{R}, \qquad \mathscr{A}_\varphi^0(\gamma) := \int_0^1 \gamma^*\lambda - f_\varphi(\gamma(0)).$$

For every $X \in T_\gamma \mathscr{L}_\varphi \widehat{W}$ we compute

$$\begin{aligned}
d\mathscr{A}_\varphi^0|_\gamma(X) &= \frac{d}{d\varepsilon}\bigg|_{\varepsilon=0} \mathscr{A}_\varphi^0(\gamma_\varepsilon)\\
&= \int_0^1 \gamma^* L_X \lambda - df_\varphi(X(0))\\
&= \int_0^1 \gamma^* i_X d\lambda + \int_0^1 \gamma^* d i_X \lambda - df_\varphi(X(0))\\
&= \int_0^1 d\lambda(X(t), \dot{\gamma}(t)) + \int_0^1 d\gamma^* i_X \lambda - df_\varphi(X(0))\\
&= \int_0^1 d\lambda(X(t), \dot{\gamma}(t)) + \lambda(X(1)) - \lambda(X(0)) - df_\varphi(X(0))\\
&= \int_0^1 d\lambda(X(t), \dot{\gamma}(t)) + \lambda(D\varphi(X(0))) - \lambda(X(0)) - df_\varphi(X(0))\\
&= \int_0^1 d\lambda(X(t), \dot{\gamma}(t)) + (\varphi^*\lambda - \lambda)(X(0)) - df_\varphi(X(0))\\
&= \int_0^1 d\lambda(X(t), \dot{\gamma}(t)).
\end{aligned}$$

Thus it follows that

$$\begin{aligned}
d\mathscr{A}_\varphi^H|_{(\gamma,\tau)}(X,\eta) &= \frac{d}{d\varepsilon}\bigg|_{\varepsilon=0} \mathscr{A}_\varphi^H(\gamma_\varepsilon, \tau + \varepsilon\eta)\\
&= d\mathscr{A}_\varphi^0|_\gamma(X) - \tau \int_0^1 dH(X(t)) - \eta \int_0^1 H(\gamma(t))
\end{aligned}$$

$$= d\mathscr{A}_\varphi^0|_\gamma(X) - \tau \int_0^1 d\lambda(X(t), X_H(\gamma(t))) - \eta \int_0^1 H(\gamma(t))$$

$$= \int_0^1 d\lambda(X(t), \dot{\gamma}(t) - \tau X_H(\gamma(t))) - \eta \int_0^1 H(\gamma(t)).$$

Let $(\gamma, \tau) \in \operatorname{Crit} \mathscr{A}_\varphi^H$. It follows immediately from (7.4) that

$$\int_0^1 H(\gamma(t))dt = 0 \qquad \text{and} \qquad \int_0^1 d\lambda(X(t), \dot{\gamma}(t) - \tau X_H(\gamma(t))) = 0$$

for all $X \in T_\gamma \mathscr{L}_\varphi \widehat{W}$. We proceed as in the proof of the fundamental lemma of the calculus of variations in Exercise 3.25. Suppose there exists $t_0 \in (0, 1)$ such that

$$\dot{\gamma}(t_0) - \tau X_H(\gamma(t_0)) \neq 0.$$

By nondegeneracy of the symplectic form $d\lambda$, there exists $v \in T_{\gamma(t_0)} \widehat{W}$ with

$$d\lambda(v, \dot{\gamma}(t_0) - \tau X_H(\gamma(t_0))) \neq 0.$$

Fix a Riemannian metric on $\widehat{W}$ and let X_v denote the unique parallel vector field along the curve $\gamma|_{[0,1]}$ such that $X_v(t_0) = v$ by [130, Theorem 4.32]. As $(0, 1)$ is open, there exists $\delta > 0$ such that $\bar{B}_\delta(t_0) \subseteq (0, 1)$. Fix a smooth bump function $\beta \in C^\infty([0, 1])$ for t_0 supported in $B_\delta(t_0)$. By shrinking δ if necessary, we may assume that

$$\int_{t_0-\delta}^{t_0+\delta} d\lambda(\beta(t)X_v(t), \dot{\gamma}(t) - \tau X_H(\gamma(t)))dt \neq 0.$$

Extending

$$(\beta X_v)(t + k) := D\varphi^k(\beta(t)X_v(t)) \qquad \forall t \in [0, 1], k \in \mathbb{Z},$$

we have that $\beta X_v \in T_\gamma \mathscr{L}_\varphi \widehat{W}$ and thus we compute

$$0 = d\mathscr{A}_\varphi^H|_{(\gamma,\tau)}(\beta X_v, 0) = \int_{t_0-\delta}^{t_0+\delta} d\lambda(\beta(t)X_v(t), \dot{\gamma}(t) - \tau X_H(\gamma(t)))dt \neq 0.$$

This is a contradiction, and we conclude

$$\dot{\gamma}(t) = \tau X_{H_t}(\gamma(t)) \qquad \forall t \in [0, 1],$$

implying

$$\begin{aligned}\dot\gamma(t+k) &= D\varphi^k(\dot\gamma(t)) && (\text{definition of } \mathscr{L}_\varphi\widehat{W})\\ &= \tau(D\varphi^k\circ X_H)(\gamma(t))\\ &= \tau(D\varphi^k\circ X_H\circ\varphi^{-k}\circ\varphi^k)(\gamma(t))\\ &= \tau\varphi^k_* X_H(\gamma(t+k))\\ &= \tau X_{\varphi^k_* H}(\gamma(t+k)) && (\text{by Lemma 2.17})\\ &= \tau X_H(\gamma(t+k))\end{aligned}$$

for all $t \in [0, 1]$ and $k \in \mathbb{Z}$. By conservation of energy (Corollary 2.28), we therefore conclude that $\gamma(\mathbb{R}) \subseteq H^{-1}(0) = \partial W$. The other direction is immediate. □

Corollary 7.27 *The differential of the perturbed twisted Rabinowitz action functional is well-defined, that is, independent of the choice of twisted variation and linear.*

Exercise 7.28 Explain why the twisted Rabinowitz action functional is not a Morse function in general and why the assumption $\operatorname{ord}(\varphi|_{\partial W}) < +\infty$ is important.

Definition 7.29 (Hessian of the Twisted Rabinowitz Action Functional) Let $(\widehat{W}, \lambda)$ be the completion of a Liouville domain (W, λ) and let $\varphi \in \operatorname{Aut}(W, \lambda)$ be a Liouville automorphism. For $H \in \mathscr{F}_\varphi(\partial W)$ and $(\gamma, \tau) \in \operatorname{Crit} \mathscr{A}^H_\varphi$, define the ***Hessian of the twisted Rabinowitz action functional***

$$\operatorname{Hess} \mathscr{A}^H_\varphi|_{(\gamma,\tau)} \colon (T_\gamma\mathscr{L}_\varphi\widehat{W} \times \mathbb{R}) \times (T_\gamma\mathscr{L}_\varphi\widehat{W} \times \mathbb{R}) \to \mathbb{R}$$

by

$$\operatorname{Hess} \mathscr{A}^H_\varphi|_{(\gamma,\tau)}((X,\eta),(Y,\sigma)) := \left.\frac{\partial^2}{\partial\varepsilon_1\partial\varepsilon_2}\right|_{\varepsilon_1=\varepsilon_2=0} \mathscr{A}^H_\varphi(\gamma_{\varepsilon_1,\varepsilon_2}, \tau+\varepsilon_1\eta+\varepsilon_2\sigma),$$

for a smooth two-parameter family $\gamma_{\varepsilon_1,\varepsilon_2}$ of twisted loops with

$$\left.\frac{\partial}{\partial\varepsilon_1}\right|_{\varepsilon_1=0}\gamma_{\varepsilon_1,0} = X \qquad\text{and}\qquad \left.\frac{\partial}{\partial\varepsilon_2}\right|_{\varepsilon_2=0}\gamma_{0,\varepsilon_2} = Y.$$

Remark 7.30 Traditionally, the differential and the Hessian of the twisted Rabinowitz action functional are called the first and second variations of the twisted Rabinowitz action functional, respectively, because it is common in Riemannian geometry [130, Theorem 6.3] and [130, Theorem 10.22].

As we have already seen in Exercise 5.1 on Morse homology, one needs a connection to define the Hessian of a smooth function. This is because the Hessian is a second derivative. However, this choice becomes irrelevant if we consider the Hessian only at critical points later on and does simplify the computations.

Definition 7.31 (Symplectic Connection) Let (M, ω) be a symplectic manifold. A ***symplectic connection on*** (M, ω) is defined to be a torsion-free connection ∇ in the tangent bundle TM such that $\nabla\omega = 0$.

Remark 7.32 Every symplectic manifold admits a symplectic connection by [99, p. 308], but in sharp contrast to the Riemannian case, a symplectic connection on a given symplectic manifold is, in general, not unique.

Lemma 7.33 *Let $(\widehat{W}, \lambda)$ be the completion of a Liouville domain (W, λ) and let $\varphi \in \mathrm{Aut}(W, \lambda)$ be a Liouville automorphism. Fix $H \in \mathscr{F}_\varphi(\partial W)$ and a symplectic connection ∇ on $(\widehat{W}, \lambda)$. If $(\gamma, \tau) \in \mathrm{Crit}\, \mathscr{A}_\varphi^H$, then*

$$\begin{aligned} \mathrm{Hess}\, \mathscr{A}_\varphi^H|_{(\gamma,\tau)}((X, \eta), (Y, \sigma)) &= \int_0^1 d\lambda(Y, \nabla_t X) \\ &\quad - \tau \int_0^1 \mathrm{Hess}^\nabla H(X, Y) - \eta \int_0^1 dH(Y) - \sigma \int_0^1 dH(X) \end{aligned} \tag{7.6}$$

for all $(X, \eta), (Y, \sigma) \in T_\gamma \mathscr{L}_\varphi M \times \mathbb{R}$.

Proof We compute

$$\begin{aligned} &\left.\frac{\partial^2}{\partial\varepsilon_1 \partial\varepsilon_2}\right|_{\varepsilon_1=\varepsilon_2=0} (\tau + \varepsilon_1\eta + \varepsilon_2\sigma) \int_0^1 H \circ \gamma_{\varepsilon_1,\varepsilon_2} \\ &= \eta \int_0^1 dH(Y) + \sigma \int_0^1 dH(X) + \tau \int_0^1 \left.\frac{\partial^2}{\partial\varepsilon_1 \partial\varepsilon_2}\right|_{\varepsilon_1=\varepsilon_2=0} H(\gamma_{\varepsilon_1,\varepsilon_2}), \end{aligned} \tag{7.7}$$

and

$$\begin{aligned} \left.\frac{\partial^2}{\partial\varepsilon_1 \partial\varepsilon_2}\right|_{\varepsilon_1=\varepsilon_2=0} H(\gamma_{\varepsilon_1,\varepsilon_2}) &= \left.\frac{\partial}{\partial\varepsilon_1}\right|_{\varepsilon_1=0} dH\left(\partial_{\varepsilon_2}|_{\varepsilon_2=0}\gamma_{\varepsilon_1,\varepsilon_2}\right) \\ &= - \left.\frac{\partial}{\partial\varepsilon_1}\right|_{\varepsilon_1=0} d\lambda\left(X_H(\gamma_{\varepsilon_1,0}), \partial_{\varepsilon_2}|_{\varepsilon_2=0}\gamma_{\varepsilon_1,\varepsilon_2}\right) \\ &= - d\lambda\left(\nabla_{\varepsilon_1}|_{\varepsilon_1=0} X_H(\gamma_{\varepsilon_1,0}), Y\right) \\ &\quad - d\lambda\left(X_H(\gamma), \nabla_{\varepsilon_1}|_{\varepsilon_1=0}\partial_{\varepsilon_2}|_{\varepsilon_2=0}\gamma_{\varepsilon_1,\varepsilon_2}\right). \end{aligned}$$

The $d\lambda$-compatibility of ∇ implies

$$
\begin{aligned}
d\lambda\left(\nabla_{\varepsilon_1}|_{\varepsilon_1=0}X_H(\gamma_{\varepsilon_1,0}), Y\right) &= d\lambda\left(\nabla_{\partial_{\varepsilon_1}|_{\varepsilon_1=0}\gamma_{\varepsilon_1,0}}X_H, Y\right) \\
&= d\lambda(\nabla_X X_H, Y) \\
&= \nabla_X d\lambda(X_H, Y) - d\lambda(X_H, \nabla_X Y) \\
&= -\nabla_X dH(Y) + dH(\nabla_X Y) \\
&= -X(Y(H)) + (\nabla_X Y)H \\
&= -\operatorname{Hess}^{\nabla} H|_{\gamma}(X, Y).
\end{aligned}
\tag{7.8}
$$

Next we compute

$$
\begin{aligned}
\frac{\partial^2}{\partial\varepsilon_1\partial\varepsilon_2}\bigg|_{\varepsilon_1=\varepsilon_2=0}\int_0^1 \gamma^*_{\varepsilon_1,\varepsilon_2}\lambda &= \int_0^1 \frac{\partial}{\partial\varepsilon_1}\bigg|_{\varepsilon_1=0}\gamma^*_{\varepsilon_1,0}i_{\partial_{\varepsilon_2}|_{\varepsilon_2=0}\gamma_{\varepsilon_1,\varepsilon_2}}(d\lambda\circ\gamma_{\varepsilon_1,0}) \\
&\quad + \int_0^1 \frac{\partial}{\partial\varepsilon_1}\bigg|_{\varepsilon_1=0} d\gamma^*_{\varepsilon_1,0}i_{\partial_{\varepsilon_2}|_{\varepsilon_2=0}\gamma_{\varepsilon_1,\varepsilon_2}}(\lambda\circ\gamma_{\varepsilon_1,0}) \\
&= \int_0^1 \frac{\partial}{\partial\varepsilon_1}\bigg|_{\varepsilon_1=0} d\lambda\left(\partial_{\varepsilon_2}|_{\varepsilon_2=0}\gamma_{\varepsilon_1,\varepsilon_2}, \dot\gamma_{\varepsilon_1,0}\right) \\
&\quad + \frac{\partial}{\partial\varepsilon_1}\bigg|_{\varepsilon_1=0} \lambda(\partial_{\varepsilon_2}|_{\varepsilon_2=0}\gamma_{\varepsilon_1,\varepsilon_2})|_0^1 \\
&= \int_0^1 d\lambda\left(\nabla_{\varepsilon_1}|_{\varepsilon_1=0}\partial_{\varepsilon_2}|_{\varepsilon_2=0}\gamma_{\varepsilon_1,\varepsilon_2}, \dot\gamma\right) \\
&\quad + \int_0^1 d\lambda\left(Y, \nabla_{\varepsilon_1}|_{\varepsilon_1=0}\partial_t\gamma_{\varepsilon_1,0}\right) \\
&\quad + \frac{\partial}{\partial\varepsilon_1}\bigg|_{\varepsilon_1=0} \lambda(\partial_{\varepsilon_2}|_{\varepsilon_2=0}\gamma_{\varepsilon_1,\varepsilon_2})|_0^1 \\
&= \tau\int_0^1 d\lambda\left(\nabla_{\varepsilon_1}|_{\varepsilon_1=0}\partial_{\varepsilon_2}|_{\varepsilon_2=0}\gamma_{\varepsilon_1,\varepsilon_2}, X_H(\gamma)\right) \\
&\quad + \int_0^1 d\lambda\left(Y, \nabla_t\partial_{\varepsilon_1}|_{\varepsilon_1=0}\gamma_{\varepsilon_1,0}\right) \\
&\quad + \frac{\partial}{\partial\varepsilon_1}\bigg|_{\varepsilon_1=0} \lambda(\partial_{\varepsilon_2}|_{\varepsilon_2=0}\gamma_{\varepsilon_1,\varepsilon_2})|_0^1
\end{aligned}
$$

$$
\begin{aligned}
&= \tau \int_0^1 d\lambda \left(\nabla_{\varepsilon_1}|_{\varepsilon_1=0} \partial_{\varepsilon_2}|_{\varepsilon_2=0} \gamma_{\varepsilon_1,\varepsilon_2}, X_H(\gamma) \right) \\
&\quad + \int_0^1 d\lambda(Y, \nabla_t X) \\
&\quad + \frac{\partial}{\partial \varepsilon_1}\bigg|_{\varepsilon_1=0} \lambda(\partial_{\varepsilon_2}|_{\varepsilon_2=0} \gamma_{\varepsilon_1,\varepsilon_2})|_0^1 .
\end{aligned} \tag{7.9}
$$

Moreover

$$
\begin{aligned}
\int_0^1 d\lambda(Y, \nabla_t X) &= \int_0^1 \frac{d}{dt} d\lambda(Y, X) - \int_0^1 d\lambda(\nabla_t Y, X) \\
&= d\lambda(Y, X)|_0^1 + \int_0^1 d\lambda(X, \nabla_t Y) \\
&= d\lambda \left(D\varphi(Y(0)), D\varphi(X(0)) \right) - d\lambda(Y(0), X(0)) + \int_0^1 d\lambda(X, \nabla_t Y) \\
&= \int_0^1 d\lambda(X, \nabla_t Y).
\end{aligned} \tag{7.10}
$$

Finally we compute

$$
\begin{aligned}
\lambda(\partial_{\varepsilon_2}|_{\varepsilon_2=0} \gamma_{\varepsilon_1,\varepsilon_2})|_0^1 &= \lambda \left(\partial_{\varepsilon_2}|_{\varepsilon_2=0} \gamma_{\varepsilon_1,\varepsilon_2}(1) \right) - \lambda \left(\partial_{\varepsilon_2}|_{\varepsilon_2=0} \gamma_{\varepsilon_1,\varepsilon_2}(0) \right) \\
&= \lambda \left(\partial_{\varepsilon_2}|_{\varepsilon_2=0} \varphi(\gamma_{\varepsilon_1,\varepsilon_2}(0)) \right) - \lambda \left(\partial_{\varepsilon_2}|_{\varepsilon_2=0} \gamma_{\varepsilon_1,\varepsilon_2}(0) \right) \\
&= \lambda \left(D\varphi \left(\partial_{\varepsilon_2}|_{\varepsilon_2=0} \gamma_{\varepsilon_1,\varepsilon_2}(0) \right) \right) - \lambda \left(\partial_{\varepsilon_2}|_{\varepsilon_2=0} \gamma_{\varepsilon_1,\varepsilon_2}(0) \right) \\
&= (\varphi^* \lambda - \lambda)(\partial_{\varepsilon_2}|_{\varepsilon_2=0} \gamma_{\varepsilon_1,\varepsilon_2}(0)) \\
&= df_\varphi(\partial_{\varepsilon_2}|_{\varepsilon_2=0} \gamma_{\varepsilon_1,\varepsilon_2}(0)).
\end{aligned} \tag{7.11}
$$

Combining (7.9), (7.7), (7.10), and (7.11) yields (7.6). □

Corollary 7.34 *The Hessian of the twisted Rabinowitz action functional is a well-defined, that is, independent of the choice of twisted two-parameter family, symmetric bilinear form.*

In the next calculation, we remove the dependence on the symplectic connection of the Hessian of the twisted Rabinowitz action functional in formula (7.6). Moreover, we provide a first description for the kernel $\ker \operatorname{Hess} \mathscr{A}_\varphi^H|_{(\gamma,\tau)}$ of the twisted Rabinowitz action function $\mathscr{A}_\varphi^H$ at a critical point $(\gamma, \tau) \in \operatorname{Crit} \mathscr{A}_\varphi^H$.

Lemma 7.35 *Let $(\widehat{W}, \lambda)$ be the completion of a Liouville domain (W, λ) and let $\varphi \in \operatorname{Aut}(W, \lambda)$ be a Liouville automorphism. Fix a twisted defining Hamiltonian* *function* $H \in \mathscr{F}_\varphi(\partial W)$. *If* $(\gamma, \tau) \in \operatorname{Crit} \mathscr{A}_\varphi^H$, *then*

$$\operatorname{Hess} \mathscr{A}_\varphi^H|_{(\gamma,\tau)}((X, \eta), (Y, \sigma)) = \int_0^1 d\lambda(Y, L_{\tau X_H} X - \eta X_H(\gamma)) - \sigma \int_0^1 dH(X) \tag{7.12}$$

for all $(X, \eta), (Y, \sigma) \in T_\gamma \mathscr{L}_\varphi \widehat{W} \times \mathbb{R}$, *where*

$$L_{\tau X_H} X(t) = \left.\frac{d}{ds}\right|_{s=0} D\phi_{-s\tau}^{X_H} X(s+t) \qquad \forall t \in \mathbb{R}.$$

In particular, the kernel of the Hessian of the twisted Rabinowitz action functional at a critical point $(\gamma, \tau) \in \operatorname{Crit} \mathscr{A}_\varphi^H$ *consists precisely of all* $(X, \eta) \in T_\gamma \mathscr{L}_\varphi \widehat{W} \times \mathbb{R}$ *satisfying*

$$L_{\tau X_H} X = \eta X_H(\gamma) \qquad \text{and} \qquad \int_0^1 dH(X) = 0.$$

Proof Inserting $\operatorname{Hess}^\nabla(X, Y) = d\lambda(Y, \nabla_X X_H)$ into (7.6) yields

$$\operatorname{Hess} \mathscr{A}_\varphi^H|_{(\gamma,\tau)}((X, \eta), (Y, \sigma)) = \int_0^1 d\lambda(Y, \nabla_t X - \tau \nabla_X X_H) - \eta \int_0^1 dH(Y) - \sigma \int_0^1 dH(X).$$

But as ∇ has no torsion by assumption, we compute

$$\nabla_t X - \tau \nabla_X X_H = \nabla_{\dot\gamma} X - \tau \nabla_X X_H = \nabla_{\tau X_H} X - \tau \nabla_X X_H = [\tau X_H, X],$$

and

$$\begin{aligned} [\tau X_H, X](t) &= L_{\tau X_H} X(t) \\ &= \left.\frac{d}{ds}\right|_{s=0} D\phi_{-s\tau}^{X_H}(X(\phi_{s\tau}^{X_H}(\gamma(t)))) \\ &= \left.\frac{d}{ds}\right|_{s=0} D\phi_{-s\tau}^{X_H}(X(\phi_{s\tau}^{X_H}(\phi_{t\tau}^{X_H}(\gamma(0))))) \\ &= \left.\frac{d}{ds}\right|_{s=0} D\phi_{-s\tau}^{X_H} X(s+t) \end{aligned}$$

for all $t \in \mathbb{R}$. □

Using the description of the kernel of the Hessian of the twisted Rabinowitz action functional provided in Lemma 7.35, we can show that this vector space is actually finite-dimensional! This ensures that the components of the critical manifolds are finite-dimensional if the twisted Rabinowitz action functional is a Morse–Bott function.

Proposition 7.36 *Let* $\varphi \in \operatorname{Aut}(W, \lambda)$ *be a Liouville automorphism and fix a twisted defining Hamiltonian function* $H \in \mathscr{F}_\varphi(\partial W)$*. For every* $(\gamma, \tau) \in \operatorname{Crit} \mathscr{A}_\varphi^H$*, there is a canonical isomorphism*

$$\ker \operatorname{Hess} \mathscr{A}_\varphi^H|_{(\gamma,\tau)} \cong \mathfrak{K}(\gamma, \tau),$$

where

$$\mathfrak{K}(\gamma, \tau) := \{(v_0, \eta) \in T_{\gamma(0)}\widehat{W} \times \mathbb{R} : \text{solution of (7.13)}\}$$

with

$$D(\phi_{-\tau}^{X_H} \circ \varphi)v_0 = v_0 + \eta X_H(\gamma(0)) \qquad \textit{and} \qquad dH(v_0) = 0. \tag{7.13}$$

Proof We follow [77, p. 99–100]. Let $(X, \eta) \in \ker \operatorname{Hess} \mathscr{A}_\varphi^H|_{(\gamma,\tau)}$ and define

$$v\colon [0, 1] \to T_{\gamma(0)}M, \qquad v(t) := D\phi_{-\tau t}^{X_H} X(t).$$

By Lemma 7.33, the assumption that (X, η) belongs to the kernel of the Hessian is equivalent to

$$\dot{v} = \eta X_H(\gamma(0)) \qquad \text{and} \qquad \int_0^1 dH(v) = 0. \tag{7.14}$$

We claim that

$$\ker \operatorname{Hess} \mathscr{A}_\varphi^H|_{(\gamma,\tau)} \to \mathfrak{K}(\gamma, \tau), \qquad (X, \eta) \mapsto (v(0), \eta) \tag{7.15}$$

is an isomorphism. First, we show that the above homomorphism is indeed well defined. Integrating the first equation in (7.14) yields

$$v(t) = v_0 + t\eta X_H(\gamma(0)) \qquad \forall t \in [0, 1],$$

with $v_0 := v(0)$. Thus $(v_0, \eta) \in \mathfrak{K}(\gamma, \tau)$ follows from

$$\begin{aligned} v(1) &= D\phi_{-\tau}^{X_H} X(1) \\ &= D\phi_{-\tau}^{X_H} D\varphi(X(0)) \end{aligned}$$

$$= D(\phi_{-\tau}^{X_H} \circ \varphi)X(0)$$

$$= D(\phi_{-\tau}^{X_H} \circ \varphi)v_0. \tag{7.16}$$

That (7.15) is an isomorphism follows by considering the inverse

$$\mathfrak{K}(\gamma, \tau) \to \ker \operatorname{Hess} \mathscr{A}_\varphi^H|_{(\gamma,\tau)}, \qquad (v_0, \eta) \mapsto (X, \eta),$$

where $X \in T_\gamma \mathscr{L}_\varphi \widehat{W}$ is defined by

$$X(t) := D\phi_{\tau t}^{X_H}(v_0 + t\eta X_H(\gamma(0))) \qquad \forall t \in \mathbb{R}.$$

This establishes the canonical isomorphism (7.15). □

Up to now, we have only made use of the assumption $H^{-1}(0) = \partial W$ for a twisted defining Hamiltonian function $H \in \mathscr{F}_\varphi(\partial W)$ in the definition of the twisted Rabinowitz action functional $\mathscr{A}_\varphi^H$. The next result uses the property $X_H|_{\partial W} = R$.

Definition 7.37 (Strict Contactomorphism) Let (Σ, α) be a contact manifold. A ***strict contactomorphism of*** (Σ, α) is a diffeomorphism $\varphi \in \operatorname{Diff}(\Sigma)$ with $\varphi^*\alpha = \alpha$.

Example 7.38 Let $\varphi \in \operatorname{Aut}(W, \lambda)$ be a Liouville automorphism for a Liouville domain (W, λ). Then the restriction $\varphi|_{\partial W} : (\partial W, \lambda|_{\partial W}) \to (\partial W, \lambda|_{\partial W})$ is a strict contactomorphism as df_φ is compactly supported in $W \setminus \partial W$.

Exercise 7.39 Let $\varphi : (\Sigma, \alpha) \to (\Sigma, \alpha)$ be a strict contactomorphism of a compact contact manifold (Σ, α). Show that $\varphi \circ \phi_t^R = \phi_t^R \circ \varphi$ holds for all $t \in \mathbb{R}$.

Definition 7.40 (Parametrised Twisted Periodic Reeb Orbit) Let $\varphi \in \operatorname{Aut}(W, \lambda)$ be a Liouville automorphism for a Liouville domain (W, λ). A ***parametrised twisted periodic Reeb orbit*** in ∂W is defined as a tuple $(\gamma, \tau) \in \mathscr{L}_\varphi W \times \mathbb{R}$ satisfying

$$\gamma(\mathbb{R}) \subseteq \partial W \qquad \text{and} \qquad \dot{\gamma}(t) = \tau R(\gamma(t)) \qquad \forall t \in \mathbb{R}.$$

We denote the set of all parametrised twisted periodic Reeb orbits in ∂W by $\mathscr{P}_\varphi(\partial W)$.

Definition 7.41 (Twisted Spectrum) Let $\varphi \in \operatorname{Aut}(W, \lambda)$ be a Liouville automorphism for a Liouville domain (W, λ). Define the ***twisted spectrum*** by

$$\operatorname{Spec}_\varphi(\partial W) := \{\tau \in \mathbb{R} : \exists \gamma \in C^\infty(\mathbb{R}, \partial W) \text{ such that } (\gamma, \tau) \in \mathscr{P}_\varphi(\partial W)\}.$$

Proposition 7.42 (Kernel of the Hessian of the Twisted Rabinowitz Action Functional) *Let $(\widehat{W}, \lambda)$ be the completion of a Liouville domain (W, λ) and let*

$\varphi \in \operatorname{Aut}(W, \lambda)$ be a Liouville automorphism. Fix a twisted defining Hamiltonian function $H \in \mathscr{F}_\varphi(\partial W)$. Then

$$\operatorname{Crit} \mathscr{A}_\varphi^H = \mathscr{P}_\varphi(\partial W)$$

and

$$\ker \operatorname{Hess} \mathscr{A}_\varphi^H|_{(\gamma,\tau)} \cong \ker\left(D(\phi_{-\tau}^R \circ \varphi)|_{\gamma(0)} - \mathrm{id}_{T_{\gamma(0)}\partial W}\right)$$

for all $(\gamma, \tau) \in \mathscr{P}_\varphi(\partial W)$. Moreover, we always have $R(\gamma(0)) \in \ker \operatorname{Hess} \mathscr{A}_\varphi^H|_{(\gamma,\tau)}$ and if $\mathscr{P}_\varphi(\partial W) \subseteq \partial W \times \mathbb{R}$ is an embedded submanifold, then $\operatorname{Spec}(\partial W)$ is discrete.

Remark 7.43 If $(\gamma, \tau) \in \mathscr{P}_\varphi(\partial W)$, we have the period-action equality

$$\mathscr{A}_\varphi^H(\gamma, \tau) = \int_0^1 \gamma^*\lambda = \int_0^1 \lambda(\dot{\gamma}) = \tau \int_0^1 \lambda(R(\gamma)) = \tau$$

for all $H \in \mathscr{F}_\varphi(\partial W)$.

Proof The identity $\operatorname{Crit} \mathscr{A}_\varphi^H = \mathscr{P}_\varphi(\partial W)$ follows from Proposition 7.26. The proof of the formula for the kernel of the Hessian of $\mathscr{A}_\varphi^H$ is inspired by [77, p. 102]. By Proposition 7.36 we have that

$$\ker \operatorname{Hess} \mathscr{A}_\varphi^H|_{(\gamma,\tau)} \cong \mathfrak{K}(\gamma, \tau),$$

where $(v_0, \eta) \in T_{\gamma(0)}\widehat{W} \times \mathbb{R}$ belongs to $\mathfrak{K}(\gamma, \tau)$ if and only if

$$D(\phi_{-\tau}^{X_H} \circ \varphi)v_0 = v_0 + \eta X_H(\gamma(0)) \qquad \text{and} \qquad dH(v_0) = 0.$$

Thus in our setting, the second condition implies $v_0 \in T_{\gamma(0)}\partial W$. Decompose

$$v_0 = v_0^\xi + aR(\gamma(0)) \qquad v_0^\xi \in \xi_{\gamma(0)}, a \in \mathbb{R},$$

where $\xi := \ker \lambda|_{\partial W}$ denotes the contact distribution. Then using Exercise 7.39 we compute

$$\begin{aligned} D(\phi_{-\tau}^R \circ \varphi)R(\gamma(0)) &= D(\phi_{-\tau}^R \circ \varphi)\left(\frac{d}{dt}\bigg|_{t=0} \phi_t^R(\gamma(0))\right) \\ &= \frac{d}{dt}\bigg|_{t=0} (\phi_{-\tau}^R \circ \varphi \circ \phi_t^R)(\gamma(0)) \\ &= \frac{d}{dt}\bigg|_{t=0} (\phi_t^R \circ \varphi \circ \phi_{-\tau}^R)(\gamma(0)) \end{aligned}$$

$$= \left.\frac{d}{dt}\right|_{t=0} \phi_t^R(\gamma(0))$$
$$= R(\gamma(0)).$$

It follows that

$$v_0 + \eta R(\gamma(0)) = D(\phi_{-\tau}^R \circ \varphi)v_0 = D^\xi(\phi_{-\tau}^R \circ \varphi)v_0^\xi + aR(\gamma(0)),$$

where

$$D^\xi(\phi_{-\tau}^R \circ \varphi) := D(\phi_{-\tau}^R \circ \varphi)|_\xi \colon \xi \to \xi,$$

implies

$$\eta = 0 \qquad \text{and} \qquad D^\xi(\phi_{-\tau}^R \circ \varphi)v_0^\xi = v_0^\xi$$

by considering the splitting $T(\partial W) = \xi \oplus \langle R \rangle$ from Exercise 4.12 implying

$$\mathfrak{K}(\gamma, \tau) = \ker\left(D(\phi_{-\tau}^R \circ \varphi)|_{\gamma(0)} - \mathrm{id}_{T_{\gamma(0)}\Sigma}\right) \times \{0\}.$$

Finally, assume that $\mathscr{P}_\varphi(\partial W) \subseteq \Sigma \times \mathbb{R}$ is an embedded submanifold via the identification of $(\gamma, \tau) \in \mathscr{P}_\varphi(\partial W)$ with $(\gamma(0), \tau) \in \partial W \times \mathbb{R}$. Fix a path (γ_s, τ_s) in the critical manifold $\mathscr{P}_\varphi(\partial W) = \operatorname{Crit} \mathscr{A}_\varphi^H$ from (γ_0, τ_0) to (γ_1, τ_1). Using Remark 7.43 we compute

$$\partial_s \tau_s = \partial_s \mathscr{A}_\varphi^H(\gamma_s, \tau_s) = d\mathscr{A}_\varphi^H|_{(\gamma_s, \tau_s)}(\partial_s \gamma_s, \partial_s \tau_s) = 0,$$

implying that τ_s is constant, and in particular $\tau_0 = \tau_1$. Consequently, $\mathscr{A}_\varphi^H$ is constant on each path-connected component of $\mathscr{P}_\varphi(\partial W)$. As $\mathscr{P}_\varphi(\partial W)$ is a submanifold of $\partial W \times \mathbb{R}$, there are only countably many connected components by definition, implying that $\operatorname{Spec}(\partial W)$ is discrete. □

7.2 Compactness of the Moduli Space of Twisted Negative Gradient Flow Lines

In Corollary 6.14 we have already seen a prototypical compactness result for the negative gradient flow lines of the Rabinowitz action functional. However, in order to achieve compactness, one needs to uniformly bound Lagrange multipliers.

Definition 7.44 (Adapted Almost Complex Structure) Let (W, λ) be a Liouville domain and let $\varphi \in \operatorname{Aut}(W, \lambda)$ be a Liouville automorphism. An ***adapted almost complex structure*** is defined to be a $d\lambda$-compatible φ-invariant almost complex structure J on (W, λ) such that J restricts to define a $d\lambda|_{\partial W}$-compatible almost complex structure on the contact distribution $\ker \lambda|_{\partial W}$ and $JR = \partial_r$ holds near the boundary ∂W.

Example 7.45 Fix a $d\lambda|_{\partial W}$ compatible almost complex structure $J^{\partial W}$ in the contact distribution $\ker \lambda|_{\partial W}$ that can be chosen to be φ-invariant by [136, Lemma 5.5.6]. Extend this almost complex structure to $((-\delta, +\infty) \times \partial W, d(e^r \lambda|_{\partial W}))$ by

$$J^{\partial W}|_{(a,x)}(b, v) := \Big(\lambda_x(v), J^{\partial W}|_x(\pi(v)) - bR(x)\Big),$$

where $\delta = \delta(W, \varphi)$ is chosen as in the proof of Lemma 7.18, and

$$\pi\colon \ker \lambda|_{\partial W} \oplus \langle R \rangle \to \ker \lambda|_{\partial W}$$

denotes the projection. Choose an additional $d\lambda$-compatible φ-invariant almost complex structure $J^{W\setminus \partial W}$ in $W \setminus \partial W$, and let $\{\beta^{\partial W}, \beta^{W\setminus\partial W}\}$ be a smooth partition of unity subordinate to $\{U_\delta, W \setminus \partial W\}$. Define a Riemannian metric on W by

$$m := \beta^{\partial W} m_{J^{\partial W}} + \beta^{W\setminus\partial W} m_{J^{W\setminus\partial W}}$$

and let J be the corresponding $d\lambda$-compatible φ-invariant almost complex structure on the completion (W, λ). This J is an adapted almost complex structure.

Definition 7.46 (Rabinowitz–Floer Data) Let $(\widehat{W}, \lambda)$ be the completion of a Liouville domain (W, λ) and let $\varphi \in \operatorname{Aut}(W, \lambda)$ be a Liouville automorphism. We define ***Rabinowitz–Floer data for φ*** to be a tuple (H, J) consisting of a twisted defining Hamiltonian function $H \in \mathscr{F}_\varphi(\partial W)$ together with an adapted almost complex structure J.

Definition 7.47 (L^2-Metric) Let (H, J) be Rabinowitz–Floer data for a Liouville automorphism $\varphi \in \operatorname{Aut}(W, \lambda)$. Define an ***$L^2$-metric*** on $\mathscr{L}_\varphi \widehat{W} \times \mathbb{R}$ by

$$\langle (X, \eta), (Y, \sigma) \rangle_J := \int_0^1 d\lambda(JX(t), Y(t))dt + \eta\sigma \tag{7.17}$$

for all $(X, \eta), (Y, \sigma) \in T_\gamma \mathscr{L}_\varphi \widehat{W} \times \mathbb{R}$ and $\gamma \in \mathscr{L}_\varphi \widehat{W}$.

Exercise 7.48 Show that the gradient $\operatorname{grad}_J \mathscr{A}_\varphi^H \in \mathfrak{X}(\mathscr{L}_\varphi \widehat{W} \times \mathbb{R})$ of the twisted Rabinowitz action functional $\mathscr{A}_\varphi^H$ with respect to the L^2-metric (7.17) is given by

$$\operatorname{grad}_J \mathscr{A}_\varphi^H|_{(\gamma,\tau)}(t) = \begin{pmatrix} J(\dot{\gamma}(t) - \tau X_H(\gamma(t))) \\ -\int_0^1 H \circ \gamma \end{pmatrix} \qquad \forall t \in \mathbb{R}$$

and every $(\gamma, \tau) \in \mathscr{L}_\varphi \widehat{W} \times \mathbb{R}$.

Definition 7.49 (Twisted Negative Gradient Flow Line) Let (H, J) be Rabinowitz–Floer data for a Liouville automorphism $\varphi \in \operatorname{Aut}(W, \lambda)$. A ***twisted negative gradient flow line*** is a tuple $(u, \tau) \in C^\infty(\mathbb{R}, \mathscr{L}_\varphi \widehat{W} \times \mathbb{R})$ such that

$$\partial_s(u, \tau) = -\operatorname{grad}_J \mathscr{A}_\varphi^H|_{(u(s),\tau(s))} \qquad \forall s \in \mathbb{R}.$$

Definition 7.50 (Energy) Let (H, J) be Rabinowitz–Floer data for a Liouville automorphism $\varphi \in \operatorname{Aut}(W, \lambda)$. The ***energy of a twisted negative gradient flow line*** (u, τ) is defined by

$$E_J(u, \tau) := \int_{-\infty}^{+\infty} \|\partial_s(u, \tau)\|_J^2 ds = \int_{-\infty}^{+\infty} \|\operatorname{grad}_J \mathscr{A}_\varphi^H|_{(u(s),\tau(s))}\|_J^2 ds.$$

In order to prove the compactness of sequences of twisted negative gradient flow lines, we need to ensure that the Lagrange multipliers stay uniformly bounded along the sequence. A first step is to show a period-action inequality: a weakened form of the period-action equality in Remark 7.43.

Lemma 7.51 (Fundamental Lemma) *Let (H, J) be Rabinowitz–Floer data for a Liouville automorphism $\varphi \in \operatorname{Aut}(W, \lambda)$. There exists a constant $C = C(\lambda, H, J) > 0$ such that*

$$\|\operatorname{grad}_J \mathscr{A}_\varphi^H|_{(\gamma,\tau)}\|_J < \frac{1}{C} \quad \Rightarrow \quad |\tau| \leq C(|\mathscr{A}_\varphi^H(\gamma, \tau)| + 1)$$

for all $(\gamma, \tau) \in \mathscr{L}_\varphi \widehat{W} \times \mathbb{R}$.

Proof We proceed in three steps.
Step 1: There exist constants $\delta > 0$ and $0 < C_\delta < +\infty$ such that if $(\gamma, \tau) \in \mathscr{L}_\varphi \widehat{W}$ with $\gamma([0, 1]) \subseteq H^{-1}((-\delta, \delta)) =: U_\delta$, then

$$|\tau| \leq 2|\mathscr{A}_\varphi^H(\gamma, \tau)| + C_\delta \|\operatorname{grad}_J \mathscr{A}_\varphi^H|_{(\gamma,\tau)}\|_J + 2\|f_\varphi\|_\infty.$$

Choose $\delta > 0$ such that $U_\delta \subseteq \operatorname{supp} X_H$ and

$$\lambda_x(X_H(x)) \geq \frac{1}{2} + \delta \qquad \forall x \in U_\delta.$$

This is possible as $X_H|_{\partial W} = R$. Moreover, set

$$C_\delta := 2\|\lambda|_{U_\delta}\|_\infty < +\infty$$

since dH is compactly supported. We estimate

$$\begin{aligned}
|\mathscr{A}_\varphi^H(\gamma,\tau)| &= \left|\int_0^1 \gamma^*\lambda - \tau\int_0^1 H(\gamma) - f_\varphi(\gamma(0))\right| \\
&= \left|\tau\int_0^1 \lambda(X_H(\gamma)) + \int_0^1 \lambda(\dot\gamma - \tau X_H(\gamma)) - \tau\int_0^1 H(\gamma) - f_\varphi(\gamma(0))\right| \\
&\geq |\tau|\left(\frac{1}{2}+\delta\right) - \left|\int_0^1 \lambda(\dot\gamma - \tau X_H(\gamma))\right| - |\tau|\delta - \|f_\varphi\|_\infty \\
&\geq \frac{|\tau|}{2} - \frac{C_\delta}{2}\int_0^1 \|\dot\gamma(t) - \tau X_H(\gamma(t))\|_J dt - \|f_\varphi\|_\infty \\
&\geq \frac{|\tau|}{2} - \frac{C_\delta}{2}\sqrt{\int_0^1 \|\dot\gamma(t) - \tau X_H(\gamma(t))\|_J^2 dt} - \|f_\varphi\|_\infty \\
&\geq \frac{|\tau|}{2} - \frac{C_\delta}{2}\|\operatorname{grad}_J \mathscr{A}_\varphi^H|_{(\gamma,\tau)}\|_J - \|f_\varphi\|_\infty
\end{aligned}$$

by Jensen's inequality.

Step 2: For all $\delta > 0$, there exists $\varepsilon = \varepsilon(\delta) > 0$ with the property that if there exists $t_0 \in [0,1]$ with $|H(\gamma(t_0))| \geq \delta$ for $(\gamma,\tau) \in \mathscr{L}_\varphi\widehat{W}$, then $\|\operatorname{grad}_J \mathscr{A}_\varphi^H|_{(\gamma,\tau)}\|_J \geq \varepsilon$.

Assume first that $\gamma(t) \in M \setminus U_{\frac{\delta}{2}}$ for all $t \in [0,1]$. In this case we estimate

$$\|\operatorname{grad}_J \mathscr{A}_\varphi^H|_{(\gamma,\tau)}\|_J \geq \left|\int_0^1 H(\gamma)\right| \geq \frac{\delta}{2}.$$

Otherwise, we may assume without loss of generality that there exists $t_1 \in [0,1]$ such that $|H(\gamma(t_1))| \leq \frac{\delta}{2}$, $t_0 < t_1$ and $\frac{\delta}{2} \leq |H(\gamma(t))| \leq \delta$ for all $t \in [t_0, t_1]$. Set

$$\kappa := \max_{x\in\bar U_\delta} \|\operatorname{grad}_J H\|_J > 0 \qquad \text{and} \qquad \varepsilon = \varepsilon(\delta) := \frac{\delta}{2\max\{1,\kappa\}},$$

as $dH \neq 0$ in a neighbourhood of ∂W. We estimate

$$\begin{aligned}
\|\operatorname{grad}_J \mathscr{A}_\varphi^H|_{(\gamma,\tau)}\|_J &\geq \sqrt{\int_0^1 \|\dot\gamma(t) - \tau X_H(\gamma(t))\|_J^2 dt} \\
&\geq \int_0^1 \|\dot\gamma(t) - \tau X_H(\gamma(t))\|_J dt
\end{aligned}$$

$$
\begin{aligned}
&\geq \int_{t_0}^{t_1} \|\dot{\gamma}(t) - \tau X_H(\gamma(t))\|_J dt \\
&\geq \frac{1}{\kappa} \int_{t_0}^{t_1} \|\operatorname{grad}_J H(\gamma(t))\|_J \|\dot{\gamma}(t) - \tau X_H(\gamma(t))\|_J dt \\
&\geq \frac{1}{\kappa} \int_{t_0}^{t_1} |m_J(\operatorname{grad}_J H(\gamma(t)), \dot{\gamma}(t) - \tau X_H(\gamma(t)))| dt \\
&= \frac{1}{\kappa} \int_{t_0}^{t_1} |dH(\dot{\gamma}(t) - \tau X_H(\gamma(t)))| dt \\
&= \frac{1}{\kappa} \int_{t_0}^{t_1} |dH(\dot{\gamma}(t))| dt \\
&= \frac{1}{\kappa} \int_{t_0}^{t_1} \left| \frac{d}{dt}(H \circ \gamma) \right| \\
&\geq \frac{1}{\kappa} \left| \int_{t_0}^{t_1} \frac{d}{dt}(H \circ \gamma) \right| \\
&= \frac{1}{\kappa} |H(\gamma(t_1)) - H(\gamma(t_0))| \\
&\geq \frac{1}{\kappa} \left(|H(\gamma(t_0))| - |H(\gamma(t_1))|\right) \\
&\geq \frac{\delta}{2\kappa}
\end{aligned}
$$

by Cauchy–Schwarz. Hence it follows that $\| \operatorname{grad}_J \mathscr{A}_\varphi^H|_{(\gamma,\tau)}\|_J \geq \varepsilon$.
Step 3: We prove the Fundamental Lemma. Choose $\delta > 0$ and $0 < C_\delta < +\infty$ as in Step 1 and $\varepsilon = \varepsilon(\delta) > 0$ as in Step 2. Set

$$C_0 := \max\left\{2, C_\delta \varepsilon + 2\|f_\varphi\|_\infty\right\}.$$

Assume $\| \operatorname{grad}_J \mathscr{A}_\varphi^H|_{(\gamma,\tau)}\|_J < \varepsilon$ for $(\gamma, \tau) \in \mathscr{L}_\varphi \widehat{W}$. Then $\gamma([0,1]) \subseteq U_\delta$, as otherwise there exists $t_0 \in [0,1]$ with $|H(\gamma(t_0))| \geq \delta$ implying $\| \operatorname{grad}_J \mathscr{A}_\varphi^H|_{(\gamma,\tau)}\|_J \geq \varepsilon$ by Step 2. Thus with Step 1 we estimate

$$|\tau| \leq C_0(|\mathscr{A}_\varphi^H(\gamma, \tau)| + 1). \tag{7.18}$$

Finally, set

$$C := \max\left\{C_0, \frac{1}{\varepsilon}\right\}.$$

This proves the Fundamental Lemma. □

We are ready to state the main theorem ensuring the compactness of sequences of twisted negative gradient flow lines.

Theorem 7.52 (Compactness) *Let (H, J) be Rabinowitz–Floer data for a Liouville automorphism $\varphi \in \operatorname{Aut}(W, \lambda)$. Suppose (u_k, τ_k) is a sequence of negative gradient flow lines of the twisted Rabinowitz action functional $\mathscr{A}_\varphi^H$ such that there exist constants $a, b \in \mathbb{R}$ with*

$$a \leq \mathscr{A}_\varphi^H (u_k(s), \tau_k(s)) \leq b \qquad \forall k \in \mathbb{N}, s \in \mathbb{R}.$$

Then for every sequence $(s_k) \subseteq \mathbb{R}$ there exists a subsequence (s_{k_j}) and a negative gradient flow line (u_∞, τ_∞) of $\mathscr{A}_\varphi^H$ such that

$$\left(u_{k_j}(\cdot + s_{k_j}), \tau_{k_j}(\cdot + s_{k_j})\right) \xrightarrow{C^\infty_{\text{loc}}} (u_\infty, \tau_\infty) \qquad \text{as } j \to \infty.$$

Proof In view of Corollary 6.14, we need to establish a uniform L^∞-bound on u_k, and a uniform L^∞-bound on τ_k. To obtain a uniform L^∞-bound on the sequence of twisted negative gradient flow lines u_k, we observe that there exists $r \in (0, +\infty)$ with

$$\operatorname{supp} X_H \cap [r, +\infty) \times \partial W = \varnothing$$

by definition of Rabinowitz–Floer data for φ, and J is adapted to the boundary of $W \cup_{\partial W} [0, r] \times \partial W$ for all $t \in \mathbb{R}$. Consequently, [136, Corollary 9.2.11] implies that every u_k remains within the compact set $W \cup_{\partial W} [0, r] \times \partial W$ as the asymptotics belong to $W \cup_{\partial W} [0, r) \times \partial W$ for all $k \in \mathbb{N}$. Indeed, this follows from

$$\begin{aligned}
E_J(u_k, \tau_k) &= \int_{-\infty}^{+\infty} \|\partial_s(u_k, \tau_k)\|_J^2 ds \\
&= \int_{-\infty}^{+\infty} \langle \partial_s(u_k, \tau_k), \partial_s(u_k, \tau_k)\rangle_J ds \\
&= -\int_{-\infty}^{+\infty} \langle \operatorname{grad}_J \mathscr{A}_\varphi^H|_{(u_k(s), \tau_k(s))}, \partial_s(u_k, \tau_k)\rangle_J ds \\
&= -\int_{-\infty}^{+\infty} d\mathscr{A}_\varphi^H(\partial_s(u_k, \tau_k)) ds \\
&= -\int_{-\infty}^{+\infty} \partial_s \mathscr{A}_\varphi^H(u_k, \tau_k) ds \\
&= \lim_{s \to -\infty} \mathscr{A}_\varphi^H(u_k(s), \tau_k(s)) - \lim_{s \to +\infty} \mathscr{A}_\varphi^H(u_k(s), \tau_k(s)) \\
&\leq b - a,
\end{aligned}$$

as this implies by the negative gradient flow equation

$$\lim_{s\to\pm\infty} \|\partial_s(u_k, \tau_k)\|_J = \lim_{s\to\pm\infty} \left\| \operatorname{grad}_J \mathscr{A}_\varphi^H|_{(u_k(s),\tau_k(s))} \right\|_J = 0.$$

The uniform L^∞-bound on the Lagrange multipliers τ_k follows from the Fundamental Lemma 7.51. Fix a twisted negative gradient flow line (u, τ) and let $C > 0$ be as in the Fundamental Lemma 7.51. For every $\sigma \in \mathbb{R}$ we can define $\zeta(\sigma) \geq 0$ by

$$\zeta(\sigma) := \inf\left\{ \zeta \geq 0 : \| \operatorname{grad}_J \mathscr{A}_\varphi^H|_{(u(\sigma+\zeta),\tau(\sigma+\zeta))} \|_J < \frac{1}{C} \right\}.$$

We estimate

$$b - a \geq E_J(u, \tau) \geq \int_\sigma^{\sigma+\zeta(\sigma)} \| \operatorname{grad}_J \mathscr{A}_\varphi^H|_{(u_s,\tau(s))} \|_J^2 ds \geq \frac{\zeta(\sigma)}{C^2}.$$

By the Fundamental Lemma 7.51 we have that

$$|\tau(\sigma + \zeta(\sigma))| < C(\max\{|a|, |b|\} + 1) \qquad \forall \sigma \in \mathbb{R},$$

and thus using the negative gradient flow equation again we estimate

$$\begin{aligned} |\tau(\sigma)| &\leq |\tau(\sigma + \zeta(\sigma))| + \int_\sigma^{\sigma+\zeta(\sigma)} |\partial_s \tau(s)| ds \\ &= |\tau(\sigma + \zeta(\sigma))| + \int_\sigma^{\sigma+\zeta(\sigma)} \left| \int_0^1 H(u(s,t))dt \right| ds \\ &\leq C(\max\{|a|, |b|\} + 1) + \zeta(\sigma)\|H\|_\infty \\ &\leq C(\max\{|a|, |b|\} + 1) + C^2(b - a)\|H\|_\infty, \end{aligned}$$

for all $\sigma \in \mathbb{R}$. Hence

$$\|\tau\|_\infty \leq C(\max\{|a|, |b|\} + 1) + C^2(b - a)\|H\|_\infty$$

is independent of the twisted negative gradient flow line (u, τ). □

Using Theorem 7.52, we can finally define twisted Rabinowitz–Floer homology! Let (H, J) be Rabinowitz–Floer data for a Liouville automorphism $\varphi \in \operatorname{Aut}(W, \lambda)$. Assume that the twisted Rabinowitz action functional $\mathscr{A}_\varphi^H$ is Morse–Bott, that is, $\operatorname{Crit} \mathscr{A}_\varphi^H \subseteq \partial W \times \mathbb{R}$ is a properly embedded submanifold by Proposition 7.42, and fix a Morse function $h \in C^\infty(\operatorname{Crit} \mathscr{A}_\varphi^H)$. For the completion $(\widehat{W}, \lambda)$ of (W, λ), define the twisted Rabinowitz–Floer chain group $\mathrm{RFC}^{\varphi}(\partial W, \widehat{W})$ to be the $\mathbb{Z}_2$-vector space consisting of all formal linear combinations

$$\zeta = \sum_{(\gamma,\tau)\in \mathrm{Crit}(h)} \zeta_{(\gamma,\tau)}(\gamma, \tau)$$

satisfying the ***Novikov finiteness condition***

$$\#\left\{(\gamma, \tau) \in \mathrm{Crit}(h) : \zeta_{(\gamma,\tau)} \neq 0, \mathscr{A}_{\varphi}^{H}(\gamma, \tau) \geq \kappa\right\} < \infty \qquad \forall \kappa \in \mathbb{R}. \tag{7.19}$$

Define a boundary operator

$$\partial \colon \mathrm{RFC}^{\varphi}(\partial W, \widehat{W}) \to \mathrm{RFC}^{\varphi}(\partial W, \widehat{W})$$

by

$$\partial(\gamma^-, \tau^-) := \sum_{(\gamma^+,\tau^+)\in \mathrm{Crit}(h)} n_{\varphi}(\gamma^{\pm}, \tau^{\pm})(\gamma^+, \tau^+),$$

where

$$n_{\varphi}(\gamma^{\pm}, \tau^{\pm}) := \#_2 \mathscr{M}_{\varphi}^{0}(\gamma^{\pm}, \tau^{\pm}) \in \mathbb{Z}_2,$$

with $\mathscr{M}_{\varphi}^{0}(\gamma^{\pm}, \tau^{\pm})$ denoting the zero-dimensional component of the moduli space of all unparametrised twisted negative gradient flow lines with cascades from (γ^-, τ^-) to (γ^+, τ^+). This is well defined by Theorem 7.52. Define the ***twisted Rabinowitz–Floer homology of ∂W and φ*** by

$$\mathrm{RFH}^{\varphi}(\partial W, \widehat{W}) := \frac{\ker \partial}{\operatorname{im} \partial}.$$

Exercise 7.53 Explain why we impose the Novikov finiteness condition (7.19) in order to define twisted Rabinowitz–Floer homology. Why did we not have to assume condition (7.19) to define Hamiltonian Floer homology?

If there are no nonconstant twisted periodic Reeb orbits on the boundary of a Liouville domain, one can easily compute the twisted Rabinowitz–Floer homology.

Proposition 7.54 *Let $\varphi \in \mathrm{Aut}(W, \lambda)$ be a Liouville automorphism. If ∂W does not admit any nonconstant twisted periodic Reeb orbit, then*

$$\mathrm{RFH}_*^{\varphi}(\partial W, \widehat{W}) \cong \mathrm{H}_*(\mathrm{Fix}(\varphi|_{\partial W}); \mathbb{Z}_2).$$

Proof If there do not exist nonconstant twisted periodic Reeb orbits, then

$$\mathrm{Crit}\,\mathscr{A}_\varphi^H = \{(c_x, 0) : x \in \mathrm{Fix}(\varphi|_{\partial W})\} \cong \mathrm{Fix}(\varphi|_{\partial W})$$

for any $H \in \mathscr{F}_\varphi(\partial W)$. Since $\mathrm{Fix}(\varphi|_{\partial W})$ is a properly embedded submanifold of ∂W by [130, Problem 8-32] or [137, Lemma 5.5.7], $\mathscr{A}_\varphi^H$ is a Morse–Bott function. Choose a Morse function h on $\mathrm{Fix}(\varphi|_{\partial W})$ and any critical point $c_x \in \mathrm{Fix}(\varphi|_{\partial W})$. Then we can define a $\mathbb{Z}$-grading of $\mathrm{RFC}^\varphi(\partial W, \widehat{W})$ by using the classical Morse index of h. Consequently, we have that

$$\mathrm{RFH}_*^\varphi(\partial W, \widehat{W}) = \mathrm{HM}_*(\mathrm{Fix}(\varphi|_{\partial W}); \mathbb{Z}_2) \cong \mathrm{H}_*(\mathrm{Fix}(\varphi|_{\partial W}); \mathbb{Z}_2)$$

as there are only twisted negative gradient flow lines with zero cascades, that is, ordinary Morse gradient flow lines of h. Indeed, suppose that there is a nonconstant twisted negative gradient flow line (u, τ) of $\mathscr{A}_\varphi^H$ such that

$$\lim_{s \to \pm\infty} (u(s), \tau(s)) = (\gamma^\pm, \tau^\pm) \in \mathrm{Crit}\,\mathscr{A}_\varphi^H.$$

Using the twisted negative gradient flow equation we estimate

$$\begin{aligned} \tau^- - \tau^+ &= \mathscr{A}_\varphi^H(\gamma^-, \tau^-) - \mathscr{A}_\varphi^H(\gamma^+, \tau^+) \\ &= \lim_{s \to -\infty} \mathscr{A}_\varphi^H(u(s), \tau(s)) - \lim_{s \to +\infty} \mathscr{A}_\varphi^H(u(s), \tau(s)) \\ &= \int_{-\infty}^{+\infty} \| \mathrm{grad}_J\, \mathscr{A}_\varphi^H|_{(u(s),\tau(s))} \|_J^2 ds > 0. \end{aligned}$$

Hence $\tau^+ < \tau^-$, contradicting $\tau^\pm = 0$. □

In Proposition 7.54, there was a natural choice to grade the twisted Rabinowitz–Floer homology $\mathrm{RFH}^\varphi(\partial W, \widehat{W})$. In general, this does not work except that one makes some assumption similar to the grading of Hamiltonian Floer homology.

Definition 7.55 (Transverse Conley–Zehnder Index) Let $\varphi \in \mathrm{Aut}(W^{2n}, \lambda)$ be a Liouville automorphism, and let $(\gamma_0, \tau_0), (\gamma_1, \tau_1) \in \mathscr{P}_\varphi(\Sigma, \lambda|_\Sigma)$ be such that there exists a path γ_σ in $\mathscr{L}_\varphi \partial W$ from γ_0 to γ_1. Define the ***transverse Conley–Zehnder index*** by

$$\mu((\gamma_0, \tau_0), (\gamma_1, \tau_1)) := \mu_{\mathrm{CZ}}(\Psi^1) - \mu_{\mathrm{CZ}}(\Psi^0) \in \mathbb{Z},$$

with

$$\Psi^0 \colon [0, 1] \to \mathrm{Sp}(2n-2), \qquad \Psi_t^0 := \Phi_{t,0}^{-1} \circ D^\xi \phi_{\tau_0 t}^R \circ \Phi_{0,0},$$

$$\Psi^1 \colon [0,1] \to \mathrm{Sp}(2n-2), \qquad \Psi^1_t := \Phi^{-1}_{t,1} \circ D^\xi \phi^R_{\tau_1 t} \circ \Phi_{0,1},$$

where $\Phi_{t,\sigma} \colon \mathbb{R}^{2n-2} \to \xi_{\gamma_\sigma(t)}$ is a symplectic trivialisation of $F^*\xi$, $\xi := \ker \lambda|_{\partial W}$ with $F \in C^\infty(\mathbb{R} \times [0,1], M)$ being defined by $F(t,\sigma) := \gamma_\sigma(t)$, satisfying

$$\Phi_{t+1,\sigma} = D\varphi \circ \Phi_{t,\sigma} \qquad \forall (t,\sigma) \in \mathbb{R} \times [0,1]. \tag{7.20}$$

Let $\varphi \in \mathrm{Aut}(W,\lambda)$ be a Liouville automorphism and denote by

$$\partial W_\varphi := \frac{\partial W \times \mathbb{R}}{(x,t) \sim (\varphi(x), t+1)}$$

the mapping torus of φ as in Example 4.7, giving rise to the fibration

$$\pi_\varphi \colon \partial W_\varphi \to \mathbb{T}, \qquad \pi_\varphi([x,t]) := [t].$$

The vertical bundle $\ker D^\xi \pi_\varphi \to \partial W_\varphi$ is a symplectic vector bundle.

Lemma 7.56 *In the setup of Definition 7.55, the transverse Conley–Zehnder index is well-defined, that is, independent of the choice of symplectic trivialisation. Moreover, if $c_1(\ker D^\xi \pi_\varphi) = 0$, then the transverse Conley–Zehnder index is independent of the choice of path in $\mathscr{L}_\varphi \partial W$ from γ_0 to γ_1.*

Proof First we need to show that one can always construct a symplectic trivialisation

$$\Phi_{t,\sigma} \colon \mathbb{R}^{2n-2} \to \xi_{F(t,\sigma)} \qquad \forall (t,\sigma) \in \mathbb{R} \times [0,1],$$

of $F^*\xi$ satisfying the twist condition (7.20). By [137, Theorem 2.1.3], there exists a linear symplectomorphism $\Phi_{0,0} \colon \mathbb{R}^{2n-2} \to \xi_{F(0,0)}$ and by [137, Lemma 2.6.6], we get a symplectic trivialisation $\Phi_{t,0} \colon \mathbb{R}^{2n-2} \to \xi_{F(t,0)}$ satisfying $\Phi_{1,0} = D\varphi \circ \Phi_{0,0}$ for all $t \in [0,1]$. Extend this trivialisation to $\mathbb{R}$ by setting

$$\Phi_{t+k,0} := D\varphi^k \circ \Phi_{t,0} \qquad \forall k \in \mathbb{Z}.$$

Next, trivialise along each ray $\sigma \mapsto F(t,\sigma)$ for fixed $t \in \mathbb{R}$. Hence, we get a symplectic trivialisation $\Phi_{t,\sigma} \colon \mathbb{R}^{2n-2} \to \xi_{F(t,\sigma)}$ of $F^*\xi$ satisfying (7.20). Now we show that the transverse Conley–Zehnder index μ_{CZ} is independent of the choice of trivialisation. Suppose that $\widetilde{\Phi}_{t,\sigma} \colon \mathbb{R}^{2n-2} \to \xi_{F(t,\sigma)}$ is another symplectic trivialisation of $F^*\xi$ satisfying

$$\widetilde{\Phi}_{t+1,\sigma} = D\varphi \circ \widetilde{\Phi}_{t,\sigma} \qquad \forall (t,\sigma) \in \mathbb{R} \times [0,1].$$

Then we have

$$\widetilde{\Psi}_{t,\sigma} = \widetilde{\Phi}_{t,\sigma}^{-1} \circ \Phi_{t,\sigma} \circ \Psi_{t,\sigma} \circ \Phi_{0,\sigma}^{-1} \circ \widetilde{\Phi}_{0,\sigma} \qquad \forall (t,\sigma) \in \mathbb{R} \times \{0,1\},$$

where

$$\Psi_{t,\sigma} = \Phi_{t,\sigma}^{-1} \circ D^{\xi} \phi_t^{\tau_\sigma R} \circ \Phi_{0,\sigma}.$$

Define

$$\eta \colon \mathbb{T} \times [0,1] \to \mathrm{Sp}(2n-2), \qquad \eta(t,\sigma) := \Phi_{0,\sigma}^{-1} \circ \widetilde{\Phi}_{0,\sigma} \circ \widetilde{\Phi}_{t,\sigma}^{-1} \circ \Phi_{t,\sigma}.$$

Indeed, we compute

$$\begin{aligned} \eta(t+1,\sigma) &= \Phi_{0,\sigma}^{-1} \circ \widetilde{\Phi}_{0,\sigma} \circ \widetilde{\Phi}_{t+1,\sigma}^{-1} \circ \Phi_{t+1,\sigma} \\ &= \Phi_{0,\sigma}^{-1} \circ \widetilde{\Phi}_{0,\sigma} \circ \widetilde{\Phi}_{t,\sigma}^{-1} \circ D\varphi^{-1} \circ D\varphi \circ \Phi_{t,\sigma} \\ &= \eta(t,\sigma), \end{aligned}$$

for all $(t,\sigma) \in \mathbb{R} \times [0,1]$. Using the naturality as well as the loop property [164, p. 20] of the Conley–Zehnder index we compute

$$\mu_{\mathrm{CZ}}(\widetilde{\Psi}^s) = \mu_{\mathrm{CZ}}(\eta_s \cdot \Psi^s) = \mu_{\mathrm{CZ}}(\Psi^s) + 2\mu_{\mathrm{M}}(\eta_s) \qquad \forall s \in \{0,1\},$$

where μ_{M} denotes the Maslov index. In particular,

$$\begin{aligned} \mu_{\mathrm{CZ}}(\widetilde{\Psi}^1) - \mu_{\mathrm{CZ}}(\widetilde{\Psi}^0) &= \mu_{\mathrm{CZ}}(\Psi^1) - \mu_{\mathrm{CZ}}(\Psi^0) + 2\left(\mu_{\mathrm{M}}(\eta_1) - \mu_{\mathrm{M}}(\eta_0)\right) \\ &= \mu_{\mathrm{CZ}}(\Psi^1) - \mu_{\mathrm{CZ}}(\Psi^0) \end{aligned}$$

by the invariance property of the Maslov index [77, p. 195]. Assume that $\widetilde{F}$ is another homotopy in $\mathscr{L}_\varphi \Sigma$ from γ_0 to γ_1. The concatenation with the reversed path F^- can be identified with the map

$$\widetilde{F}\#F^- \colon \mathbb{T}^2 \to \partial W_\varphi, \qquad (t,\sigma) \mapsto [(\widetilde{F}\#F^-(t,\sigma), t)].$$

Hence using the concatenation property of the Conley–Zehnder index [77, p. 195] as well as the functoriality of the Chern number [137, p. 85], we compute

$$\begin{aligned} \mu_{\mathrm{CZ}}(\widetilde{\Psi}^s) - \mu_{\mathrm{CZ}}(\Psi^s) &= \mu_{\mathrm{CZ}}\left(\widetilde{\Psi}^s\#(\Psi^s)^-\right) \\ &= 2\mu_{\mathrm{M}}\left(\widetilde{\Psi}^s\#(\Psi^s)^-\right) \\ &= 2c_1\left((\widetilde{F}\#F^-)^* \ker D^{\xi}\pi_\varphi\right) \\ &= 2c_1(\ker D^{\xi}\pi_\varphi) \end{aligned}$$

for $s \in \{0, 1\}$. Thus, if the transverse Conley–Zehnder index is viewed as a number in $\mathbb{Z}_2$ or $c_1(\ker D^{\xi}\pi_{\varphi}) = 0$, then it additionally does not depend on the choice of path in $\mathscr{L}_{\varphi}\partial W$. □

Twisted Rabinowitz–Floer homology is stable under twisted homotopies of Liouville domains.

Theorem 7.57 (Invariance of Twisted Rabinowitz–Floer Homology) *Let $(\widehat{W}, \lambda)$ be the completion of a Liouville domain (W, λ) and let $\varphi \in \mathrm{Aut}(W, \lambda)$ be a Liouville automorphism. If $H \in C^{\infty}(\widehat{W} \times [0, 1])$ is a twisted homotopy of Liouville domains such that both $\mathscr{A}_{\varphi}^{H_0}$ and $\mathscr{A}_{\varphi}^{H_1}$ are Morse–Bott, then there is a canonical isomorphism*

$$\mathrm{RFH}^{\varphi}(\partial W_0, \widehat{W}) \cong \mathrm{RFH}^{\varphi}(\partial W_1, \widehat{W}).$$

Sketch of Proof The proof follows from the same adiabatic argument as in [51, p. 275–277] reviewed in Problem 5.47. Crucial is that [51, Theorem 3.6] remains true in our setting, as well as the genericity of the Morse–Bott condition. In fact, if (M, λ) is an exact symplectic manifold and $\varphi \in \mathrm{Diff}(M)$ is of finite order such that $\varphi^*\lambda = \lambda$, then we have the following generalisation of [51, Theorem B.1]. One can show that there exists a subset $\mathscr{U} \subseteq \{H \in C^{\infty}_{\varphi}(M) : \mathrm{supp}\, dH \text{ compact}\}$, of the second category such that for every $H \in \mathscr{U}$, $\mathscr{A}_{\varphi}^{H}$ is Morse–Bott with the critical manifold being $\mathrm{Fix}(\varphi|_{H^{-1}(0)})$ together with a disjoint union of circles. Again, this works only since the contact condition is an open condition in contrast, for example, to the stability condition [55]. □

Remark 7.58 The invariance of the twisted Rabinowitz–Floer homology allows us to define the twisted Rabinowitz–Floer homology also in the case where $\mathscr{A}_{\varphi}^{H}$ is not a Morse–Bott function. Indeed, as the proof of Theorem 7.57 shows, we can perturb the hypersurface ∂W slightly to make the twisted Rabinowitz action functional $\mathscr{A}_{\varphi}^{H}$ a Morse–Bott function. Thus we can define the twisted Rabinowitz–Floer homology of such a hypersurface to be the twisted Rabinowitz–Floer homology of any Morse–Bott perturbation which is well defined by Theorem 7.57.

An important implication of the invariance established in Theorem 7.57 is the following result, showing the independence of the Rabinowitz–Floer homology of the twisted defining Hamiltonian function.

Corollary 7.59 (Independence) *Let $\varphi \in \mathrm{Aut}(W, \lambda)$ a Liouville automorphism and let $H_0, H_1 \in \mathscr{F}_{\varphi}(\partial W)$ be twisted defining Hamiltonian functions such that $\mathscr{A}_{\varphi}^{H_0}$ or $\mathscr{A}_{\varphi}^{H_1}$ is a Morse–Bott function. Then the definition of twisted Rabinowitz–Floer homology $\mathrm{RFH}^{\varphi}(\partial W, \widehat{W})$ is independent of the choice of a twisted defining Hamiltonian function for ∂W.*

Proof Note that $\mathscr{F}_\varphi(\partial W)$ is a convex space. Indeed, set

$$H_\sigma := (1-\sigma)H_0 + \sigma H_1 \qquad \sigma \in [0,1].$$

Then $\varphi^* H_\sigma = H_\sigma$, dH_σ has a compact support, and $X_{H_\sigma}|_{\partial W} = R$ for all $\sigma \in [0,1]$. Moreover, for the Liouville vector field $X \in \mathfrak{X}(\widehat{W})$ we compute

$$\left.\frac{d}{dt}\right|_{t=0} H \circ \phi_t^X|_{\partial W} = dH(X)|_{\partial W} = d\lambda(X, X_H)|_{\partial W} = \lambda(X_H)|_{\partial W} = \lambda(R) = 1,$$

for any $H \in \mathscr{F}_\varphi(\partial W)$, and thus $H < 0$ on $\operatorname{Int} W$ and $H > 0$ on $\widehat{W} \setminus W$. Consequently, $H_\sigma^{-1}(0) = \partial W$ and so $H_\sigma \in \mathscr{F}_\varphi(\partial W)$ for all $\sigma \in [0,1]$. Hence $(H_\sigma)_{\sigma\in[0,1]}$ is a twisted homotopy of Liouville domains in $\widehat{W}$ and Theorem 7.57 implies the claim. □

Problems

7.60 (Brieskorn Manifolds, [70]) Let $a = (a_0, a_1, \dots, a_n) \in \mathbb{N}_{\geq 2}^{n+1}$ be a vector of natural numbers. Define the smooth function

$$f_a \colon \mathbb{C}^{n+1} \to \mathbb{C}, \qquad f_a(z_0, z_1, \dots, z_n) := \sum_{j=0}^{n} z_j^{a_j}.$$

A ***Brieskorn manifold*** is defined as the intersection $\Sigma_a := f_a^{-1}(0) \cap \mathbb{S}^{2n+1}$.

1. Can you give two examples of Brieskorn manifolds? *Hint:* See [123, Section 2.2].
2. Show that Σ_a and $f_a^{-1}(r)$ are smooth manifolds for all $r \in \mathbb{R} \setminus \{0\}$.
3. Show that $\lambda_a \in \Omega^1(\mathbb{C}^{n+1})$ defined by

$$\lambda_a := \frac{i}{2} \sum_{j=0}^{n} a_j (z_j d\bar{z}_j - \bar{z}_j dz_j)$$

   restricts to Σ_a giving rise to a contact form with associated Reeb vector field

$$R_a = i\left(\frac{z_0}{a_0}, \frac{z_1}{a_1}, \dots, \frac{z_n}{a_n}\right) \in \mathfrak{X}(\Sigma_a).$$

4. Choose a smooth cutoff function $\rho \in C^\infty(\mathbb{R})$, that is, ρ is monotone decreasing and satisfies

$$\rho(r) = \begin{cases} 1 & x \leq \frac{1}{4}, \\ 0 & x \geq \frac{3}{4}. \end{cases}$$

For $\varepsilon \in \mathbb{R}$ define

$$V_\varepsilon(a) := \left\{ z \in \mathbb{C}^{n+1} : f(z) = \varepsilon\rho(|z|^2) \right\}.$$

Show that $(V_\varepsilon(a), \lambda_a)$ is a Liouville domain with boundary $\partial V_\varepsilon(a) = \Sigma_a$ with vanishing first Chern class $c_1(TV_\varepsilon(a))$, provided ε is sufficiently small.

5. Pick a defining Hamiltonian function for Σ_a and show that the corresponding Rabinowitz action functional is a Morse–Bott function. *Hint:* See [70].

7.61 (Symplectic Homology) Let $(\widehat{W}, \lambda)$ be the completion of a Liouville domain (W, λ) satisfying $c_1(T\widehat{W}) = 0$. For a general time-dependent Hamiltonian function $H \in C^\infty(\widehat{W} \times \mathbb{T})$ we consider the symplectic action functional

$$\mathscr{A}_H : \mathscr{L}\widehat{W} \to \mathbb{R}, \qquad \mathscr{A}_H(\gamma) := \int_0^1 \gamma^*\lambda - \int_0^1 H_t(\gamma(t))dt.$$

Denote by $\operatorname{Spec}(\partial W, \lambda|_{\partial W})$ the spectrum of ∂W, that is, the set of all periods of periodic Reeb orbits. For $a \in \mathbb{R} \cup \{\pm\infty\}$ with $a \notin \operatorname{Spec}(\partial W, \lambda|_{\partial W})$ we define the free $\mathbb{Z}_2$-vector space

$$\mathrm{CF}_k^{<a} := \mathrm{Crit}_k^a(\mathscr{A}_H) \otimes \mathbb{Z}_2,$$

where

$$\mathrm{Crit}_k^a(H) := \{\gamma \in \mathrm{Crit}(\mathscr{A}_H) : \mu_{\mathrm{CZ}}(\gamma) = k \text{ and } \mathscr{A}_H(\gamma) < a\}$$

for all $k \in \mathbb{Z}$.

1. Show that the usual boundary operator $\partial_k : \mathrm{CF}_k(H) \to \mathrm{CF}_{k-1}(H)$ in Floer homology defined by counting negative gradient flow lines of $\mathscr{A}_H$ descends to a boundary operator $\partial_*^{(a,b)}$ on the quotient

$$\mathrm{CF}_*^{(a,b)}(H) := \frac{\mathrm{CF}_*^{<b}(H)}{\mathrm{CF}_*^{<a}(H)} \qquad \forall -\infty \le a < b \le +\infty.$$

2. A time-dependent Hamiltonian function $H \in C^\infty(\widehat{W} \times \mathbb{T})$ is said to be ***admissible***, if it satisfies

   - $H \le 0$ on W.
   - H only admits nondegenerate periodic orbits.
   - H is linear at infinity, that is, $H(r, x) = ar + b$ for r sufficiently large and where $0 < a \notin \operatorname{Spec}(\partial W, \lambda|_{\partial W})$.

   A typical admissible Hamiltonian function is depicted in Fig. 7.5a. Draw an admissible Hamiltonian function for defining Rabinowitz–Floer homology. What are the similarities and differences?

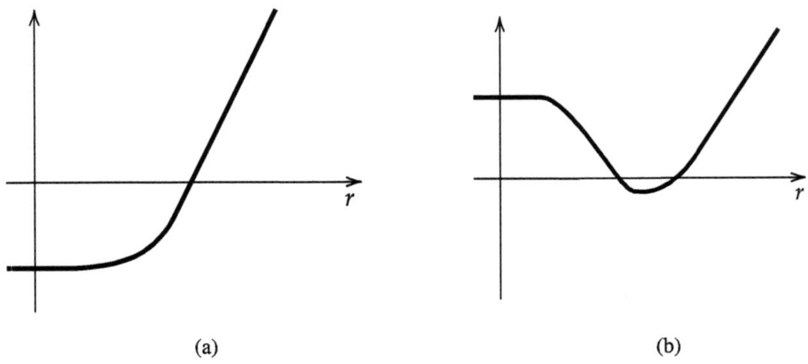

Fig. 7.5 (**a**) A typical admissible Hamiltonian function in symplectic homology. (**b**) A typical admissible $\bigvee$-shaped Hamiltonian function

3. One can define a partial order $\preccurlyeq$ on the set of all admissible Hamiltonian functions by setting

$$H \preccurlyeq K \qquad :\Leftrightarrow \qquad H_t(x) \leq K_t(x) \qquad \forall (x,t) \in \widehat{W} \times \mathbb{T}.$$

Let $H_- \preccurlyeq H_+$. Show that a monotone increasing homotopy from H_- to H_+ induces chain maps

$$\sigma_k^{(a,b)} \colon \mathrm{CF}_k^{(a,b)}(H_-) \to \mathrm{CF}_k^{(a,b)}(H_+) \qquad \forall k \in \mathbb{Z}.$$

Moreover, show that these maps descend to define unique maps on ***filtered Floer homology***

$$\mathrm{HF}_k^{(a,b)}(H_\pm) := \frac{\ker \partial_k \colon \mathrm{CF}_k^{(a,b)}(H_\pm) \to \mathrm{CF}_{k-1}^{(a,b)}(H_\pm)}{\operatorname{im} \partial_{k+1}^{(a,b)} \colon \mathrm{CF}_{k+1}^{(a,b)}(H_\pm) \to \mathrm{CF}_k^{(a,b)}(H_\pm)}.$$

4. Define the ***symplectic homology groups of W*** by the direct limit

$$\mathrm{SH}_*^{(a,b)}(W) := \varinjlim \mathrm{HF}_*^{(a,b)}(H), \qquad H \to \infty. \tag{7.21}$$

Moreover, define

$$\mathrm{SH}_*^-(W) := \lim_{b \to 0} \mathrm{SH}_*^{(-\infty,b)}(W).$$

As in [54, Lemma 2.1], show that

$$\mathrm{SH}_*^-(W) \cong \mathrm{H}^{n-*}(W;\mathbb{Z}_2) \cong \mathrm{H}_{*+n}(W,\partial W;\mathbb{Z}_2), \qquad n = \frac{1}{2}\dim W.$$

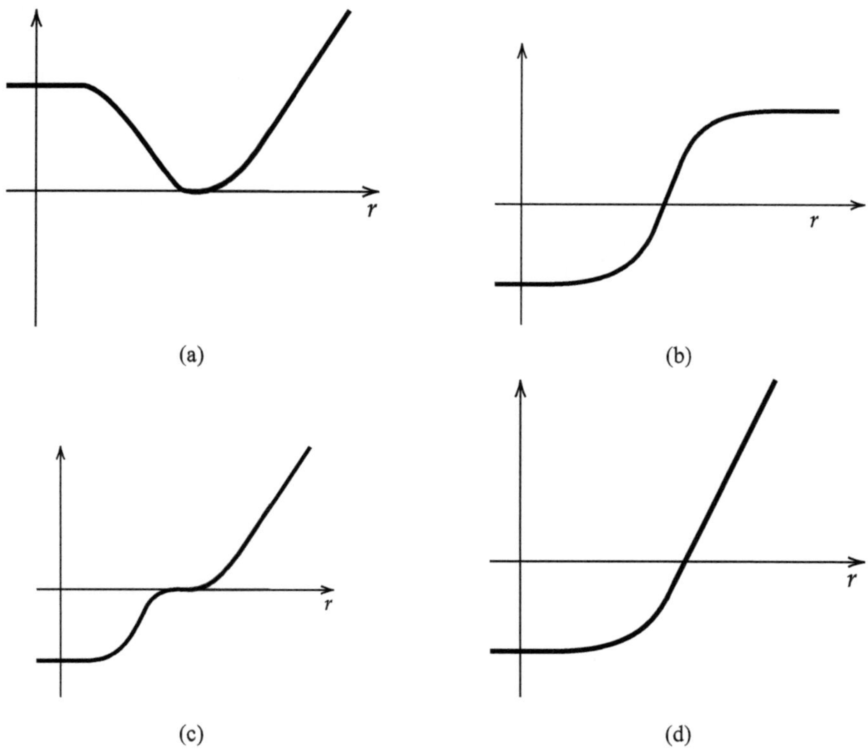

Fig. 7.6 A proof-by-pictures of the isomorphism between Rabinowitz–Floer homology and the symplectic homology of $\bigvee$-shaped Hamiltonian functions

5. Give a proof-by-pictures of the isomorphism

$$\mathrm{RFH}_*(W) \cong \check{\mathrm{SH}}_*(W)$$

established in [54, Theorem 1.5], where $\check{\mathrm{SH}}_*(W)$ denotes symplectic homology (7.21) for $\bigvee$-shaped Hamiltonian functions depicted in Fig. 7.5b. To achieve this, bring the different radial Hamiltonian functions depicted in Fig. 7.6 in the correct order and describe what exactly happens in each step. Does this "proof" also show

$$\text{“}\mathrm{RFH}_*(W) \cong \mathrm{SH}_*(W)\text{”},$$

where the full symplectic homology of W is defined by

$$\mathrm{SH}_*(W) := \mathrm{SH}_*^{(-\infty,+\infty)}(W)?$$

7.62 (A Maximum Principle) In order to extend the theory of J-holomorphic curves to noncompact symplectic manifolds like the completion of a Liouville domain, one needs a hypothesis to ensure that J-holomorphic curves cannot escape a suitable compact subset such that Gromov compactness still applies as in Corollary 6.14.

1. Let (W, λ) be a Liouville domain. Show that there exists a pair (f, J), where J is a $d\lambda$-compatible almost complex structure and $f\colon W \to (-\infty, 0]$ is a smooth function, such that $f^{-1}(0) = \partial W$ and for all $v \in TW$ we have that
$$\omega_f(Jv, v) \geq 0, \qquad \omega_f := -d(df \circ J).$$
2. Let (W, λ) be a Liouville domain and let (f, J) be the pair from the first part of this exercise. Show that for any J-holomorphic curve $u \in C^\infty(U, W)$ for some open subset $U \subseteq \mathbb{C}$, we have that $f \circ u$ is subharmonic, that is, we have
$$\Delta(f \circ u) = \omega_f(J\partial_s u, \partial_s u) \geq 0.$$
3. Let (W, λ) be a Liouville domain and let Σ be a connected Riemann surface without boundary. Moreover, let $u \in C^\infty(\Sigma, W)$ be a smooth map such that the restriction of u to $u^{-1}(U)$ is J-holomorphic for some open neighbourhood $U \subseteq W$ of ∂W. Show that then
$$u(\Sigma) \cap \partial W \neq \varnothing \qquad \Leftrightarrow \qquad u(\Sigma) \subseteq \partial W.$$

7.63 ($\mathbb{Z}$-Coefficients) Let (M, ω) be a compact symplectically aspherical symplectic manifold satisfying $c_1(TM)|_{\pi_2(M)} = 0$ and $H \in C^\infty(M \times \mathbb{T})$ such that the symplectic action functional $\mathscr{A}_H$ is a Morse function. Choose a coherent orientation σ on M as in Problem 5.46 or [72]. Set
$$\operatorname{Crit}_k(\mathscr{A}_H) := \{x \in \operatorname{Crit}(\mathscr{A}_H) : \mu_{\mathrm{CZ}}(x) = k\} \qquad \forall k \in \mathbb{Z}$$
and
$$\operatorname{CF}_k(H) := \operatorname{Crit}_k(\mathscr{A}_H) \otimes \mathbb{Z}.$$
For every $k \in \mathbb{Z}$ define a map $\partial_k^\sigma\colon \operatorname{CM}_k(H) \to \operatorname{CM}_{k-1}(H)$ by
$$\partial_k^\sigma \gamma^- := \sum_{\gamma^+ \in \operatorname{Crit}_{k-1}(\mathscr{A}_H)} \sum_{u \in \mathscr{M}(\gamma^-, \gamma^+)} \tau_\sigma(u)\gamma^+,$$
where $\tau_\sigma(u) = \pm 1$ denotes the characteristic sign induced by the coherent orientation σ by comparing the coherent orientation with the flow orientation.

1. Show that Floer chain complex $(\mathrm{CF}_*(H), \partial_*^\sigma)$ is a chain complex, that is, there holds

$$\partial_{k-1}^\sigma \circ \partial_k^\sigma = 0 \qquad \forall k \in \mathbb{Z}.$$

Hint: See [164, Theorem 3.5].

2. Denote the induced homology by $\mathrm{HF}_*^\sigma(H)$. Show that there is a natural isomorphism

$$\mathrm{HF}_*^\sigma(H) \cong \mathrm{HM}_{*+n}(H; \mathbb{Z})$$

for a sufficiently C^2-small Hamiltonian function $H \in C^\infty(M)$. *Hint:* See [24, Theorem 10.1.1].

3. Let (Q, m) be a compact Riemannian manifold without boundary and fix a potential function $V \in C^\infty(Q \times \mathbb{T})$. Consider the mechanical Lagrangian function

$$L\colon \mathscr{L}Q \times \mathbb{T} \to \mathbb{R}, \qquad L(q, \dot{q}, t) = \frac{1}{2}\|\dot{q}\|_m^2 - V(q, t).$$

Assume that the time-dependent Lagrangian action functional

$$\mathscr{E}_L\colon \mathscr{L}Q \to \mathbb{R}, \qquad \mathscr{E}_L(\gamma) := \int_0^1 L(\gamma, \dot{\gamma}, t)dt$$

is a Morse function. This is by [191, Example 3.3.33] generically true for potentials in $C^\infty(Q \times \mathbb{T})$. Show that there is a natural isomorphism

$$\mathrm{HM}_*^\sigma(\mathscr{E}_L) \cong \mathrm{H}_*(\mathscr{L}Q; \mathbb{Z}).$$

Hint: See [190, Theorem A].

Applications of Twisted Rabinowitz–Floer Homology

8

Following [14], consider the restricted three-body problem given by a (massless) satellite, Jupiter, and its moon Europa. Imagine the plausible scenario of a comet passing by, interfering with the planned trajectory of the satellite. During this time, the satellite is no longer confined to its periodic orbit about Europa as energy is no longer conserved. It would be desirable that the satellite returns to its initial periodic orbit after the perturbation by a comet. To analytically investigate this phenomenon, one can introduce the notion of a leaf-wise intersection point. In this chapter, we will prove the third part of Theorem 7.22 using a clever argument introduced in [12].

8.1 Twisted Leaf-Wise Intersection Points

Definition 8.1 (Twisted Leaf-Wise Intersection Point) Let $(\widehat{W}, \lambda)$ be the completion of a Liouville domain (W, λ) and let $\varphi \in \operatorname{Aut}(W, \lambda)$ be a Liouville automorphism. A point $x \in \partial W$ is a ***twisted leaf-wise intersection point*** for a Hamiltonian symplectomorphism $\varphi_F \in \operatorname{Ham}(\widehat{W}, \lambda)$, if

$$\varphi_F(x) \in L_{\varphi(x)} = \{\phi_t^R(\varphi(x)) : t \in \mathbb{R}\} \subseteq \partial W$$

for the Reeb vector field $R \in \mathfrak{X}(\partial W)$.

Exercise 8.2 Explain why the notion of a twisted leaf-wise intersection point formally captures the physical scenario of an orbit of a satellite perturbed by a flying-by comet depicted in Fig. 8.1.

Y. Bähni, *Lectures on Twisted Rabinowitz-Floer Homology*, Pathways in Mathematics, https://doi.org/10.1007/978-3-032-10673-5_8

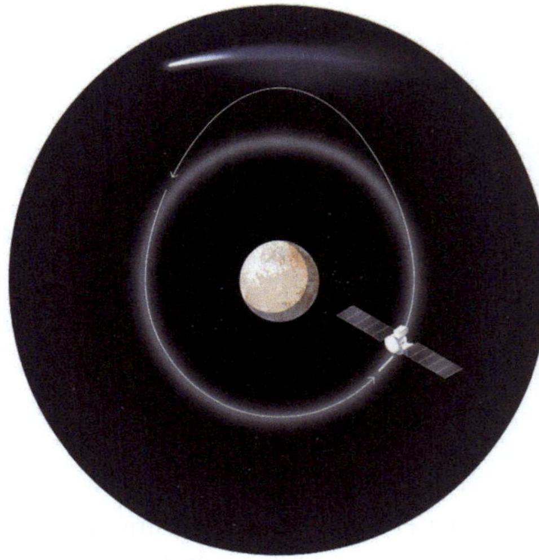

Fig. 8.1 The Europa–Jupiter–Satellite system perturbed by a comet

Definition 8.3 (Twisted Moser Pair) Let $\varphi \in \mathrm{Aut}(W, \lambda)$ be a Liouville automorphism. A ***twisted Moser pair*** is defined to be a tuple $\mathfrak{M} := (\chi H, F)$, where

1. $H \in C^\infty_\varphi(\widehat{W})$, $F \in C^\infty_\varphi(\widehat{W} \times \mathbb{R})$ and $\chi \in C^\infty(\mathbb{T}, [0, 1])$ with $\int_0^1 \chi(t)dt = 1$. Any time-dependent Hamiltonian function χH is said to be *weakly time-dependent*.
2. $\operatorname{supp} \chi \subseteq (0, \frac{1}{2})$ and $F_t = 0$ for all $t \in [0, \frac{1}{2}]$.

Lemma 8.4 *Let $\varphi \in \mathrm{Aut}(W, \lambda)$ be a Liouville automorphism. For all $H \in \mathscr{F}_\varphi(\partial W)$ and $\varphi_F \in \mathrm{Ham}(\widehat{W}, \lambda)$ there exists a corresponding twisted Moser pair $\mathfrak{M}$ such that the flow of χX_H is a time-reparametrisation of the flow of X_H.*

Proof For constructing the Hamiltonian perturbation $\widetilde{F}$, fix $\rho \in C^\infty([0, 1])$ as in Fig. 8.2a such that

$$0 \leq \rho \leq 1 \qquad \text{and} \qquad \rho(t) = \begin{cases} 0 & t \in [0, \frac{1}{2}], \\ 1 & t \in [\frac{2}{3}, 1]. \end{cases}$$

Then define $\widetilde{F} \in C^\infty_\varphi(M \times \mathbb{R})$ by

$$\widetilde{F}(x, t) := \dot{\rho}(t-k) F(\varphi^{-k}(x), \rho(t-k)) \qquad \forall t \in [k, k+1],$$

for $k \in \mathbb{Z}$ with $\dot{\rho}$ as in Fig. 8.2b. Then $\widetilde{F} \in C^\infty_\varphi(\widehat{W} \times \mathbb{R})$ and $\widetilde{F}_t = 0$ for all $t \in [0, \frac{1}{2}]$. Moreover, we claim that

$$\phi_t^{X_{\widetilde{F}}} = \phi_{\rho(t)}^{X_F} \qquad \forall t \in [0, 1].$$

(a) (b)

Fig. 8.2 (**a**) The smooth function ρ. (**b**) The derivative $\dot{\rho}$ of ρ

Indeed, we compute

$$\frac{d}{dt}\phi^{X_F}_{\rho(t)} = \dot{\rho}(t)\frac{d}{d\rho}\phi^{X_F}_{\rho(t)} = \dot{\rho}(t)\left(X_{F_{\rho(t)}} \circ \phi^{X_F}_{\rho(t)}\right) = X_{\widetilde{F}_t} \circ \phi^{X_F}_{\rho(t)}.$$

In particular

$$\varphi_{\widetilde{F}} = \phi_1^{X_{\widetilde{F}}} = \phi^{X_F}_{\rho(1)} = \phi_1^{X_F} = \varphi_F,$$

and thus the Hamiltonian symplectomorphisms $\varphi_{\widetilde{F}}$ and φ_F coincide. Finally, we have that

$$\phi_t^{\chi X_H} = \phi^{X_H}_{\tau(t)} \qquad \text{with} \qquad \tau(t) := \int_0^t \chi.$$

Indeed, we compute

$$\frac{d}{dt}\phi^{X_H}_{\tau(t)} = \chi(t)\frac{d}{d\tau}\phi^{X_H}_{\tau(t)} = \chi(t)X_H \circ \phi^{X_H}_{\tau(t)},$$

and conclude by the uniqueness of integral curves. □

Twisted leaf-wise intersection points can be detected variationally by a suitable perturbation of the twisted Rabinowitz action functional associated with a twisted Moser pair.

Definition 8.5 (Perturbed Twisted Rabinowitz Action Functional) Let $(\widehat{W}, \lambda)$ be the completion of a Liouville domain and let $\varphi \in \mathrm{Aut}(W, \lambda)$ be a Liouville automorphism. For a twisted Moser pair $\mathfrak{M} = (\chi H, F)$ we define the ***perturbed***

twisted Rabinowitz action functional

$$\mathscr{A}_{\varphi}^{\mathfrak{M}} : \mathscr{L}_{\varphi}\widehat{W} \times \mathbb{R} \to \mathbb{R}, \qquad \mathscr{A}_{\varphi}^{\mathfrak{M}}(\gamma, \tau) := \mathscr{A}_{\varphi}^{\chi H}(\gamma, \tau) - \int_0^1 F_t(\gamma(t))dt,$$

where $\mathscr{A}_{\varphi}^{\chi H}$ denotes the twisted Rabinowitz action functional.

A computation similar to the one in Proposition 7.26 shows

$$(\gamma, \tau) \in \operatorname{Crit} \mathscr{A}_{\varphi}^{\mathfrak{M}} \qquad \Leftrightarrow \qquad \begin{cases} \dot{\gamma}(t) = \tau\chi(t)X_H(\gamma(t)) + F_t(\gamma(t)), \\ \int_0^1 \chi(t)H(\gamma(t))dt = 0, \end{cases} \tag{8.1}$$

for all $t \in \mathbb{R}$. Note that energy is no longer conserved, in general!

Proposition 8.6 (Variational Characterisation of Twisted Leaf-Wise Intersection Points) *Let $\varphi \in \operatorname{Aut}(W, \lambda)$ be a Liouville automorphism, and let $\varphi_F \in \operatorname{Ham}(\widehat{W}, \lambda)$ be a Hamiltonian symplectomorphism. If $(\gamma, \tau) \in \operatorname{Crit} \mathscr{A}_{\varphi}^{\mathfrak{M}}$, then $x := \gamma(\frac{1}{2})$ is a twisted leaf-wise intersection point for φ_F, where $\mathfrak{M}$ is the associated twisted Moser pair.*

Proof Let $\mathfrak{M} = (\chi H, F)$ denote the twisted Moser pair from Lemma 8.4. Then we compute for all $t \in [0, \frac{1}{2}]$

$$\frac{d}{dt}H(\gamma(t)) = dH(\tau\chi(t)X_H(\gamma(t)) + X_{F_t}(\gamma(t))) = \tau\chi(t)dH(X_H(\gamma(t))) = 0.$$

Thus, $H(\gamma(t)) = c \in \mathbb{R}$ for $t \in [0, \frac{1}{2}]$ with

$$c = c\int_0^1 \chi(t)dt = \int_0^{\frac{1}{2}} \chi(t)H(\gamma(t))dt = \int_0^1 \chi(t)H(\gamma(t))dt = 0,$$

since $(\gamma, \tau) \in \operatorname{Crit}(\mathscr{A}_{\varphi}^{\mathfrak{M}})$. Consequently, $\gamma(0) \in L_x$ and $x \in \partial W$. Moreover, we also have $\gamma(1) = \varphi(\gamma(0)) \in \partial W$ by the φ-invariance of H. For $t \in [\frac{1}{2}, 1]$, $\dot{\gamma} = X_{F_t}(\gamma)$ and so $\gamma(1) = \varphi_F(x) \in \partial W$. Exercise 7.39 now implies

$$L_{\varphi(x)} = \{\phi_t^R(\varphi(x)) : t \in \mathbb{R}\} = \{\varphi(\phi_t^R(x)) : t \in \mathbb{R}\} = \varphi(L_x),$$

and so $\varphi_F(x) = \gamma(1) = \varphi(\gamma(0)) \in L_{\varphi(x)}$. □

Exercise 8.7 Let $\varphi \in \operatorname{Aut}(W, \lambda)$ be a Liouville automorphism. Show that if ∂W is displaceable in $(\widehat{W}, \lambda)$, then $\operatorname{Crit}(\mathscr{A}_{\varphi}^{\mathfrak{M}}) = \varnothing$ for any displacing Hamiltonian symplectomorphism $\varphi_F \in \operatorname{Ham}_c(\widehat{W}, \lambda)$.

We are now in a position to give a proof of the third part of Theorem 7.22.

Theorem 8.8 *Let $\varphi \in \operatorname{Aut}(W, \lambda)$ be a Liouville automorphism. Assume that ∂W is displaceable in the completion $(\widehat{W}, \lambda)$. Then* $\mathrm{RFH}^{\varphi}(\partial W, \widehat{W}) \cong 0$.

Proof Let $\varphi_F \in \operatorname{Ham}_c(\widehat{W}, \lambda)$ be a displacing Hamiltonian symplectomorphism for the boundary ∂W and denote by $\mathfrak{M} = (\chi H, F)$ the associated twisted Moser pair from Lemma 8.4. Then $\operatorname{Crit} \mathscr{A}_{\varphi}^{\mathfrak{M}} = \varnothing$ by Exercise 8.7. Consequently, the perturbed twisted Rabinowitz action functional $\mathscr{A}_{\varphi}^{\mathfrak{M}}$ is a Morse function. By adapting the Fundamental Lemma 7.51 to the current setting as in [12, Theorem 2.9], the Floer homology $\mathrm{HF}(\mathscr{A}_{\varphi}^{\mathfrak{M}})$ is well-defined and satisfies

$$
\begin{aligned}
0 &= \mathrm{HF}(\mathscr{A}_{\varphi}^{\mathfrak{M}}) && (\operatorname{Crit} \mathscr{A}_{\varphi}^{\mathfrak{M}} = \varnothing) \\
&\cong \mathrm{HF}(\mathscr{A}_{\varphi}^{(\chi H, 0)}) && \text{(continuation homomorphisms)} \\
&\cong \mathrm{RFH}^{\varphi}(\partial W, \widehat{W}) && \text{(period-action equality)},
\end{aligned}
$$

where the last equation is the observation that twisted Rabinowitz–Floer homology in the autonomous case extends to the weakly time-dependent case without any issues. In fact, crucial is that the period-action equality in Remark 7.43 is still valid in this case. Indeed, we compute

$$
\mathscr{A}_{\varphi}^{(\chi H, 0)}(\gamma, \tau) = \int_0^1 \gamma^* \lambda = \int_0^1 \lambda(\dot{\gamma}) = \tau \int_0^1 \chi \lambda(R(\gamma)) = \tau \int_0^1 \chi = \tau
$$

for all $(\gamma, \tau) \in \operatorname{Crit} \mathscr{A}_{\varphi}^{(\chi H, 0)}$. □

Corollary 8.9 *Let $\Sigma \subseteq \mathbb{R}^{2n}$ be a star-shaped hypersurface. Then Σ admits a periodic Reeb orbit.*

Proof As any star-shaped hypersurface is displaceable in $\mathbb{R}^{2n}$, Theorem 8.8 implies that $\mathrm{RFH}(\Sigma, \mathbb{R}^{2n}) \cong 0$. However, if Σ does not admit any non-constant periodic Reeb orbit, then $\mathrm{RFH}_*(\Sigma, \mathbb{R}^{2n}) \cong \mathrm{H}_*(\Sigma; \mathbb{Z}_2)$ by Proposition 7.54. Thus, there must exist at least one periodic Reeb orbit on Σ. □

Exercise 8.10 Let $(\widehat{W}, \lambda)$ be the completion of a Liouville domain (W, λ). Explain how one can interpret the critical points of the Rabinowitz action functional $\mathscr{A}^H$ for some $H \in \mathscr{F}(\partial W)$ and why it is important that $\partial W \subseteq \operatorname{Crit}(\mathscr{A}^H)$!

Corollary 8.9 shows the Weinstein conjecture for any star-shaped hypersurface. What can we say about lens spaces? If we can prove the existence of a twisted periodic Reeb orbit on Σ, then we automatically get a noncontractible periodic Reeb orbit on the quotient $\Sigma/\mathbb{Z}_m$. Unfortunately, we have $\operatorname{Fix}(\varphi|_{\Sigma}) = \varnothing$ for the rotation φ and thus the proof of Corollary 8.9 no longer works! But there is a way to bypass this issue. We can define an equivariant version of the twisted Rabinowitz–Floer homology for the discrete group $\mathbb{Z}_m$ following [15, p. 487]. In general, suppose that a topological manifold M admits a group action by a topological group G. Then there exists a principal G-bundle $EG \to BG$, where $BG = EG/G$ is the classifying space of G and EG is weakly contractible. Then G acts freely on $EG \times M$ through the diagonal action. Thus we can define the ***G-equivariant homology of M*** by

$$\mathrm{H}_*^G(M; R) := \mathrm{H}_*(EG \times M/G; R),$$

for any coefficient ring R and where $EG \times M/G$ is the homotopy quotient of M by G. See [183, pp. 30–31]. For example, if $G = \mathbb{Z}_m$, then $E\mathbb{Z}_m = \mathbb{S}^\infty$ and $B\mathbb{Z}_m = \mathbb{S}^\infty/\mathbb{Z}_m$ is a lens space. Since $\mathbb{Z}_m$ acts freely on $\mathbb{S}^{2n-1}$, there is a fibre bundle

$$EG \to EG \times M/G \to M/G,$$

inducing an isomorphism $\mathrm{H}_*^G(M) = \mathrm{H}_*(M/G)$ by [183, Corollary 9.6] and [183, Theorem 3.3]. Fix a φ-invariant $d\lambda_0$-compatible almost complex structure on $(\mathbb{R}^{2n}, \lambda_0)$. Then the rotation φ induces a free $\mathbb{Z}_m$-action on $\operatorname{Crit} \mathscr{A}_\varphi^{H_\Sigma}$ and on the moduli space of twisted negative gradient flow lines with cascades of $\mathscr{A}_\varphi^{H_\Sigma}$. Thus, we can define ***$\mathbb{Z}_m$-sequivariant twisted Rabinowitz-Floer homology***

$$\overline{\mathrm{RFH}}_k^\varphi(\Sigma/\mathbb{Z}_m) := \frac{\ker \bar{\partial}_k}{\operatorname{im} \bar{\partial}_{k+1}} \qquad \forall k \in \mathbb{Z},$$

as the homology of the $\mathbb{Z}$-graded chain complex

$$\bar{\partial}_k \colon \mathrm{RFC}_k^\varphi(\Sigma, \mathbb{R}^{2n})/\mathbb{Z}_m \to \mathrm{RFC}_{k-1}^\varphi(\Sigma, \mathbb{R}^{2n})/\mathbb{Z}_m$$

given by

$$\bar{\partial}_k[(\gamma, \tau)] := [\partial_k(\gamma, \tau)] \qquad \forall (\gamma, \tau) \in \operatorname{Crit}(h),$$

for some φ-invariant Morse function h on $\operatorname{Crit} \mathscr{A}_\varphi^{H_\Sigma}$. The main computation we need is the $\mathbb{Z}_m$-equivariant Rabinowitz–Floer homology of the sphere $\mathbb{S}^{2n-1}$.

Theorem 8.11 *Let $n \geq 2$. For $m \geq 1$ consider the rotation*

$$\varphi \colon \mathbb{C}^n \to \mathbb{C}^n, \quad \varphi(z_1, \dots, z_n) := \left(e^{2\pi i k_1/m} z_1, \dots, e^{2\pi i k_n/m} z_n\right)$$

for $k_1, \dots, k_n \in \mathbb{Z}$ coprime to m. Then

$$\overline{\mathrm{RFH}}_k^\varphi(\mathbb{S}^{2n-1}/\mathbb{Z}_m) \cong \begin{cases} \mathbb{Z}_2 & m \text{ even}, \\ 0 & m \text{ odd}, \end{cases} \qquad \forall k \in \mathbb{Z}.$$

If m is even, then $\overline{\mathrm{RFH}}_k^{\varphi}(\mathbb{S}^{2n-1}/\mathbb{Z}_m)$ is generated by a noncontractible periodic Reeb orbit in the lens space $\mathbb{S}^{2n-1}/\mathbb{Z}_m$ for all $k \in \mathbb{Z}$.

Proof First, we consider the special case

$$\varphi\colon \mathbb{C}^n \to \mathbb{C}^n, \qquad \varphi(z) = e^{2\pi i/m} z.$$

The Reeb vector field on $\mathbb{S}^{2n-1}$ is given by

$$R = 2\sum_{j=1}^{n}\left(y_j\frac{\partial}{\partial x_j} - x_j\frac{\partial}{\partial y_j}\right)\bigg|_{\mathbb{S}^{2n-1}} = 2i\sum_{j=1}^{n}\left(\bar{z}_j\frac{\partial}{\partial \bar{z}_j} - z_j\frac{\partial}{\partial z_j}\right)\bigg|_{\mathbb{S}^{2n-1}}.$$

Suppose that $(\gamma, \tau) \in \operatorname{Crit} \mathscr{A}_{\varphi}^H$ for some $H \in \mathscr{F}_{\varphi}(\mathbb{S}^{2n-1})$. If $\tau = 0$, then γ is constant. This cannot happen since $\operatorname{Fix}(\varphi|_{\mathbb{S}^{2n-1}}) = \varnothing$. So, assume $\tau \neq 0$. Define a reparametrisation

$$\gamma_\tau\colon \mathbb{R} \to \mathbb{S}^{2n-1}, \qquad \gamma_\tau(t) := \gamma(t/\tau).$$

Then γ_τ is the unique integral curve of R starting at $z := \gamma(0)$ and thus

$$\gamma_\tau(t) = e^{-2it} z \qquad \forall t \in \mathbb{R}.$$

From $\gamma(t) = \gamma_\tau(\tau t)$ and the requirement

$$e^{-2i\tau} z = \gamma(1) = \varphi(\gamma(0)) = \varphi(z) = e^{2\pi i/m} z,$$

we conclude $\tau \in \frac{\pi}{m}(m\mathbb{Z} - 1)$. Hence

$$\operatorname{Crit} \mathscr{A}_{\varphi}^H = \{(\phi^{\tau_k R}(z), \tau_k) : k \in \mathbb{Z}, z \in \mathbb{S}^{2n-1}\} \cong \mathbb{S}^{2n-1} \times \mathbb{Z},$$

where

$$\tau_k := \frac{\pi}{m}(mk - 1).$$

By Proposition 7.42, $(z_0, \eta) \in T_z\mathbb{S}^{2n-1} \times \mathbb{R}$ belongs to the kernel of the Hessian at the critical point $(z, k) \in \operatorname{Crit} \mathscr{A}_{\varphi}^H$ if and only if $\eta = 0$ and

$$z_0 \in \ker(D(\phi_{-\tau_k}^R \circ \varphi)|_z - \mathrm{id}_{T_z\mathbb{S}^{2n-1}}).$$

A direct computation yields $D(\phi_{-\tau_k}^R \circ \varphi)|_z = \mathrm{id}_{T_z\mathbb{S}^{2n-1}}$ and thus

$$T_{(z,k)} \operatorname{Crit} \mathscr{A}_{\varphi}^H = T_z\mathbb{S}^{2n-1} \times \{0\} \cong \ker \operatorname{Hess} \mathscr{A}_{\varphi}^H|_{(z,k)}.$$

So the twisted Rabinowitz action functional $\mathscr{A}_\varphi^H$ is Morse–Bott with spheres as critical manifolds. The full Conley–Zehnder index [77, Definition 10.4.1] gives rise to a locally constant function

$$\widehat{\mu}_{CZ}\colon \operatorname{Crit}\mathscr{A}_\varphi^H \to \mathbb{Z}, \qquad \widehat{\mu}(z,k) = (2k-1)n.$$

Note that the definition of the Conley–Zehnder index also applies in this degenerate case; compare [77, Remark 10.4.2]. By the adapted proof of the Hofer–Wysocki–Zehnder Theorem [77, Theorem 12.2.1] to the n-dimensional setting revisited in Problem 8.51, the full Conley–Zehnder index coincides with the transverse Conley–Zehnder index μ_{CZ}. Indeed, for the critical point $(\gamma, \tau) \in \operatorname{Crit}\mathscr{A}_\varphi^H$ define

$$\Psi\colon I \to \operatorname{Sp}(2n), \qquad \Psi_t := D\phi^H_{\tau t}|_{\gamma(0)}\colon \mathbb{C}^n \to \mathbb{C}^n.$$

Adapting the proof of [77, Lemma 12.2.3 (iii)], we get that

$$\Psi_1(R(\gamma(0))) = R(\gamma(1)) \qquad \text{and} \qquad \Psi_1(\gamma(0)) = \gamma(1).$$

Arguing as in [77, pp. 235–236] we conclude $\mu_{CZ}(\gamma,\tau) = \widehat{\mu}_{CZ}(\gamma,\tau)$. Fix $z_0 \in \mathbb{S}^{2n-1}$ and define $\eta := \phi^{\tau_0 R}(z_0)$. Note that $\phi^{\tau_k R}(z)$ belongs to the same equivalence class in $\pi_0 \mathscr{L}_\varphi \mathbb{S}^{2n-1}$ as η for all $z \in \mathbb{S}^{2n-1}$ and $k \in \mathbb{Z}$ because $\mathbb{S}^{2n-1}$ is simply connected for $n \geq 2$. Let $h \in C^\infty(\mathbb{S}^{2n-1})$ be the standard height function. The twisted Rabinowitz–Floer homology $\mathrm{RFH}^\varphi(\mathbb{S}^{2n-1}, \mathbb{R}^{2n})$ carries the $\mathbb{Z}$-grading

$$\mu((z,k),(z_0,0)) + \operatorname{ind}_h(z) = 2kn + \operatorname{ind}_h(z) \qquad \forall (z,k) \in \mathbb{S}^{2n-1} \times \mathbb{Z}.$$

We claim that the number of twisted negative gradient flow lines between the minimum of $\mathbb{S}^{2n-1} \times \{k+1\}$ and the maximum of $\mathbb{S}^{2n-1} \times \{k\}$ must be odd, so that the critical manifold $\operatorname{Crit}\mathscr{A}_\varphi^H$ looks like a string of pearls, see Fig. 8.3. In fact, if there is an even number of such negative gradient flow lines, then $\mathrm{RFH}^\varphi_*(\mathbb{S}^{2n-1}, \mathbb{R}^{2n}) \neq 0$, contradicting Theorem 8.8 as $\mathbb{S}^{2n-1}$ is displaceable in the completion $\mathbb{R}^{2n}$. To compute the $\mathbb{Z}_m$-equivariant twisted Rabinowitz–Floer homology, choose Bott's perfect Morse function $f \in C^\infty(\mathbb{S}^{2n-1})$ from Exercise 5.9 on each connected component of $\operatorname{Crit}\mathscr{A}_\varphi^H$. Additionally, choose a $\mathbb{Z}_m$-invariant Morse function on $\operatorname{Crit}(f)$. For example, one can take

$$h\colon \mathbb{T} \to \mathbb{R}, \qquad h(t) := \cos(2\pi m t).$$

The resulting chain complex is given by

$$\cdots \longrightarrow \mathbb{Z}_2^m \xrightarrow{\mathbb{1}} \mathbb{Z}_2^m \xrightarrow{A} \mathbb{Z}_2^m \xrightarrow{\mathbb{1}} \mathbb{Z}_2^m \xrightarrow{A} \mathbb{Z}_2^m \xrightarrow{\mathbb{1}} \mathbb{Z}_2^m \longrightarrow \cdots$$

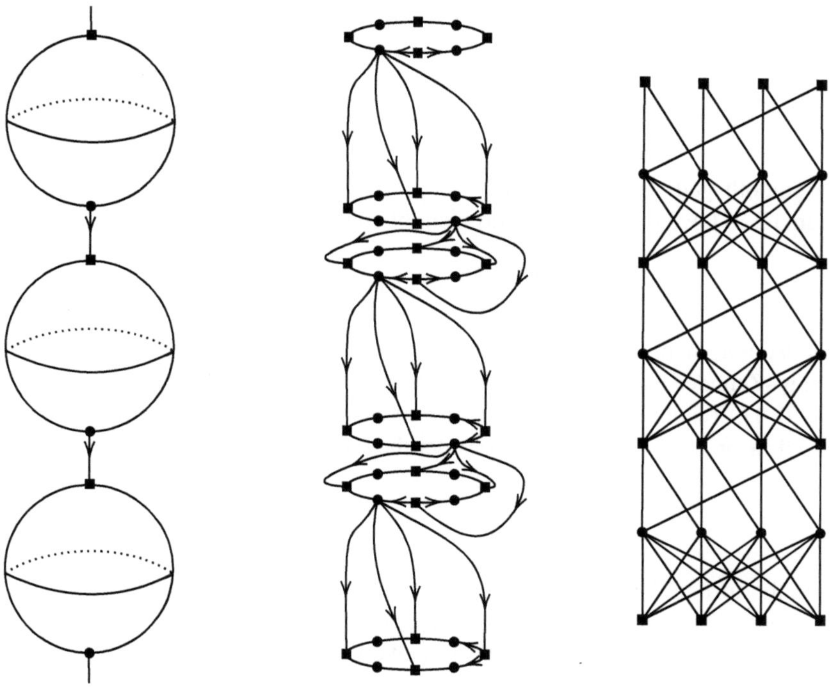

Fig. 8.3 The critical manifold $\mathbb{S}^{2n-1} \times \mathbb{Z}$ with the standard height function, the Morse–Bott function f and the resulting chain complex

where $\mathbb{1} \in \mathbb{Z}_2^{m\times m}$ has every entry equal to 1 and $A \in \mathbb{Z}_2^{m\times m}$ is defined by

$$A := I_m + \sum_{j=1}^{m-1} e_{(j+1)j} + e_{1m},$$

where $e_{ij} \in \mathbb{Z}_2^{m\times m}$ satisfies $(e_{ij})_{kl} = \delta_{ik}\delta_{jl}$. Thus, the resulting chain complex looks like a rope ladder. Compare Fig. 8.3. Passing to the quotient via the free $\mathbb{Z}_m$-action, we get the acyclic chain complex

$$\dots \longrightarrow \mathbb{Z}_2 \xrightarrow{0} \mathbb{Z}_2 \xrightarrow{0} \mathbb{Z}_2 \xrightarrow{0} \mathbb{Z}_2 \longrightarrow \dots$$

if m is even and the alternating chain complex

$$\dots \longrightarrow \mathbb{Z}_2 \xrightarrow{\mathbb{1}} \mathbb{Z}_2 \xrightarrow{0} \mathbb{Z}_2 \xrightarrow{\mathbb{1}} \mathbb{Z}_2 \xrightarrow{0} \mathbb{Z}_2 \xrightarrow{\mathbb{1}} \mathbb{Z}_2 \longrightarrow \dots$$

if m is odd. From this, the statement follows in the special case. For the general case, we note that

$$\mathbb{C}^n \times [0,1] \to \mathbb{C}^n, \qquad \varphi_\sigma(z_1,\ldots,z_n) := \left(e^{2\pi i\sigma k_1/m} z_1, \ldots, e^{2\pi i\sigma k_n/m} z_n\right)$$

is a smooth path from $\varphi_0 = \mathrm{id}_{\mathbb{C}^n}$ to $\varphi_1 = \varphi$. By adapting the proof of [185, Lemma 2.27], we obtain an isomorphism of chain complexes

$$\mathrm{RFC}(\mathbb{S}^{2n-1}, \mathbb{R}^{2n}) \cong \mathrm{RFC}^{\varphi}(\mathbb{S}^{2n-1}, \mathbb{R}^{2n}).$$

This isomorphism does not necessarily preserve the grading, but the relative Conley–Zehnder index is preserved. Note also that f is invariant under φ_σ for all $\sigma \in [0,1]$. It is no problem to allow for twists φ_σ of infinite order as the standard Reeb flow on $\mathbb{S}^{2n-1}$ is periodic. Consider the torus action

$$\mathbb{T}^n \times \mathbb{C}^n \to \mathbb{C}^n, \quad (\theta_1,\ldots,\theta_n)\cdot(z_1,\ldots,z_n) := \left(e^{2\pi i\theta_1} z_1, \ldots, e^{2\pi i\theta_n} z_n\right).$$

Since the torus $\mathbb{T}^n$ is abelian, we have that the $\mathbb{Z}_m$-action induced by φ and the different twists along the path $(\varphi_\sigma)_{\sigma\in[0,1]}$ commute. Thus, we get an isomorphism of the $\mathbb{Z}_m$-equivariant chain complexes, and consequently

$$\overline{\mathrm{RFH}}_*^{\varphi}(\mathbb{S}^{2n-1}/\mathbb{Z}_m) \cong \mathrm{RFH}_*^{\mathbb{Z}_m}(\mathbb{S}^{2n-1}, \mathbb{R}^{2n}),$$

where $\mathrm{RFH}_*^{\mathbb{Z}_m}$ denotes the $\mathbb{Z}_m$-equivariant Rabinowitz–Floer homology constructed in [15, p. 487]. Performing the same computation of the latter homology as before in the special case yields

$$\mathrm{RFH}_k^{\mathbb{Z}_m}(\mathbb{S}^{2n-1}, \mathbb{R}^{2n}) \cong \begin{cases} \mathbb{Z}_2 & m \text{ even,} \\ 0 & m \text{ odd,} \end{cases} \qquad \forall k \in \mathbb{Z}.$$

Finally, $\overline{\mathrm{RFH}}_k^{\varphi}(\mathbb{S}^{2n-1}/\mathbb{Z}_m)$ is generated by a noncontractible periodic Reeb orbit in $\mathbb{S}^{2n-1}/\mathbb{Z}_m$ for all $k \in \mathbb{Z}$ by Exercise 4.37. □

Exercise 8.12 Explain the computation performed in the proof of Theorem 8.11 explicitly for $m = 2$ and $m = 3$ in the special case and describe the differences!

For an algebraic corollary we define a special group cohomology called Tate cohomology as in [192, Definition 6.2.4] or Tate homology defined in [44, p. 135]. For a more general result, see [162, Theorem 5.6]. Following [192, Chapter 6], we introduce the Tate (co)homology groups.

Definition 8.13 (Tate Homology) Let G be a group. A ***(left) G-module*** is an abelian group A on which G acts by additive maps on the left. A ***trivial G-module***

is an abelian group A where the G-action is defined by

$$ga = a \qquad \forall g \in G, a \in A.$$

For any cyclic group $\mathbb{Z}_m$ and any $\mathbb{Z}_m$-module define the ***Tate cohomology groups*** by

$$\widehat{\mathrm{H}}^k(\mathbb{Z}_m; A) := \begin{cases} A^G/NA & k \text{ even}, \\ \{a \in A : Na = 0\} & k \text{ odd}, \end{cases}$$

where

$$A^G := \{a \in A : ga = a \, \forall g \in G\} \tag{8.2}$$

is the invariant subgroup and

$$N := 1 + 2 + 3 + \cdots + m \in \mathbb{Z}\mathbb{Z}_m$$

is the norm element in the group ring $\mathbb{Z}\mathbb{Z}_m$. We define the ***Tate homology groups*** via duality by

$$\widehat{\mathrm{H}}_k(\mathbb{Z}_m; A) := \widehat{\mathrm{H}}^{-k-1}(\mathbb{Z}_m; A) \qquad \forall k \in \mathbb{Z}.$$

Exercise 8.14 Compute the Tate homology groups for the trivial $\mathbb{Z}_m$-modules $\mathbb{Z}_2$ and $\mathbb{Z}$.

Corollary 8.15 (Tate Homology) *Let $m \geq 1$ be an integer. Then for the trivial left $\mathbb{Z}_m$-module $\mathbb{Z}_2$ we have that*

$$\overline{\mathrm{RFH}}_k^{\varphi}(\mathbb{S}^{2n-1}/\mathbb{Z}_m) \cong \widehat{\mathrm{H}}_k(\mathbb{Z}_m; \mathbb{Z}_2) \qquad \forall k \in \mathbb{Z}.$$

Remark 8.16 By Corollary 8.15, one can also prove Theorem 8.11 more abstractly by showing that the $\mathbb{Z}_m$-equivariant twisted Rabinowitz–Floer homology coincides with the Tate homology similar to [162, Theorem 5.6].

Having computed the $\mathbb{Z}_m$-equivariant twisted Rabinowitz–Floer homology of the sphere $\mathbb{S}^{2n-1}$, we are able to prove the existence result for noncontractible periodic Reeb orbits on lens spaces.

Theorem 8.17 ([27, Theorem 1.2]) *Let $\Sigma \subseteq \mathbb{R}^{2n}$, $n \geq 2$, be a star-shaped hypersurface invariant under the rotation*

$$\varphi \colon \mathbb{C}^n \to \mathbb{C}^n, \quad \varphi(z_1, \ldots, z_n) := \left(e^{2\pi i k_1/m} z_1, \ldots, e^{2\pi i k_n/m} z_n\right)$$

for some even $m \geq 2$ *and* $k_1, \dots, k_n \in \mathbb{Z}$ *coprime to* m. *Then* $\Sigma/\mathbb{Z}_m$ *admits a noncontractible periodic Reeb orbit that generates* $\pi_1(\Sigma/\mathbb{Z}_m) \cong \mathbb{Z}_m$.

Remark 8.18 Theorem 8.17 has been shown to hold in full generality without restricting the parity of m in [10, Theorem 1.1].

Proof By Theorem 7.57 we have that

$$\mathrm{RFH}_*^{\varphi}(\Sigma, \mathbb{R}^{2n}) \cong \mathrm{RFH}_*^{\varphi}(\mathbb{S}^{2n-1}, \mathbb{R}^{2n}),$$

giving rise to a canonical isomorphism of the associated $\mathbb{Z}_m$-equivariant twisted Rabinowitz–Floer homology

$$\overline{\mathrm{RFH}}_*^{\varphi}(\Sigma/\mathbb{Z}_m) \cong \overline{\mathrm{RFH}}_*^{\varphi}(\mathbb{S}^{2n-1}/\mathbb{Z}_m). \tag{8.3}$$

Indeed, this follows from observing that the standard continuation homomorphism [51, p. 276] is $\mathbb{Z}_m$-invariant and thus descends to the quotient. In fact, extend the twisted homotopy $(H_\sigma)_{\sigma \in [0,1]}$ smoothly to a smooth twisted homotopy $(H_s)_{s \in \mathbb{R}}$ with

$$H_s = \begin{cases} H_0 & s \leq 0, \\ H_1 & s \geq 1. \end{cases}$$

Pick $\mathbb{Z}_m$-invariant Morse functions $h^\pm \in \mathrm{Crit}(\mathscr{A}_\varphi^{H_\pm})$ and for $c^\pm \in \mathrm{Crit}(h^\pm)$ consider the moduli space of twisted negative gradient flow lines with cascades $\mathscr{M}_\varphi(c^-, c^+)$ for the s-dependent action functional $\mathscr{A}_\varphi^{H_s}$. By proving the Compactness Theorem 7.52 for such s-dependent functionals by adapting [54, Theorem 3.6] to the twisted case, we get that the zero-dimensional component $\mathscr{M}_\varphi^0(c^-, c^+)$ is a finite set. Thus we can define a linear map

$$\Phi\colon \mathrm{RFC}_k^{\varphi}(\Sigma^-, \mathbb{R}^{2n}; h^-) \to \mathrm{RFC}_k^{\varphi}(\Sigma^+, \mathbb{R}^{2n}; h^+)$$

by defining

$$\Phi(c^-) := \sum_{c^+ \in \mathrm{Crit}_k(h^+)} \#_2 \mathscr{M}_\varphi^0(c^-, c^+) c^+$$

on generators. Then we can show that $\partial^+ \circ \Phi = \Phi \circ \partial^-$ holds, and thus Φ is a chain map that descends to homology. This homomorphism is actually an isomorphism between $\mathrm{RFC}_*^{\varphi}(\Sigma^-, \mathbb{R}^{2n}; h^-)$ and $\mathrm{RFC}_*^{\varphi}(\Sigma^+, \mathbb{R}^{2n}; h^+)$. Now we simply consider

$$\overline{\Phi}\colon \mathrm{RFC}_k^{\varphi}(\Sigma^-, \mathbb{R}^{2n}; h^-)/\mathbb{Z}_m \to \mathrm{RFC}_k^{\varphi}(\Sigma^+, \mathbb{R}^{2n}; h^+)/\mathbb{Z}_m$$

by defining

$$\overline{\Phi}(c^-) := \sum_{c^+ \in \mathrm{Crit}_k(h^+/\mathbb{Z}_m)} \#_2 \left(\mathscr{M}^0_\varphi(c^-, c^+)/\mathbb{Z}_m \right) c^+.$$

Establishing the isomorphism (8.3). But by Theorem 8.11, the latter homology in (8.3) does not vanish as $m \geq 2$ is even by assumption, and the statement follows. □

Exercise 8.19 How could one modify the proof of Theorem 8.17 to remove the restriction of m being even?

Remark 8.20 We put Theorem 8.17 in context. If $\Sigma^{2n-1}/\mathbb{Z}_m$ satisfies the index condition

$$\mu_{\mathrm{CZ}}(\gamma) > 4 - n \tag{8.4}$$

for all contractible Reeb orbits γ, the $\bigvee$-shaped symplectic homology $\check{\mathrm{SH}}(\Sigma)$ can be defined in the positive cylindrical end $[0, +\infty) \times \Sigma$ by [184, Corollary 3.7]. If $\Sigma/\mathbb{Z}_m$ admits a Liouville filling W, then we have that

$$\check{\mathrm{SH}}_*(\Sigma/\mathbb{Z}_m, \widehat{W}) \cong \mathrm{RFH}_*(\Sigma/\mathbb{Z}_m, \widehat{W}).$$

Note that even in the case of projective spaces, this need not be the case, as, for example, $\mathbb{RP}^{2n-1}$ is not Liouville fillable for any odd $n \geq 2$ by [85, Theorem 1.1]. As the index condition (8.4) is only required for contractible Reeb orbits and they come from the universal covering manifold Σ, we can say something in the case where Σ is strictly convex. Indeed, the Hofer–Wysocki–Zehnder Theorem [77, Theorem 12.2.1] then implies that Σ is dynamically convex, that is,

$$\mu_{\mathrm{CZ}}(\gamma) \geq n + 1$$

holds for all periodic Reeb orbits γ. Thus, for $n \geq 2$, the index condition is satisfied and we actually compute $\check{\mathrm{SH}}_*(\mathbb{S}^{2n-1}/\mathbb{Z}_m)$ via the $\mathbb{Z}_m$-equivariant version of the symplectic homology $\check{\mathrm{SH}}_*(\mathbb{S}^{2n-1})$.

In the remainder of this section, we give a further application of Theorem 8.11 and an alternative proof of Theorem 8.17.

Definition 8.21 (Spectral Value) Let $(\widehat{W}, \lambda)$ be the completion of a Liouville domain (W, λ) and let $\varphi \in \mathrm{Aut}(W, \lambda)$ be a Liouville automorphism. Assume that $\mathfrak{M}$ is a twisted Moser pair such that the perturbed twisted Rabinowitz action functional

$\mathscr{A}_\varphi^{\mathfrak{M}}$ is a Morse function. For a homology class $X \in \mathrm{RFH}^\varphi(\partial W, \widehat{W}) \setminus \{0\}$ we define its ***spectral value*** by

$$\sigma_{\mathfrak{M}}(X) := \inf_{[\zeta]=X} \max_{\zeta_{(\gamma,\tau)} \neq 0} \mathscr{A}_\varphi^{\mathfrak{M}}(\gamma, \tau) \in \mathbb{R} \cup \{-\infty\},$$

where

$$\zeta = \sum_{(\gamma,\tau)\in \mathrm{Crit}(\mathscr{A}_\varphi^{\mathfrak{M}})} \zeta_{(\gamma,\tau)}(\gamma, \tau) \in \mathrm{RFC}^\varphi(\partial W, \widehat{W}) \setminus \{0\}.$$

Exercise 8.22 Show that if $\sigma_{\mathfrak{M}}(X) \in \mathbb{R}$, then $\sigma_{\mathfrak{M}}(X)$ is a critical value, meaning

$$\sigma_{\mathfrak{M}}(X) \in \mathscr{A}_\varphi^{\mathfrak{M}} \left(\mathrm{Crit}(\mathscr{A}_\varphi^{\mathfrak{M}})\right).$$

Hint: Use the fact that the Fundamental Lemma 7.51 is also valid for the perturbed twisted Rabinowitz action functional by adapting [12, Lemma 2.11].

Let $(\widehat{W}, \lambda)$ be the completion of a Liouville domain (W, λ) and let $\varphi \in \mathrm{Aut}(W, \lambda)$ be a Liouville automorphism. Assume that $\mathfrak{M}_0$ is a twisted Moser pair such that the perturbed twisted Rabinowitz action functional $\mathscr{A}_\varphi^{\mathfrak{M}_0}$ is a Morse function. It follows from [13, Corollary 5.13], that if the set

$$\{\sigma_{\mathfrak{M}_0}(X) : X \in \mathrm{RFH}^\varphi(\partial W, \widehat{W}) \setminus \{0\}\} \subseteq \mathbb{R} \cup \{+\infty\}$$

is unbounded from above, then also the set

$$\{\sigma_{\mathfrak{M}}(X) : X \in \mathfrak{B}(\overline{\mathfrak{M}}, \mathfrak{M}_0)\}$$

is unbounded from above for all Moser pairs $\overline{\mathfrak{M}}$, not necessarily such that the associated perturbed Rabinowitz action functional $\mathscr{A}^{\overline{\mathfrak{M}}}$ is a Morse function! Here, the set $\mathfrak{B}(\overline{\mathfrak{M}}, \mathfrak{M}_0)$ is defined in such a way that the map $\mathfrak{M} \to \sigma_{\mathfrak{M}}(X)$ is locally Lipschitz continuous around $\overline{\mathfrak{M}}$ for all $X \in \mathfrak{B}(\overline{\mathfrak{M}}, \mathfrak{M}_0)$, where the precise definition of this set also depends on the choice of $\mathfrak{M}_0$. It is worth pointing out that this result holds for all Moser pairs sufficiently C^0-close to $\overline{\mathfrak{M}}$, so the regularity assumption is very weak. It is also worth mentioning that this result holds for the unperturbed twisted Rabinowitz action functional, if we assume that $\mathscr{A}_\varphi^{H_0}$ is a Morse–Bott function. In this case, one has to slightly adapt the definition of the spectral value provided in Definition 8.21 to incorporate the additional Morse function on $\mathrm{Crit}(\mathscr{A}_\varphi^{H_0})$. Generally speaking, spectral invariants are currently used in persistence theory [156], which will not be further explored in this book but is an active area of research; for example, see the paper [60]. However, using spectral values, we can also prove the existence of twisted leaf-wise intersection points similar to [15, Theorem 1.2]. Let $\Sigma \subseteq \mathbb{R}^{2n}$ be a φ-symmetric star-shaped hypersurface for the rotation

$$\varphi\colon \mathbb{C}^n \to \mathbb{C}^n, \qquad \varphi(z_1, \ldots, z_n) := \left(e^{2\pi i k_1/m} z_1, \ldots, e^{2\pi i k_n/m} z_n\right)$$

for $m \geq 1$ an integer and $k_1, \ldots, k_n \in \mathbb{Z}$ coprime to m. Define the set of ***φ*-invariant Hamiltonian symplectomorphisms** by

$$\mathrm{Ham}^{\varphi}(\mathbb{R}^{2n}, \lambda_0) := \{\varphi_F \in \mathrm{Ham}(\mathbb{R}^{2n}, \lambda_0) : F_t(x) = F_t(\varphi(x)) \forall (x, t) \in \mathbb{R}^{2n} \times [0, 1]\}.$$

Exercise 8.23 Show that no $\varphi_F \in \mathrm{Ham}^{\varphi}(\mathbb{R}^{2n}, \lambda_0)$ displaces Σ.

Theorem 8.24 *Let $\Sigma \subseteq \mathbb{R}^{2n}$ be a star-shaped hypersurface invariant under the rotation φ for some $m \geq 1$ even. Then every element in $\mathrm{Ham}^{\varphi}(\mathbb{R}^{2n}, \lambda_0)$ has infinitely many leaf-wise intersection points or there exists a leaf-wise intersection point on a closed leaf.*

Proof We first show the statement for compactly supported Hamiltonian symplectomorphisms in $\mathrm{Ham}^{\varphi}(\mathbb{R}^{2n}, \lambda_0)$. The general case follows from this one by making use of cut-off functions. Let $\varphi_F \in \mathrm{Ham}^{\varphi}_c(\mathbb{R}^{2n}, \lambda_0)$ and for $\sigma \in [0, 1]$ consider the smooth family of perturbed Rabinowitz action functionals $\mathscr{A}_\sigma \colon \mathscr{L}\mathbb{R}^{2n} \times \mathbb{R} \to \mathbb{R}$ defined by

$$\mathscr{A}_\sigma(\gamma, \tau) := \int_0^1 \gamma^*\lambda_0 - \tau \int_0^1 H_\sigma(\gamma(t))dt - \sigma \int_0^1 F_t(\gamma(t))dt,$$

for a twisted homotopy $(H_\sigma)_{\sigma \in [0,1]}$ from $\mathbb{S}^{2n-1}$ to Σ. Clearly, every $\mathscr{A}_\sigma$ is φ-invariant. From the proof of Theorem 8.11 it follows that the set

$$\{\sigma(X) : X \in \mathrm{RFH}^{\mathbb{Z}_m}_*(\Sigma, \mathbb{R}^{2n}) \setminus \{0\}\} = 2\pi\mathbb{Z}$$

is unbounded from above as m is even by assumption. Those are precisely the critical values of $\mathscr{A}_0$. Consequently, also $\mathscr{A}_1$ has critical points of arbitrarily large critical values. In particular, $\mathscr{A}_1$ admits infinitely many critical points. By Proposition 8.6, every critical point of the perturbed Rabinowitz action functional $\mathscr{A}_1$ gives rise to a leaf-wise intersection point. By [16, Proposition 1.2], the map

$$\mathrm{Crit}\,\mathscr{A}_1 \to \{\text{leaf-wise intersection points}\}$$

is injective unless there exists a leaf-wise intersection point on a closed leaf.

Let $\varphi_F \in \mathrm{Ham}^{\varphi}(\mathbb{R}^{2n}, \lambda_0)$ be a φ-invariant Hamiltonian symplectomorphism that is not necessarily compactly supported. Observe that the set

$$K := \{\varphi_t^{X_F}(x) : x \in \Sigma, t \in [0, 1]\} \subseteq \mathbb{R}^{2n}$$

is compact since Σ is compact. In particular, all critical points of $\mathscr{A}_\sigma$ are contained in K for all $\sigma \in [0, 1]$. Thus, we can cut off F outside of K to make it compactly supported without changing the critical points of $\mathscr{A}_\sigma$, and the general statement follows. □

Exercise 8.25 Give an alternative proof of Theorem 8.17 using spectral invariants.

8.2 On a Theorem by Schlenk

In this section, we leave our setup of a completion of a Liouville domain in favour of a more general setting. Still, the main geometrical picture to consider is that of star-shaped hypersurfaces in $\mathbb{R}^{2n}$.

Definition 8.26 (Stable Hypersurface, [56, p. 1774]) Let (M, ω) be a connected symplectic manifold. A ***stable hypersurface*** in (M, ω) is a compact and connected hypersurface $\Sigma \subseteq M$ such that the following conditions hold:

1. Σ is separating, that is, $M \setminus \Sigma$ consists of two connected components $M^\pm$, where M^- is bounded and M^+ is unbounded.
2. There exists a vector field X in a neighbourhood of Σ such that X is outward-pointing to $\Sigma \cup M^-$ and $\ker \omega|_\Sigma \subseteq \ker L_X\omega|_\Sigma$.

We write $(\Sigma, \omega|_\Sigma, \lambda)$ for a stable hypersurface, where the stabilising form $\lambda \in \Omega^1(\Sigma)$ is defined by $\lambda := i_X\omega|_\Sigma$.

Definition 8.27 (Symmetric Stable Hypersurface) Let $(\Sigma, \omega|_\Sigma, \lambda)$ be a stable hypersurface in a connected symplectic manifold (M, ω) and let $\varphi \in \mathrm{Symp}(M, \omega)$ be a symplectomorphism. We say that Σ is ***φ-symmetric***, if $\varphi(\Sigma) = \Sigma$, φ is of finite order and $\varphi^*X = X$.

Example 8.28 (Liouville Automorphism) Let $(\widehat{W}, \lambda)$ be the completion of a Liouville domain (W, λ) and let $\varphi \in \mathrm{Aut}(W, \lambda)$ be a Liouville automorphism. Then the contact manifold $(\partial W, \lambda|_{\partial W})$ is a φ-symmetric hypersurface in $(\widehat{W}, \lambda)$.

Example 8.29 (Magnetic Torus) Consider a magnetic torus as in Example 4.6 with the energy hypersurfaces $\Sigma_k := H^{-1}(k)$, $k \in \mathbb{R}$, for the mechanical Hamiltonian function

$$H(q, p) := \frac{1}{2}\|p\|^2 \qquad \forall (q, p) \in \mathbb{T}^n \times \mathbb{R}^n.$$

Let $\varphi \in \mathrm{Diff}(\mathbb{T}^n)$ be an isometry of finite order satisfying

$$D\varphi \circ J = J \circ D\varphi \tag{8.5}$$

and consider the cotangent lift

$$D_*\varphi\colon \mathbb{T}^n \times \mathbb{R}^n \to \mathbb{T}^n \times \mathbb{R}^n, \qquad D_*\varphi(q,p) = \left(\varphi(q), \left(D\varphi^{-1}(q)\right)^T p\right).$$

Then clearly $\varphi(\Sigma_k) = \Sigma_k$ as φ is an isometry and $D_*\varphi$ is of finite order as φ is. Also $D_*\varphi \in \operatorname{Symp}(T^*\mathbb{T}^n, \omega_\rho)$, because $D_*\varphi \in \operatorname{Symp}(T^*\mathbb{T}^n, dp \wedge dq)$ and $D_*\varphi$ preserves $\rho(\cdot,\cdot) := \langle \cdot, J\cdot\rangle$ by assumption (8.5). Lastly, we see that $\varphi^*\lambda = \lambda$ by considering formula (4.1) together with assumption (8.5), and thus also $\varphi^*X = X$.

As is the case in Hamiltonian Floer homology, one cannot define a symplectic area functional in general unless the symplectic form is exact.

Definition 8.30 (Contractible Twisted Loop Space) Let (M,ω) be a symplectic manifold and let $\varphi \in \operatorname{Symp}(M,\omega)$ be of finite order $m := \operatorname{ord}(\varphi)$. A contractible loop $v \in \Lambda M$ is said to be a ***contractible twisted periodic loop***, if there exists $\gamma \in \mathscr{L}_\varphi M$ such that $v = \iota(\gamma)$, where

$$\iota\colon \mathscr{L}_\varphi M \hookrightarrow \mathscr{L}M, \qquad \iota(\gamma)(t) := \gamma(mt)\ \forall t \in \mathbb{T}.$$

We denote the space of all contractible twisted periodic loops of M and φ by $\Lambda_\varphi M$.

Exercise 8.31 Explain the definition of the space of all contractible twisted periodic loops $\Lambda_\varphi M$ and sketch a typical element.

Definition 8.32 (Twisted Rabinowitz Action Functional) Let $(\Sigma, \omega|_\Sigma, \lambda)$ be a φ-symmetric stable hypersurface in a symplectically aspherical symplectic manifold (M,ω). For a defining Hamiltonian function $H \in \mathscr{F}_\varphi(\Sigma)$, we define the ***twisted Rabinowitz action functional***

$$\mathscr{A}_\varphi^H\colon \Lambda_\varphi M \times \mathbb{R} \to \mathbb{R}, \qquad \mathscr{A}_\varphi^H(v,\tau) := \frac{1}{\operatorname{ord}(\varphi)} \int_{\mathbb{D}} \bar{v}^*\omega - \tau \int_0^1 H(v(t))dt.$$

Exercise 8.33 Show that if $\omega = d\lambda$ is exact, then Definition 8.32 coincides with the Rabinowitz action functional in Definition 4.23.

Remark 8.34 A routine computation as we have performed in several exercises and results shows that

$$(v,\tau) \in \operatorname{Crit}(\mathscr{A}_\varphi^H) \qquad \Leftrightarrow \qquad \begin{cases} v(\mathbb{T}) \subseteq \Sigma, \\ \dot{\gamma}(t) = \tau R(\gamma(t))\ \forall t \in \mathbb{R}, \end{cases}$$

where $R \in \mathfrak{X}(\Sigma)$ is the stable Reeb vector field. If J is a φ-invariant almost complex structure compatible with ω, then the gradient $\operatorname{grad}_J \mathscr{A}_\varphi^H \in \mathfrak{X}(\Lambda_\varphi M \times \mathbb{R})$ with respect to the standard L^2-metric

$$m\left((X,\eta),(Y,\sigma)\right)=\int_0^1 \omega(JX(t),Y(t))dt+\eta\sigma$$

for all $(X,\eta),(Y,\sigma)\in T_{(v,\tau)}\Lambda_\varphi M\times\mathbb{R}$ and $(v,\tau)\in\Lambda_\varphi M\times\mathbb{R}$ is given by

$$\operatorname{grad}_J \mathscr{A}_\varphi^H|_{(v,\tau)}(t)=\begin{pmatrix} J(\dot{\gamma}(\operatorname{ord}(\varphi)t)-\tau X_H(v(t))) \\ -\int_0^1 H\circ v \end{pmatrix} \qquad \forall t\in\mathbb{T}.$$

Hence $(u,\tau)\in C^\infty(\mathbb{R},\Lambda_\varphi M\times\mathbb{R})$ is a twisted negative gradient flow line if the Rabinowitz–Floer equations in Corollary 6.14 are satisfied. Consequently, most of the analysis developed so far is also applicable in the current more general setting.

Definition 8.35 (Morse–Bott Component, [12, p. 86]) Let $(\Sigma,\omega|_\Sigma,\lambda)$ be a φ-symmetric stable hypersurface in a symplectically aspherical symplectic manifold (M,ω). A subset $C\subseteq\operatorname{Crit}\mathscr{A}_\varphi^H$ is called a ***Morse–Bott component***, if

1. C is an action-constant submanifold of $M\times\mathbb{R}$.
2. $T_xC=\ker\operatorname{Hess}\mathscr{A}_\varphi^H(x)$ for all $x\in C$ for the Hessian $\operatorname{Hess}\mathscr{A}_\varphi^H$ of $\mathscr{A}_\varphi^H$.

Exercise 8.36 Show that $\operatorname{Fix}(\varphi|_\Sigma)$ is a Morse–Bott component of $\mathscr{A}_\varphi^H$.

The following condition guarantees that a similar maximum principle as in the case of the completion of a Liouville domain holds in Problem 7.62. This assumption was used to define Floer theory for noncompact manifolds in [94].

Definition 8.37 ([56, p. 1768]) A symplectic manifold (M,ω) is called ***geometrically bounded***, if there exists an ω-compatible almost complex structure J and a complete Riemannian metric such that the following conditions hold.

1. There are constants $C_0,C_1>0$ with

   $$\omega(Jv,v)\geq C_0\|v\|^2 \qquad \text{and} \qquad |\omega(u,v)|\leq C_1\|u\|\|v\|$$

   for all $u,v\in T_xM$ and $x\in M$.
2. The sectional curvature of the metric is bounded above and its injectivity radius is bounded away from zero.

Example 8.38 Let (T^*Q,ω_σ) be a magnetic cotangent bundle as in Example 2.11 for a compact smooth manifold Q and $\sigma\in\Omega^2(Q)$ closed. By [53, Proposition 2.2], any such magnetic cotangent bundle is geometrically bounded, and a similar argument also shows that any completion $(\widehat{W}, \lambda)$ of a Liouville domain (W, λ) is geometrically bounded.

Let $(\widehat{W}, \lambda)$ be the completion of a Liouville domain (W, λ) and let $\varphi \in \operatorname{Aut}(W, \lambda)$ be a Liouville automorphism. In this setup, the kernel of the twisted Rabinowitz action functional $\mathscr{A}_\varphi^H$ admits the canonical description

$$\ker \operatorname{Hess} \mathscr{A}_\varphi^H|_{(v,\tau)} \cong \ker \left(D^\xi(\phi_{-\tau}^R \circ \varphi)|_{v(0)} - \operatorname{id}|_{\xi_{v(0)}} \right) \oplus \langle R(v(0)) \rangle$$

by Proposition 7.42, where $\xi := \ker \lambda|_{\partial W}$ denotes the contact distribution.

Definition 8.39 (Transversal Nondegeneracy, [77, Definition 7.3.1]) Let $(\widehat{W}, \lambda)$ be the completion of a Liouville domain (W, λ) and let $\varphi \in \operatorname{Aut}(W, \lambda)$ be a Liouville automorphism. A contractible twisted periodic Reeb orbit $(v, \tau) \in \operatorname{Crit}(\mathscr{A}_\varphi^H)$ is said to be ***transversally nondegenerate***, if

$$\ker \left(D^\xi(\phi_{-\tau}^R \circ \varphi)|_{v(0)} - \operatorname{id}|_{\xi_{v(0)}} \right) = \{0\}.$$

Under some circumstances, one can improve Theorem 8.17, forcing the existence of another twisted Reeb orbit, given a nondegenerate periodic Reeb orbit!

Theorem 8.40 ([28, Theorem 1]) *Let $\Sigma \subseteq \mathbb{R}^{2n}$, $n \geq 2$, be a star-shaped hypersurface invariant under the rotation*

$$\varphi\colon \mathbb{C}^n \to \mathbb{C}^n, \quad \varphi(z_1, \ldots, z_n) := \left(e^{2\pi i k_1/m} z_1, \ldots, e^{2\pi i k_n/m} z_n \right)$$

for some even $m \geq 2$ and $k_1, \ldots, k_n \in \mathbb{Z}$ coprime to m. Assume that there exists a nondegenerate twisted Reeb orbit (γ_0, τ_0) on Σ. Then there exists a twisted Reeb orbit (γ, τ) on Σ with

$$0 < |\tau - \tau_0| \leq e(\Sigma). \tag{8.6}$$

Example 8.41 (Ellipsoids) For real numbers $0 < a_1 \leq \ldots \leq a_n$ we consider the ellipsoid

$$E(a_1, \ldots, a_n) = \left\{ (z_1, \ldots, z_n) \in \mathbb{C}^n : \sum_{j=1}^n \frac{\pi}{a_j} |z_j|^2 = 1 \right\}$$

as in Example 1.7 with the corresponding Reeb vector field R_E given by

$$R_E = X_{H_E}|_E = 2 \sum_{j=1}^n \frac{\pi}{a_j} \left(y_j \frac{\partial}{\partial x_j} - x_j \frac{\partial}{\partial y_j} \right) \bigg|_E .$$

Hence the Reeb flow $\phi_t \colon E(a_1, \dots, a_n) \to E(a_1, \dots, a_n)$ is given by

$$\phi_t(z_1, \dots, z_n) = (e^{-2\pi i t/a_1} z_1, \dots, e^{-2\pi i t/a_n} z_n) \qquad \forall t \in \mathbb{R}.$$

If we assume that $a_1, \dots, a_n$ are linearly independent over $\mathbb{Z}$, all periodic orbits are transversally nondegenerate as $z_j = 0$ except for one coordinate $z_1, \dots, z_n$. Moreover, these periodic orbits are clearly invariant under φ and the twisted periodic Reeb orbits on $E(a_1, \dots, a_n)$ are given by

$$\left(t \mapsto e^{-2\pi i \tau_j t/a_j} z, \tau_j\right), \qquad |z|^2 = \frac{a_j}{\pi} \quad \text{and} \quad \tau_j \in a_j\mathbb{Z} - \frac{a_j k_j}{m},$$

for $j = 1, \dots, n$. Consider the twisted Reeb orbit (γ_0, τ_0) defined by

$$\gamma_0(t) := e^{-2\pi i \tau_0 t/a_1} z, \qquad \tau_0 := a_1 \left(k - \frac{k_1}{m}\right),$$

for $\pi |z|^2 = a_1$ and $k \in \mathbb{Z}$. Then for $\gamma(t) := e^{-2\pi i t} \gamma_0(t)$ we compute

$$\tau - \tau_0 = a_1 + \tau_0 - \tau_0 = a_1 = e(E(a_1, \dots, a_n))$$

by [137, Example 12.1.7]. Thus, the bound in (8.6) is sharp.

Remark 8.42 As Example 8.41 shows, one cannot conclude the existence of two geometrically distinct simple symmetric periodic Reeb orbits as in [10, Theorem 1.2] from Theorem 8.40. Indeed, even under the additional assumption that Σ is dynamically convex, the estimate (8.6) is not only satisfied for geometrically distinct closed orbits as the ellipsoid $\Sigma = E(a_1, \dots, a_n)$ is dynamically convex by the Hofer–Wysocki–Zehnder Theorem [77, Theorem 12.2.1], revisited in Problem 8.51. As we shall see, the strength of the estimate (8.6) is to provide an upper bound for symmetric systoles.

Theorem 8.40 follows immediately from a much more general result.

Theorem 8.43 (Forcing, [28, Theorem 2]) *Let Σ be a φ-symmetric stable displaceable hypersurface in a symplectically aspherical, geometrically bounded, symplectic manifold (M, ω) for a symplectomorphism $\varphi \in \operatorname{Symp}(M, \omega)$ of finite order $\operatorname{ord}(\varphi)$ and suppose that v_0 is a contractible twisted periodic Reeb orbit on Σ belonging to a Morse–Bott component C of $\mathscr{A}_\varphi^H$. Then there exists a contractible twisted periodic Reeb orbit $v \notin C$ such that*

$$0 \le \left| \int_{\mathbb{D}} \bar{v}^* \omega - \int_{\mathbb{D}} \bar{v}_0^* \omega \right| \le \operatorname{ord}(\varphi) e(\Sigma).$$

Remark 8.44 The case $(\Sigma, M) = (\mathbb{S}^{2n-1}, \mathbb{R}^{2n})$ or $(\Sigma, M) = (E(a_1, \ldots, a_n), \mathbb{R}^{2n})$ with the ellipsoid from Example 1.7 and the rotation

$$\varphi \colon \mathbb{C}^n \to \mathbb{C}^n, \quad \varphi(z_1, \ldots, z_n) = \left(e^{2\pi i k_1/m} z_1, \ldots, e^{2\pi i k_n/m} z_n\right)$$

shows that the estimate in Theorem 8.43 is sharp.

Exercise 8.45 Explain why Theorem 8.43 is called the "Forcing Theorem".

An immediate corollary of Theorem 8.43 is provided in the following exercise.

Exercise 8.46 Let Σ be a φ-symmetric stable displaceable hypersurface for a symplectomorphism $\varphi \in \operatorname{Symp}(M, \omega)$ with $\operatorname{Fix}(\varphi|_\Sigma) \neq \varnothing$ in a symplectically aspherical, geometrically bounded, symplectic manifold (M, ω). Describe in detail why there exists a nonconstant contractible twisted periodic Reeb orbit v such that

$$0 < \left| \int_{\mathbb{D}} \bar{v}^* \omega \right| \leq \operatorname{ord}(\varphi) e(\Sigma).$$

Remark 8.47 (Schlenk's Theorem) In particular, if we take $\varphi = \mathrm{id}_M$ in Exercise 8.46, we recover Schlenk's theorem as stated in [56, Theorem 4.9] about the existence of contractible closed characteristics on stable, displaceable hypersurfaces with energy less than or equal to the displacement energy of the hypersurface. Schlenk proved this result in [167, Theorem 1.1] using quite different methods.

Before embarking on the proof of Theorem 8.43, we show some existence results for stable regular energy hypersurfaces in magnetic Hamiltonian systems.

Exercise 8.48 (Magnetic Hamiltonian Systems) Let (T^*Q, ω_σ) be a magnetic cotangent bundle for some closed $\sigma \in \Omega^2(Q)$ on a compact smooth manifold Q. As in Problem 3.35, define

$$e_0(H) := \inf\left\{k \in \mathbb{R} : \pi_{T^*Q}\left(H^{-1}(k)\right) = Q\right\},$$

where $\pi_{T^*Q} \colon T^*Q \to Q$ denotes the projection.

1. If $\sigma = d\theta$ is exact, show that the map

$$\varphi_\theta \colon (T^*Q, dp \wedge dq) \to (T^*Q, \omega_{d\theta}), \qquad \varphi_\theta(q, p) := (q, p - \theta_q)$$

is an exact symplectomorphism. Conclude that the associated Hamiltonian function $H_\theta := \varphi_\theta^* H$ of a mechanical Hamiltonian function

$$H \colon (T^*Q, \omega_{d\theta}) \to \mathbb{R}, \qquad H(q, p) = \frac{1}{2}\|p\|_{m^*}^2 + V(q)$$

for some potential function $V \in C^\infty(Q)$ and some Riemannian metric m on TQ is given by

$$H_\theta(q,p) = \frac{1}{2}\|p - \theta_q\|^2_{m^*} + V(q).$$

2. Consider the magnetic torus as in Example 8.29 with $V = 0$ and

$$\theta_q(v) := \frac{1}{2}\langle q, Jv\rangle \qquad \forall (q,v) \in \mathbb{T}^n \times \mathbb{R}^n.$$

Show that $e_0 = 0$ and that a point $(q,p) \in \Sigma_k = H^{-1}(k)$ gives rise to a twisted periodic Reeb orbit if and only if

$$\int_0^\tau e^{sJ} p ds + q = \varphi(q), \quad e^{\tau J} p = \left(D\varphi^{-1}(q)\right)^T p, \quad \text{and} \quad \|p\|^2 = 2k.$$

3. Let $(T^*Q, dp \wedge dq, H)$ be a Hamiltonian system for a compact configuration space Q and suppose that $\Sigma_k := H^{-1}(k)$ with $k < e_0(H)$ is a φ-symmetric stable regular energy hypersurface admitting a contractible twisted periodic Reeb orbit (q_0, p_0) belonging to a Morse–Bott component C. Show that then there exists a contractible twisted periodic Reeb orbit $(q,p) \notin C$ with

$$\int_0^1 p(t)\dot{q}(t)dt - \int_0^1 p_0(t)\dot{q}_0(t)dt \leq \operatorname{ord}(\varphi)e(\Sigma_k).$$

Does this result apply to the magnetic torus from before? Explain your answer!
4. Let $H \in C^\infty(T^*Q)$ be a Tonelli Hamiltonian function. Define the Mañé critical value

$$c := \inf_\theta \sup_{q \in \widetilde{Q}} \widetilde{H}(q, \theta_q),$$

where the infimum is over all 1-forms θ on the universal covering manifold $\widetilde{Q}$ such that $d\theta = \pi^*\sigma$, and $\widetilde{H} \in C^\infty(T^*\widetilde{Q})$ denotes the lift of H. We always have that $c \geq e_0(H)$ as in Problem 3.35. Compute the Mañé critical value c for the magnetic torus. If $k > c$, then the Rabinowitz–Floer homology $\mathrm{RFH}_*(\Sigma_k, T^*M)$ of [56] is well-defined and does not vanish. Can we apply the result of the third part for $k > c$?
5. If H is a Tonelli Hamiltonian function, that is, H is strictly fibrewise convex and superlinear, then any stable energy level of H does admit a periodic Reeb orbit by [134]. See also [2, Theorem (iv)]. Explain why this is a more general statement than the one provided in the third part!

In the remainder of this section, we give a proof of the Forcing Theorem 8.43.

Proof of Theorem 8.43 The proof of Theorem 8.43 uses a method called a "homotopy of homotopies argument". Fix $\varepsilon > 0$ and choose a Hamiltonian function $F \in C_c^\infty(M \times [0, 1])$ satisfying

$$F_t = 0\,\forall t \in [0, \tfrac{1}{2}], \qquad \|F\| < e(\Sigma) + \varepsilon \qquad \text{and} \qquad \varphi_F(\Sigma) \cap \Sigma = \varnothing.$$

This is possible by the definition of the displacement energy. Next, we need to carefully choose a twisted defining Hamiltonian function H for the stable hypersurface Σ. We postpone the construction of this Hamiltonian function, which is similar to the proof of Theorem 4.30 and explain the main idea of the proof. Choose a smooth family $(\beta_r)_{r\in[0,+\infty)}$ of cutoff functions $\beta_r \in C^\infty(\mathbb{R}, [0, 1])$ such that

$$\begin{cases} \beta_r(s) = 0 & |s| \geq r, \\ \beta_r(s) = 1 & |s| \leq r - 1, \\ s\beta_r'(s) \leq 0 & \forall s \in \mathbb{R}, \end{cases}$$

for all $r \in [0, +\infty)$. Define a family of twisted Rabinowitz action functionals

$$\mathscr{A}_r \colon \Lambda_\varphi M \times \mathbb{R} \times \mathbb{R} \to \mathbb{R}$$

by

$$\mathscr{A}_r(v, \tau, s) := \mathscr{A}_\varphi^H(v, \tau) - \beta_r(s) \int_0^1 F_t(v(t))dt$$

for all $r \in [0, +\infty)$. Note that $\mathscr{A}_0 = \mathscr{A}_\varphi^H$. For a suitable φ-invariant ω-compatible almost complex structure J, we consider the moduli space (Fig. 8.4)

$$\mathscr{M} := \{(u, \tau, r) \in C^\infty(\mathbb{R}, \mathscr{L}_\varphi M \times \mathbb{R}) \times [0, +\infty) : (u, \tau, r) \text{ solution of } (8.7)\},$$

where

$$\begin{cases} \partial_s(u, \tau) = \operatorname{grad} \mathscr{A}_r|_{(u(s),\tau(s),s)} & \forall s \in \mathbb{R}, \\ \lim\limits_{s\to-\infty} (u(s), \tau(s)) = (v_0, \tau_0), \\ \lim\limits_{s\to+\infty} (u(s), \tau(s)) \in C. \end{cases} \tag{8.7}$$

Note that always $(v_0, \tau_0, 0) \in \mathscr{M}$ and that such a φ-invariant ω-compatible almost complex structure always exists by [137, Lemma 5.5.6]. The gradient $\operatorname{grad} \mathscr{A}_r$ of $\mathscr{A}_r$ is taken with respect to the metric

$$m\left((X, \eta), (Y, \sigma)\right) := \int_0^1 \omega(JX(t), Y(t))dt + \eta\sigma.$$

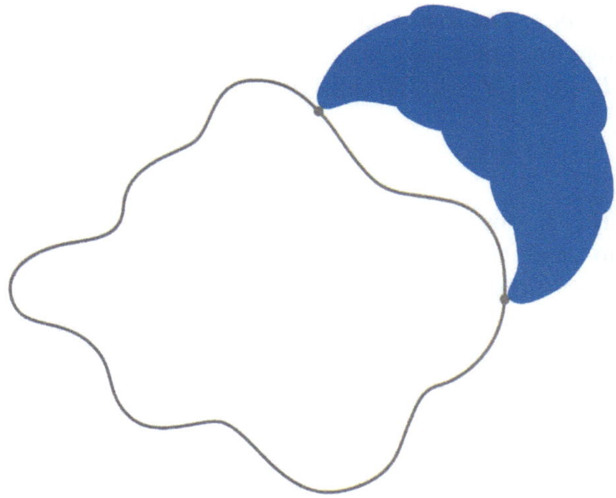

Fig. 8.4 A typical element of the moduli space $\mathscr{M}$ looks like a "croissant"

Lemma 8.49 *If*

$$\mathscr{A}_0(v, \tau) > \|F\| + \mathscr{A}_0(v_0, \tau_0) \qquad \forall (v, \tau) \in \mathrm{Crit}(\mathscr{A}_0) \setminus C, \tag{8.8}$$

then $\mathscr{M}$ is compact.

As a corollary of Lemma 8.49 we get Theorem 8.43. Indeed, the moduli space $\mathscr{M}$ is the zero level set of a Fredholm section of a bundle over a Banach manifold. As v_0 belongs to a Morse–Bott component, the Fredholm section is regular at the point v_0, that is, the linearisation of the gradient flow equation is surjective there. By compactness, we can therefore perturb the Fredholm section to make it transverse. Hence $\mathscr{M}$ is a compact smooth manifold with a boundary consisting precisely of the point v_0. See [136, Appendix A] for details. This is absurd, and we conclude that there exists a critical point $(v, \tau) \in \mathrm{Crit}(\mathscr{A}_0) \setminus C$ such that

$$\mathscr{A}_0(v, \tau) - \mathscr{A}_0(v_0, \tau_0) \leq \|F\| < e(\Sigma) + \varepsilon.$$

As $\varepsilon > 0$ was arbitrary, the statement follows since

$$\mathscr{A}_0(v, \tau) = \frac{1}{\mathrm{ord}(\varphi)} \int_{\mathbb{D}} \bar{v}^* \omega \qquad \forall (v, \tau) \in \mathrm{Crit}(\mathscr{A}_0).$$

We prove Lemma 8.49 in four steps.
Step 1: If $(u, \tau, r) \in \mathscr{M}$, then $E(u, \tau) \leq \|F\|$ for the energy

$$E(u, \tau) := \int_{-\infty}^{+\infty} \|\partial_s(u, \tau)\|_J^2 ds.$$

We estimate

$$\begin{aligned}
E(u,\tau) &= \int_{-\infty}^{+\infty} \|\partial_s(u,\tau)\|_J^2 ds \\
&= \int_{-\infty}^{+\infty} d\mathscr{A}_r(\partial_s(u,\tau),s)ds \\
&= \int_{-\infty}^{+\infty} \frac{d}{ds}\mathscr{A}_r(u,\tau,s)ds - \int_{-\infty}^{+\infty} (\partial_s\mathscr{A}_r)(u,\tau,s)ds \\
&= \lim_{s\to+\infty} \mathscr{A}_r(u,\tau,s) - \lim_{s\to-\infty} \mathscr{A}_r(u,\tau,s) - \int_{-\infty}^{+\infty} (\partial_s\mathscr{A}_r)(u,\tau,s)ds \\
&= \mathscr{A}_0(v,\tau) - \mathscr{A}_0(v_0,\tau_0) - \int_{-\infty}^{+\infty} (\partial_s\mathscr{A}_r)(u,\tau,s)ds \\
&= -\int_{-\infty}^{+\infty} (\partial_s\mathscr{A}_r)(u,\tau,s)ds \\
&= \int_{-\infty}^{+\infty} \dot\beta_r(s)\int_0^1 F_t(u(s,t))dtds \\
&\le \|F\|_+\int_{-\infty}^{0} \dot\beta_r(s)ds - \|F\|_-\int_0^{+\infty} \dot\beta_r(s)ds \\
&= \beta_r(0)(\|F\|_- + \|F\|_+) \\
&= \beta_r(0)\|F\| \\
&\le \|F\|,
\end{aligned}$$

as $\mathscr{A}_0(v,\tau) = \mathscr{A}_0(v_0,\tau_0)$ since C is action-constant by definition of a Morse–Bott component.

Step 2: There exists $r_0 \in \mathbb{R}$ such that $r \le r_0$ for all $(u,\tau,r) \in \mathscr{M}$. Crucial is the existence of a constant $\delta > 0$ such that

$$\|\operatorname{grad}\mathscr{A}_r|_{(v,\tau,s)}\|_J \ge \delta \qquad \forall (v,\tau,s) \in \Lambda_\varphi M \times \mathbb{R} \times \mathbb{R}.$$

This is proven along the lines of [51, Lemma 3.9]. With this inequality and Step 1, we estimate

$$\|F\| \ge E(u,\tau) \ge \int_{-r}^{r} \|\operatorname{grad}\mathscr{A}_r|_{(u(s),\tau(s),s)}\|_J^2 ds \ge 2r\delta^2,$$

and thus we can set

$$r_0 := \frac{\|F\|}{2\delta^2}.$$

Step 3: There exists a constant $C > 0$ such that $\|\tau\|_\infty \leq C$ for all $(u, \tau, r) \in \mathscr{M}$. This is a delicate estimate based on the explicit construction of the defining Hamiltonian H for Σ as well as an extension of the stabilising form λ. The bound on the Lagrange multiplier is derived by comparing the twisted Rabinowitz action functional to a different action functional. This modified version of the twisted Rabinowitz action functional is obtained using a suitable extension of the φ-invariant stabilising form $\lambda \in \Omega^1(\Sigma)$ to a compactly supported form $\beta_\lambda \in \Omega^1(M)$. The precise constructions can be found in [56, Section 4.2.2]. Given β_λ, we can define the auxiliary action functional

$$\widehat{\mathscr{A}_0}\colon \Lambda_\varphi M \times \mathbb{R} \to \mathbb{R}, \qquad \widehat{\mathscr{A}_0}(v, \tau) := \frac{1}{\operatorname{ord}(\varphi)} \int_0^1 v^*\beta_\lambda - \tau \int_0^1 H(v(t))dt.$$

Moreover, we consider the bilinear form on the tangent bundle $T\Lambda_\varphi M \times \mathbb{R}$

$$\widehat{m}\left((X, \eta), (Y, \sigma)\right) := \int_0^1 d\beta_\lambda(JX(t), Y(t))dt + \eta\sigma.$$

The main point in the choice of the φ-invariant $H \in C^\infty(M)$, $\beta_\lambda \in \Omega^1(M)$ and the ω-compatible φ-invariant almost complex structure J is to ensure that the following properties hold:

1. $d\widehat{\mathscr{A}_0}(v, \tau)(X, \eta) = \widehat{m}\left(\operatorname{grad} \mathscr{A}_0(v, \tau), (X, \eta)\right)$,
2. $(m - \widehat{m})\left((X, \eta), (X, \eta)\right) \leq 0$,

are true for all $(v, \tau) \in \Lambda_\varphi M \times \mathbb{R}$ and $(X, \eta) \in T_{(v,\tau)}\Lambda_\varphi M \times \mathbb{R}$. These two conditions ensure that the difference $\mathscr{A}_0 - \widehat{\mathscr{A}_0}$ is a Liapunov function for the negative gradient flow lines of the twisted Rabinowitz action functional $\mathscr{A}_0$. The uniform bound on the Lagrange multiplier τ now follows from Steps 1 and 2. For details, see [56, p. 1808]. The only subtle difference in our case is that everything needs to be φ-invariant. However, this is no problem as we explain now. The construction of H, β_λ and J is based on the existence of a stable tubular neighbourhood of Σ, that is, a pair (ρ_0, ψ) with $\rho_0 > 0$ and $\psi\colon (-\rho_0, \rho_0) \times \Sigma \hookrightarrow M$ an embedding such that

$$\psi|_{\{0\}\times\Sigma} = \iota_\Sigma\colon \Sigma \hookrightarrow M \qquad \text{and} \qquad \psi_\rho^*\omega = \omega|_\Sigma + d(\rho\lambda).$$

By [56, Proposition 2.6 (a)], the space of stable tubular neighbourhoods of (Σ, λ) is nonempty. Using the equivariant Darboux–Weinstein Theorem [98, Theorem 22.1], we get the existence of a stable tubular neighbourhood (ρ_0, ψ), satisfying

$$\varphi(\psi(\rho, x)) = \psi(\rho, \varphi(x)) \qquad \forall (\rho, x) \in (-\rho_0, \rho_0) \times \Sigma. \tag{8.9}$$

Compare also [27, Equation (3.2)]. Hence, the constructions [56, pp. 1791–1793] yield φ-invariant data H, β_λ and J due to (8.9).

Step 4: Proof of Lemma 8.49. Let (u_k, τ_k, r_k) be a sequence in the moduli space $\mathscr{M}$. By Step 2 and Step 3, the sequences (r_k) and (τ_k) are uniformly bounded. Thus, (u_k, τ_k, r_k) admits a C^∞_{loc}-convergent subsequence by standard arguments [136, Theorem B.4.2] or Corollary 6.14. Indeed, the uniform L^∞-bound on the sequence (u_k) follows from the assumption that (M, ω) is geometrically bounded and the uniform L^∞-bound on the derivatives (Du_k) follows from the absence of bubbling, as (M, ω) is symplectically aspherical. In particular, there cannot exist a nonconstant J-holomorphic sphere when the sequence of derivatives is unbounded [136, Section 4.2]. Denote the limit of this subsequence by (u, τ, r). This limit clearly satisfies the first equation in (8.7), thus one only needs to check the asymptotic conditions in (8.7). Again, by compactness, (u, τ) converges to critical points $(w_\pm, \tau_\pm)$ of $\mathscr{A}_0$ at its asymptotic ends. We claim that

$$\mathscr{A}_r(u(s), \tau(s), s) \in [-\|F\| + \mathscr{A}_0(v_0, \tau_0), \|F\| + \mathscr{A}_0(v_0, \tau_0)] \qquad \forall s \in \mathbb{R}. \tag{8.10}$$

In particular, $\mathscr{A}_0(w_\pm, \tau_\pm) \in [-\|F\| + \mathscr{A}_0(v_0, \tau_0), \|F\| + \mathscr{A}_0(v_0, \tau_0)]$. So, if the equation (8.10) holds, then by assumption (8.8) we conclude that $(w_\pm, \tau_\pm) \in C$ and $\mathscr{M}$ is compact. It remains to prove (8.10). It is enough to show

$$\mathscr{A}_r(u_k(s), \tau_k(s), s) \in [-\|F\| + \mathscr{A}_0(v_0, \tau_0), \|F\| + \mathscr{A}_0(v_0, \tau_0)] \qquad \forall s \in \mathbb{R}$$

for every $k \in \mathbb{N}$. As in the proof of [12, Lemma 2.8] we estimate

$$\begin{aligned}
0 &\leq \int_{s_0}^{+\infty} d\mathscr{A}_r(\partial_s(u_k, \tau_k), s)ds \\
&= \int_{s_0}^{+\infty} \frac{d}{ds}\mathscr{A}_r(u_k, \tau_k, s)ds - \int_{s_0}^{+\infty} (\partial_s \mathscr{A}_r)(u_k, \tau_k, s)ds \\
&= \lim_{s\to+\infty} \mathscr{A}_r(u_k, \tau_k, s) - \mathscr{A}_r(u_k(s_0), \tau_k(s_0), s_0) - \int_{s_0}^{+\infty} (\partial_s \mathscr{A}_r)(u_k, \tau_k, s)ds \\
&= \mathscr{A}_0(v, \tau) - \mathscr{A}_r(u_k(s_0), \tau_k(s_0), s_0) + \int_{s_0}^{+\infty} \dot{\beta}_r(s) \int_0^1 F_t(u_k(s, t))dtds \\
&\leq \mathscr{A}_0(v, \tau) - \mathscr{A}_r(u_k(s_0), \tau_k(s_0), s_0) + \int_{-\infty}^{+\infty} \|\dot{\beta}_r(s)F\|_+ ds \\
&\leq \mathscr{A}_0(v, \tau) - \mathscr{A}_r(u_k(s_0), \tau_k(s_0), s_0) + \|F\| \\
&= \mathscr{A}_0(v_0, \tau_0) - \mathscr{A}_r(u_k(s_0), \tau_k(s_0), s_0) + \|F\|
\end{aligned}$$

for all $s_0 \in \mathbb{R}$. Similarly, we compute

$$
\begin{aligned}
0 &\leq \int_{-\infty}^{s_0} d\mathscr{A}_r(\partial_s(u_k, \tau_k), s)ds \\
&= \int_{-\infty}^{s_0} \frac{d}{ds}\mathscr{A}_r(u_k, \tau_k, s)ds - \int_{-\infty}^{s_0} (\partial_s \mathscr{A}_r)(u_k, \tau_k, s)ds \\
&= \mathscr{A}_r(u_k(s_0), \tau_k(s_0), s_0) - \lim_{s \to -\infty} \mathscr{A}_r(u_k, \tau_k, s) - \int_{-\infty}^{s_0} (\partial_s \mathscr{A}_r)(u_k, \tau_k, s)ds \\
&= \mathscr{A}_r(u_k(s_0), \tau_k(s_0), s_0) - \mathscr{A}_0(v_0, \tau_0) + \int_{-\infty}^{s_0} \dot{\beta}_r(s) \int_0^1 F_t(u_k(s,t))dtds,
\end{aligned}
$$

and thus we estimate

$$
\begin{aligned}
\mathscr{A}_r(u_k(s_0), \tau_k(s_0), s_0) &\geq \mathscr{A}_0(v_0, \tau_0) - \int_{-\infty}^{s_0} \dot{\beta}_r(s) \int_0^1 F_t(u_k(s,t))dtds \\
&\geq \mathscr{A}_0(v_0, \tau_0) - \int_{-\infty}^{+\infty} \|\dot{\beta}(s)F\|_+ ds \\
&\geq \mathscr{A}_0(v_0, \tau_0) - \|F\|.
\end{aligned}
$$

This shows the estimate (8.10) and so the proof of Lemma 8.49 and Theorem 8.43 is complete. □

Exercise 8.50 Summarise the results from this chapter and give an outline for further steps in twisted Rabinowitz–Floer homology.

In the following last chapter of the book, we conclude our study of twisted periodic Reeb orbits by considering the ones of least "length". In symplectic geometry, length will always mean the smallest (positive) period on a fixed energy hypersurface inside a symplectic manifold.

Problems

8.51 (Hofer–Wysocki–Zehnder) A star-shaped hypersurface $\Sigma \subseteq \mathbb{R}^{2n}$, $n \geq 2$, is called ***dynamically convex***, if $\mu_{CZ}(\gamma) \geq n+1$ for all $(\gamma, \tau) \in \mathscr{P}(\Sigma, \lambda_0|_\Sigma)$ such that $\tau > 0$. The aim of this problem is to understand an important result saying that every strictly convex hypersurface is dynamically convex. Since the Conley–Zehnder index is invariant under symplectomorphisms, dynamical convexity is a much more natural assumption on a star-shaped hypersurface in symplectic geometry than mere convexity. Here a star-shaped hypersurface is said to be ***strictly convex***, if $dN(x) > 0$ holds for all points $x \in \Sigma$, where

$$N\colon \Sigma \mapsto \mathbb{S}^{2n-1}, \qquad N(x) := \frac{\nabla H_\Sigma}{\|\nabla H_\Sigma\|}.$$

By [77, Lemma 12.1.2], a star-shaped hypersurface is strictly convex, if and only if the endomorphism $dN(x)\colon T_x\Sigma \to T_x\mathbb{S}^{2n-1} \cong T_x\Sigma$ is positive-definite for all $x \in \Sigma$. Moreover, by [77, 12.1.3], the bounded part $\{x \in \mathbb{R}^{2n} : H_\Sigma(x) < 0\}$ is convex.

1. Assume that $(\gamma, \tau) \in \mathscr{P}(\Sigma, \lambda_0|_\Sigma)$ with $\tau > 0$ is a periodic Reeb orbit. Consider the smooth path of symplectic matrices

$$\Psi_\gamma\colon [0,1] \to \mathrm{Sp}(2n), \qquad \Psi_\gamma^t := D\phi_{\tau t}^{R_\Sigma}(\gamma(0)),$$

where $R_\Sigma \in \mathfrak{X}(\Sigma)$ denotes the Reeb vector field. Show that Ψ_γ satisfies the following properties:
(a) $\Psi_\gamma^0 = \mathrm{id}_{\mathbb{R}^{2n}}$.
(b) It holds that $\partial_t \Psi_\gamma^t = \tau J_0 \operatorname{Hess}_{H_\Sigma}(\gamma(t))\Psi_\gamma^t$ for all $t \in [0,1]$.
(c) $\Psi_\gamma^1(R(\gamma(0))) = R(\gamma(0))$ and $\Psi_\gamma^1(\gamma(0)) = \gamma(0)$.
2. Fix a filling disc $\bar{\gamma} \in C^\infty(\mathbb{D}, \mathbb{R}^{2n})$. Then the vector bundle $\bar{\gamma}^*\mathbb{R}^{2n} \to \mathbb{D}$ splits as

$$\bar{\gamma}^*\mathbb{R}^{2n} = \bar{\gamma}^* \ker(\lambda_0|_\Sigma) \oplus \bar{\gamma}^*\eta, \qquad \eta := \mathrm{span}(X, R),$$

where $X \in \mathfrak{X}(\mathbb{R}^{2n})$ denotes the Liouville vector field. Fix a symplectic trivialisation $\Phi_\xi : \bar{\gamma}^*\xi \to \mathbb{D} \times \mathbb{R}^{2n-2}$, and observe that there is a canonical trivialisation

$$\Phi_\eta\colon \bar{\gamma}^*\eta \to \mathbb{D} \times \mathbb{R}^2, \qquad \Phi_\eta(aX + bR) := (a, b).$$

Show that Ψ_γ preserves the splitting, meaning that

$$\Psi_\gamma^t : \xi_{\gamma(0)} \to \xi_{\gamma(t)} \qquad \text{and} \qquad \Psi_\gamma^t : \eta_{\gamma(0)} \to \eta_{\gamma(t)}$$

holds for all $t \in [0,1]$. Moreover, show that

$$\Phi_{\eta,\gamma(t)} \circ \Psi_\gamma^t \circ \Phi_{\eta,\gamma(0)}^{-1} = \mathrm{id}_{\mathbb{R}^{2n}} \qquad \forall t \in [0,1].$$

3. Explain why the full Conley–Zehnder index coincides with the transverse one.

8.52 Let $(\widehat{W}, \lambda)$ be the completion of a Liouville domain and let $\varphi \in \mathrm{Aut}(W, \lambda)$ be a Liouville automorphism. Assume that the boundary ∂W is displaceable in the completion $(\widehat{W}, \lambda)$. Choose $\varphi_F \in \mathrm{Ham}_c(\widehat{W}, \lambda)$ displacing ∂W and denote by $\mathfrak{M}$ the associated twisted Moser pair such that $\varphi_F(\mathrm{supp}(X_H)) \cap \mathrm{supp}(X_H) = \varnothing$. The aim of this problem is to show that then there exists a constant $\delta > 0$ such that

$$\| \operatorname{grad}_J \mathscr{A}_\varphi^{\mathfrak{M}}(\gamma, \tau)\|_J \geq \delta \tag{8.11}$$

holds for all $(\gamma, \tau) \in \mathscr{L}_\varphi \widehat{W} \times \mathbb{R}$, where J is some φ-invariant $d\lambda$-compatible almost complex structure. The uniform lower bound in (8.11) was originally used to prove that $\mathrm{RFH}^\varphi(\partial W, \widehat{W}) \cong 0$ in the paper [51].

1. Explain why (8.11) implies $\mathrm{Crit}(\mathscr{A}_\varphi^{\mathfrak{M}}) = \varnothing$.
2. We prove (8.11) in three steps following [51, Lemma 3.9]. For all $x \in \widehat{W}$ define a norm on the tangent space $T_x\widehat{W}$ by

$$\|v\|_{J,\mathfrak{M}} := \min_{t\in[0,1]} \|D\phi_t^{F_t}|_x(v)\|_J \qquad \forall (x, v) \in T_x\widehat{W}.$$

Denote by $d^1_{J,\mathfrak{M}} \colon \widehat{W} \times \widehat{W} \to [0, +\infty)$ the induced metric, and define a second metric by

$$d^2_{J,\mathfrak{M}} := (\varphi_F)_* d^1_{J,\mathfrak{M}}$$

(a) Show that there exists $\varepsilon_0 = \varepsilon_0(J)$ such that if $(\gamma, \tau) \in \mathscr{L}_\varphi\widehat{W} \times \mathbb{R}$ satisfies

$$\|\dot{\gamma} - \tau X_{\chi H}(\gamma) - X_F(\gamma)\|_{d^2} < \varepsilon_0,$$

then

$$(\gamma(0), \gamma(1/2)) \notin \mathrm{supp}(X_H) \times \mathrm{supp}(X_H).$$

(b) Show that there exist $\varepsilon_1 > 0$ and $\kappa > 0$ such that if

$$\|\dot{\gamma} - \tau X_{\chi H}(\gamma) - X_F(\gamma)\|_{d^2} < \varepsilon_1,$$

then

$$|H(\gamma(t))| \geq \frac{\kappa}{2} \qquad \forall t \in [0, 1/2].$$

(c) Show that there exists $\delta > 0$ as in (8.11).

8.53 ([56, Theorem 1.5]) Let Q be a compact connected smooth manifold without boundary and consider a magnetic Hamiltonian system (T^*Q, ω_σ, H), for some $\sigma \in \Omega^2(Q)$ and a mechanical Hamiltonian function H.

1. Show that if σ vanishes on $\pi_2(Q)$, then the pullback $\pi^*\sigma$ to the universal cover $\pi\colon \widetilde{Q} \to Q$ is exact.
2. Assume that at the Mañé critical value

$$c := \inf_{\theta} \sup_{q \in \widetilde{Q}} \widetilde{H}(q, \theta_q),$$

where the infimum is taken over all 1-forms θ on the universal covering manifold $\widetilde{Q}$ with $d\theta = \widetilde{\rho}$, the regular energy hypersurface $\Sigma_c = H^{-1}(c)$ has no contractible periodic orbits. Show that then Σ_c is not stable, provided that all levels Σ_k with $k < c$ are displaceable.

8.54 (Coisotropic Submanifolds, [114]) Let Σ be a compact submanifold of codimension k of a symplectic manifold (M^n, ω). We say that Σ is ***coisotropic***, if the vector space $(T_x\Sigma)^\perp$ is a k-dimensional subvectorspace of $T_x\Sigma$ for all $x \in \Sigma$. A coisotropic submanifold Σ is said to be of ***contact type***, if there exist differential forms $\alpha_1, \dots, \alpha_k \in \Omega^1(M)$ such that $d\alpha_i = \omega$ for all $i = 1, \dots, k$ and with

$$\alpha_1 \wedge \dots \wedge \alpha_k \wedge \omega^{n-k}|_\Sigma \in \Omega^{2n-k}(\Sigma)$$

is a volume form on Σ.

1. Let $H_1, \dots, H_k \in C^\infty(M)$. Show that if

$$\{H_i, H_j\} = 0 \qquad \forall i, j = 1, \dots, k$$

and 0 is a regular value of

$$\mathscr{H} := (H_1, \dots, H_k)\colon M \to \mathbb{R}^k,$$

then $\mathscr{H}^{-1}(0)$ is a smooth coisotropic submanifold of codimension k in (M, ω) and the symplectic orthogonal complement $(T\mathscr{H}^{-1}(0))^\perp$ is spanned by the restrictions of the Hamiltonian vector fields $X_{H_1}, \dots, X_{H_k} \in \mathfrak{X}(M)$ to $\mathscr{H}^{-1}(0)$.
2. Let Σ be a coisotropic submanifold of contact type in (M, ω). Show that there exist unique vector fields $R_1, \dots, R_k \in \mathfrak{X}(\Sigma)$ satisfying

$$\alpha_i(R_j) = \delta_{ij} \qquad \text{and} \qquad R_i \in \bigcap_{l=1}^{k} \ker d\alpha_l|_\Sigma$$

for all $i, j = 1, \dots, k$. These vector fields are called the ***Reeb vector fields*** associated with the coisotropic submanifold of contact type.

3. Assume that $\mathscr{H} \in C^\infty(M, \mathbb{R}^k)$ has Poisson-commuting components in a symplectically aspherical symplectic manifold (M, ω). Define the ***generalised Rabinowitz action functional***

$$\mathscr{A}^{\mathscr{H}} : \Lambda M \times \mathbb{R}^k \to \mathbb{R}, \qquad \mathscr{A}^{\mathscr{H}}(\gamma, \tau) := \int_{\mathbb{D}} \bar{\gamma}^* \omega - \sum_{j=1}^{k} \tau_j \int_0^1 H_j(\gamma(t))dt.$$

Show that $(\gamma, \tau) \in \operatorname{Crit}(\mathscr{A}^{\mathscr{H}})$ if and only if

$$\dot{\gamma}(t) = \sum_{j=1}^{k} \tau_j X_{H_j}(\gamma(t)) \qquad \text{and} \qquad \int_0^1 H_j(\gamma(t))dt = 0$$

for all $t \in \mathbb{T}$ and $j = 1, \dots, k$. Moreover, show that $\gamma(\mathbb{T}) \subseteq \mathscr{H}^{-1}(0)$ for all critical points $(\gamma, \tau) \in \operatorname{Crit}(\mathscr{A}^{\mathscr{H}})$.

Using the notion of the generalised Rabinowitz action functional for coisotropic submanifolds, one can generalise some results from standard Rabinowitz–Floer homology:

- Let Σ be a compact and connected coisotropic submanifold of contact type in a geometrically bounded symplectic manifold (M, ω). Let $\varphi \in \operatorname{Ham}_c(M, \omega)$ be a Hamiltonian symplectomorphism such that the Hofer norm $\|\varphi\|$ is strictly smaller than the minimal period of a contractible periodic Reeb orbit in $\operatorname{Crit}(\mathscr{A}^{\mathscr{H}})$. Then by [114, Theorem A], there exists a leafwise intersection point for φ.
- If one replaces contact type by a suitable generalisation of the stability condition for coisotropic submanifolds, one gets another generalisation of Schlenk's Theorem as opposed to the Forcing Theorem 8.43. Assume that Σ is a displaceable stable coisotropic submanifold. Then by [114, Theorem D], there exists a contractible periodic Reeb orbit $v \in C^\infty(\mathbb{T}, \Sigma)$ satisfying

$$0 < \left| \int_{\mathbb{D}} \bar{\gamma}^* \omega \right| \leq e(\Sigma).$$

9 Systolic Geometry

> *"After talking with Misha (Mikhail Gromov) for a few weeks, I said to Dennis Sullivan, I have the impression that half of what is known in Riemannian geometry is known only to Gromov."*
>
> —*Jeff Cheeger, quoted in* [180]

Systolic geometry is concerned with the study of loops of least length in a topological space in some suitable sense. For example, in Riemannian geometry one is interested in non-contractible geodesic loops. In this setting, the notion of a systole has a long history and produced beautiful results relating volume and the length of curves, such as the isoperimetric inequality described in [37] or [115]. It should be mentioned that Gromov also contributed substantially to this field. In this final chapter, we investigate a fairly recent introduction of systoles in symplectic geometry, as in [77, Section 1.3]. For a nice introduction to the Riemannian and symplectic viewpoints of systolic geometry, see [36]. For us, the main motivation will be that the systoles in contact geometry give rise to fuel-efficient space mission design.

9.1 Symplectic Capacities

We start with the notion of a symplectic capacity. A symplectic capacity can be thought of as measuring the 2-dimensional area of a subset of a symplectic manifold. For a short introduction to the heuristics behind this notion, see Appendix A, and Problem 9.35 comparing the following definition with Definition A.16.

Definition 9.1 (Symplectic Capacity) A ***symplectic capacity on*** $\mathbb{R}^{2n}$ is a map c which assigns to each subset $A \subseteq \mathbb{R}^{2n}$ a number $c(A) \in [0, +\infty]$ such that the following three properties hold.

Y. Bähni, *Lectures on Twisted Rabinowitz-Floer Homology*, Pathways in Mathematics, https://doi.org/10.1007/978-3-032-10673-5_9

1. (***Monotonicity***) If there exists a symplectomorphism $\varphi \in \mathrm{Symp}(\mathbb{R}^{2n}, \omega_0)$ such that $\varphi(A) \subseteq B$, then $c(A) \leq c(B)$.
2. (***Conformality***) $c(\lambda A) = \lambda^2 c(A)$ for all $\lambda \in \mathbb{R}$.
3. (***Normalisation***) It holds that

$$c(B^{2n}(r)) = c(Z^{2n}(r)) = \pi r^2 \qquad \forall r > 0.$$

Exercise 9.2 Consider the function $c(A) := \mathrm{Vol}(A)^{1/n}$ for all $A \subseteq \mathbb{R}^{2n}$ measureable. Explain why this function can never be a symplectic capacity for $n \geq 2$.

Exercise 9.3 Let c be a symplectic capacity. Show that $c(\{x\}) = 0$ for every point $x \in \mathbb{R}^{2n}$.

The next example provides two fundamental examples of a symplectic capacity. In practice, they are usually notoriously difficult to compute explicitly.

Example 9.4 (Ball and Cylindrical Capacity) For any subset $A \subseteq \mathbb{R}^{2n}$ define

$$c_B(A) := \sup\{\pi r^2 > 0 : \exists \varphi \in \mathrm{Symp}(\mathbb{R}^{2n}, \omega_0) \text{ such that } \varphi(B^{2n}(r)) \subseteq A\},$$

called the ***Ball capacity*** of A, and

$$c_Z(A) := \inf\{\pi R^2 > 0 : \exists \varphi \in \mathrm{Symp}(\mathbb{R}^{2n}, \omega_0) \text{ such that } \varphi(A) \subseteq Z^{2n}(R)\},$$

called the ***cylindrical capacity*** of A. Then $c_B(A), c_Z(A) \in [0, +\infty]$ and monotonicity, as well as conformality, can be directly checked by means of the definition of c_B and c_Z. Showing normalisation is considerably harder and involves a deep theorem from symplectic topology, namely Gromov's Nonsqueezing Theorem A.15.

Exercise 9.5 Using Gromov's Nonsqueezing Theorem A.15, show that both the ball capacity c_B and the cylindrical capacity c_Z are symplectic capacities, and that

$$c_B(A) \leq c(A) \leq c_Z(A) \qquad \forall A \subseteq \mathbb{R}^{2n}.$$

The next important example of a symplectic capacity is the displacement energy from Definition 2.44. With the theory we developed so far, we are able to give a very dynamical proof of this fact. By definition of the displacement energy, the hardest part is computing the displacement energy of the ball and proving normalisation. An alternative proof is provided in [137, Theorem 12.3.9].

Exercise 9.6 Show that $e(B^{2n}(r)) \leq e(Z^{2n}(r)) \leq c_B(Z^{2n}(r)) = \pi r^2$ for all $r > 0$.

Proposition 9.7 *The displacement energy e is a symplectic capacity on $\mathbb{R}^{2n}$.*

Proof Monotonicity and conformality follow directly from the definition of the displacement energy (Definition 2.44). So, it only remains to show the normalisation property. For the closed ball $B^{2n}(r)$ of radius r we estimate

$$e(\partial B^{2n}(r)) \leq e(B^{2n}(r)) \leq e(Z^{2n}(r)) \leq c_B(Z^{2n}(r)) = \pi r^2 \qquad \forall r > 0,$$

by monotonicity and Exercise 9.6. The periodic Reeb flow on $\partial B^{2n}(r)$ is given by

$$\phi_t(z) = e^{-2it/r^2} z \qquad \forall z \in \partial B^{2n}(r).$$

Hence we have that $\tau \in \pi r^2 \mathbb{Z}$ for the period τ of a periodic Reeb orbit. By Schlenk's Theorem 8.46, the only possible choice satisfying $0 < \tau \leq e(\partial B^{2n}(r))$ is $\tau = \pi r^2$. Hence, we obtain

$$\pi r^2 = e(\partial B^{2n}(r)) = e(B^{2n}(r)) \qquad \forall r > 0.$$

This shows the normalisation property of the displacement energy. □

A corollary of Proposition 9.7 is that there is a metric on the group of Hamiltonian symplectomorphisms $\mathrm{Ham}_c(\mathbb{R}^{2n}, \omega_0)$. This is the starting point of what is known as Hofer geometry [155]. But first, we justify that $\mathrm{Ham}_c(\mathbb{R}^{2n}, \omega_0)$ is indeed a group!

Exercise 9.8 Let $\varphi_F, \varphi_G \in \mathrm{Ham}_c(\mathbb{R}^{2n}, \omega_0)$. Show that the composition $\varphi_F \circ \varphi_G$ again belongs to $\mathrm{Ham}_c(\mathbb{R}^{2n}, \omega_0)$, and is generated by

$$(F\#G)_t := F_t + G_t \circ \phi_t^{-F_t}.$$

Exercise 9.9 (Hofer Metric) Following [137, Section 12.3] define the ***Hofer metric*** on $\mathrm{Ham}_c(\mathbb{R}^{2n}, \omega_0)$ by

$$d(\varphi_0, \varphi_1) := \inf_{\varphi_F = \varphi_1 \circ \varphi_0^{-1}} \|F\|$$

for all $\varphi_0, \varphi_1 \in \mathrm{Ham}_c(\mathbb{R}^{2n}, \omega_0)$.

1. Show that the Hofer distance is ***symmetric***, meaning that

$$d(\varphi_1, \varphi_0) = d(\varphi_0, \varphi_1)$$

for all $\varphi_0, \varphi_1 \in \mathrm{Ham}_c(\mathbb{R}^{2n}, \omega_0)$.
2. Show that the Hofer distance satisfies the ***triangle inequality***, meaning that

$$d(\varphi_0, \varphi_1) \leq d(\varphi_0, \varphi_2) + d(\varphi_2, \varphi_1)$$

for all $\varphi_0, \varphi_1, \varphi_2 \in \mathrm{Ham}_c(\mathbb{R}^{2n}, \omega_0)$.

3. Show that the Hofer distance is ***bi-invariant***, meaning that

$$d(\psi \circ \varphi_0, \psi \circ \varphi_1) = d(\varphi_0 \circ \psi, \varphi_1 \circ \psi) = d(\varphi_0, \varphi_1)$$

for all $\varphi_0, \varphi_1, \psi \in \mathrm{Ham}_c(\mathbb{R}^{2n}, \omega_0)$.
4. Show that the Hofer norm is ***nondegenerate***, meaning that

$$d(\varphi_0, \varphi_1) = 0 \quad \Rightarrow \quad \varphi_0 = \varphi_1.$$

for all $\varphi_0, \varphi_1 \in \mathrm{Ham}_c(\mathbb{R}^{2n}, \omega_0)$.

In [124, Corollary 1.2], the results of Exercise 9.9 are generalised to arbitrary symplectic manifolds. However, we do not have the technical tools for proving that the Hofer metric is nondegenerate for arbitrary symplectic manifolds! What we want to elaborate on is the observation that in the proof of Proposition 9.7, we used the fact that there exists a parametrised periodic Reeb orbit (γ, τ) on $\partial B^{2n}(r)$, $r > 0$, satisfying $\tau = e(\partial B^{2n}(r))$. In fact, this periodic orbit has the smallest period among all Reeb orbits on $\partial B^{2n}(r)$.

Definition 9.10 (Systole) Let $\Sigma \subseteq \mathbb{R}^{2n}$ be a star-shaped hypersurface. Then we define the ***systole of*** $\boldsymbol{\Sigma}$ to be the number

$$\mathrm{sys}(\Sigma) := \inf\{\tau \in (0, +\infty) : (\gamma, \tau) \in \mathscr{P}(\Sigma)\} \in [0, +\infty].$$

Proposition 9.11 *Let $\Sigma \subseteq \mathbb{R}^{2n}$ be a star-shaped hypersurface. Then there exists a parametrised periodic Reeb orbit $(\gamma, \tau) \in \mathscr{P}(\Sigma)$ such that $\tau = \mathrm{sys}(\Sigma)$ holds. In particular, we have $\mathrm{sys}(\Sigma) > 0$.*

Proof The proof is similar to Theorem 4.21. First of all, observe that $\mathscr{P}(\Sigma) \neq \varnothing$ by Corollary 8.9. This implies that $\mathrm{sys}(\Sigma) < +\infty$. Let (τ_j) be a monotone decreasing sequence in $(0, +\infty)$ such that $\tau_j \to \tau$ as $j \to \infty$ for $\tau := \mathrm{sys}(\Sigma)$. By definition of the systole of Σ, there exists a sequence $(\gamma_j) \subseteq C^\infty(\mathbb{T}, \Sigma)$ satisfying

$$\dot{\gamma}_j(t) = \tau_j R_\Sigma(\gamma_j(t)) \qquad \forall t \in \mathbb{T}$$

for every $j \in \mathbb{N}$. Define a family of continuous functions

$$\mathscr{F} \subseteq C^0(\mathbb{T}, \Sigma) \qquad \text{by} \qquad \mathscr{F} := \{\gamma_j : j \in \mathbb{N}\}.$$

Fix a Riemannian metric m on Σ, for example the standard Euclidean inner product of $\mathbb{R}^{2n}$. Then $\mathscr{F}$ is pointwise bounded since the set $\mathscr{F}_t := \{\gamma_j(t) : j \in \mathbb{N}\}$ is contained in the compact, thus bounded, space Σ for all $t \in \mathbb{T}$. Moreover, we claim that the family $\mathscr{F}$ is equicontinuous. Indeed, for $t, t_0 \in \mathbb{T}$ we estimate

$$d_m(\gamma_j(t), \gamma_j(t_0)) \leq \int_{t_0}^{t} |\dot{\gamma}_j(s)| ds = \tau_j \int_{t_0}^{t} |R_\Sigma(\gamma_j(s))| ds \leq C|t - t_0|,$$

as the convergent sequence (τ_j) and the Reeb vector field $R_\Sigma \colon \Sigma \mapsto T\Sigma$ are bounded. By Ascoli's theorem [145, Theorem 45.4], the closure of $\mathscr{F}$ in $C^0(\mathbb{T}, \Sigma)$ is compact. Hence, there exists $\gamma \in C^0(\mathbb{T}, \Sigma)$ such that $\gamma_j \to \gamma$ uniformly as $j \to \infty$ modulo subsequences. We observe

$$\dot{\gamma}(t) = \lim_{j\to\infty} \dot{\gamma}_j = \lim_{j\to\infty} \tau_j R_\Sigma(\gamma_j(t)) = \tau R_\Sigma(\gamma(t)) \qquad \forall t \in \mathbb{T},$$

and conclude that $\gamma \in C^1(\mathbb{T}, \Sigma)$. Hence, we can inductively improve the regularity of the continuous loop γ to be of class C^∞ via bootstrapping. So left to show is that we have $\tau > 0$ for $\gamma \in C^\infty(\mathbb{T}, \Sigma)$. Towards a contradiction, assume that $\tau = 0$. This implies that $y(t) = x$ for all $t \in \mathbb{T}$ is constant for some $x \in \Sigma$. By an argument along the lines of [200, Proposition VIII.3], we estimate in local coordinates

$$\frac{1}{\tau_j} \langle \dot{\gamma}_j(t), R_\Sigma(x) \rangle = \langle R_\Sigma(\gamma_j(t), R_\Sigma(x)) \geq (1 - \varepsilon) \| R_\Sigma(x) \|^2 \qquad \forall t \in \mathbb{T},$$

for $\varepsilon > 0$ sufficiently small and j sufficiently large. Thus we get

$$(1 - \varepsilon) \| R_\Sigma(x) \|^2 \leq \frac{1}{\tau_j} \int_0^1 \langle \dot{\gamma}_j(t), R_\Sigma(x) \rangle dt = 0 \qquad \Rightarrow \qquad \| R_\Sigma(x) \| = 0.$$

This contradicts the fact that $R_\Sigma(x) \neq 0$ as the Reeb vector field is nowhere vanishing and we conclude that $(\gamma, \tau) \in \mathscr{P}(\Sigma)$. This finishes the proof. □

The next exercise generalises our discussion to arbitrary contact manifolds.

Exercise 9.12 (Geodesics) Let (Σ, ω, α) be a compact stable Hamiltonian manifold.

1. Define the notion of the systole $\operatorname{sys}(\Sigma, \omega, \alpha)$ generalising the one already given in Definition 9.10 for star-shaped hypersurfaces $\Sigma \subseteq \mathbb{R}^{2n}$.
2. Formulate a suitable statement of Proposition 9.11 in the more general case and prove it! What is the main difference from the case of star-shaped hypersurfaces?
3. Let (Q, m) be a compact Riemannian manifold and consider the ***unit cotangent bundle*** defined by $ST^*Q := \{(q, p) \in T^*Q : m_*(p, p) = 1\}$, where m_* denotes the dual metric of m. Let $V \in \mathfrak{X}(TQ)$ denote the geodesic vector field [130, p. 129] defined by

$$Vf(q, \dot{q}) := \left.\frac{d}{dt}\right|_{t=0} f(\exp(t\dot{q}), \exp'(t\dot{q}))$$

for all $f \in C^\infty(TQ)$, where $\exp\colon TQ \to Q$ denotes the exponential map induced by the Levi–Civita connection associated with m. Consider the mechanical Hamiltonian function

$$H\colon T^*Q \to \mathbb{R}, \qquad H(q,p) := \frac{1}{2}\|p\|^2_{m^*}.$$

Show that $(ST^*Q, \lambda|_{ST^*Q})$ is a contact manifold, where $\lambda \in \Omega^1(T^*Q)$ denotes the canonical Liouville form (2.3) on T^*Q, and that $R = X_H|_{ST^*Q} \in \mathfrak{X}(ST^*Q)$, where R denotes the Reeb vector field of the contact manifold $(ST^*Q, \lambda|_{ST^*Q})$. Explain why a periodic Reeb orbit in the unit cotangent bundle ST^*Q projects down to a closed geodesic in Q, whose length coincides with the period. In particular, $\operatorname{sys}(ST^*Q)$ is the length of the shortest nonconstant closed geodesic in Q. Moreover, explain why we cannot make use of Schlenk's theorem 8.46 to find closed geodesics in Q!

The notion of a systole in contact geometry is fairly new but has a long tradition in Riemannian geometry [77, Section 16.3]. For us, the main motivation for studying systoles in classical mechanics comes from the search for fuel-efficient space missions. Thus, from now on, we will focus on star-shaped hypersurfaces and lens spaces. Let $A \subseteq \mathbb{R}^{2n}$ be a smooth star-shaped domain, that is, A is star-shaped with respect to the origin and ∂A is a smooth star-shaped hypersurface. An immediate consequence of Exercise 8.46 is the inequality

$$0 < \operatorname{sys}(\partial A) \leq e(\partial A) \leq e(A), \tag{9.1}$$

giving a nontrivial lower bound for the displacement energy of any star-shaped domain A. A natural question is whether we can quantify the gap between $\operatorname{sys}(\partial A)$ and $e(A)$. We start with the simplest case $n = 1$. Let $A \subseteq \mathbb{R}^2$ be a star-shaped domain. By definition, there exists a positive function $\rho \in C^\infty(\mathbb{T})$ such that

$$\partial A = \{\rho(\theta)e^{2\pi i\theta} : \theta \in \mathbb{T}\} \subseteq \mathbb{R}^2.$$

Consider the smooth parametrisation Γ of the boundary $\partial A = \partial A_\rho$ given by

$$\Gamma\colon \mathbb{T} \to \mathbb{R}^2, \qquad \Gamma(\theta) := \begin{pmatrix} x(\theta) \\ y(\theta) \end{pmatrix} = \begin{pmatrix} \rho(\theta)\cos(2\pi\theta) \\ \rho(\theta)\sin(2\pi\theta) \end{pmatrix}.$$

Lemma 9.13 *Let $A \subseteq \mathbb{R}^2$ be a star-shaped domain. Then* $\operatorname{sys}(\partial A) = \operatorname{Area}(A)$.

Proof For the smooth parametrisation Γ of the boundary ∂A we compute

$$\Gamma^*\lambda|_{\partial A} = \frac{1}{2}\left(x(\theta)y'(\theta) - y(\theta)x'(\theta)\right)d\theta = \pi\rho^2(\theta)d\theta \in \Omega^1(\mathbb{T}).$$

Consequently, the Reeb vector field $R \in \mathfrak{X}(\partial A)$ pulls back under Γ to

$$\Gamma^* R = \frac{2}{x(\theta)y'(\theta) - y(\theta)x'(\theta)} \frac{\partial}{\partial \theta} = \frac{1}{\pi \rho^2(\theta)} \frac{\partial}{\partial \theta} \in \mathfrak{X}(\mathbb{T}).$$

Indeed, by assumption, $\rho \in C^\infty(\mathbb{R}/2\pi\mathbb{Z})$ is a positive function and thus the pullback $\Gamma^* R$ is well defined. Using the separation of variables for the ordinary differential equation of Reeb flow $\dot{\theta}(t) = \Gamma^* R(\theta(t))$ for all $t \in \mathbb{R}$, we obtain for the period τ the formula

$$\tau = \frac{1}{2} \int_0^1 \big(x(\theta)y'(\theta) - y(\theta)x'(\theta)\big)\, d\theta.$$

Green's Theorem [129, Theorem 16.17] now states

$$\int_A \left(\frac{\partial Q}{\partial x} - \frac{\partial P}{\partial y} \right) dxdy = \int_{\partial A} Pdx + Qdy$$

for smooth functions $P, Q \in C^\infty(A)$. By choosing

$$Q(x, y) := x \qquad \text{and} \qquad P(x, y) := -y,$$

we conclude

$$\begin{aligned} \mathrm{Area}(A) &= \frac{1}{2} \int_{\partial A} xdy - ydx \\ &= \frac{1}{2} \int_0^1 \Gamma^*(xdy - ydx)d\theta \\ &= \frac{1}{2} \int_0^1 \big(x(\theta)y'(\theta) - y(\theta)x'(\theta)\big)\, d\theta \\ &= \tau. \end{aligned}$$

Therefore, the period τ of the Reeb flow along ∂A is equal to the area of the star-shaped domain A. This shows the lemma. □

Theorem 9.14 *Let $A \subseteq \mathbb{R}^2$ be a star-shaped domain. Then its displacement energy and the systole of the boundary ∂A coincide. More precisely, we have*

$$\mathrm{sys}(\partial A) = \mathrm{Area}(A) = e(A).$$

Proof The first equality follows from Lemma 9.13. By constructing an explicit displacing Hamiltonian function, we see $e(A) \leq \mathrm{Area}(A)$. That the reverse inequality $\mathrm{Area}(A) \leq e(A)$ holds follows from (9.1) combined with Lemma 9.13.

□

Exercise 9.15 Show that $e(A) = \text{Area}(A)$ for every star-shaped domain $A \subseteq \mathbb{R}^2$ by writing the proof of the Forcing Theorem 8.43 explicitly in that case.

Remark 9.16 (Symplectic Capacities in Dimension Two) The equality between the displacement energy and the area for star-shaped domains in Theorem 9.14 also follows from a more general observation. By [174, Theorem 4.2] or [166, Corollary C.10 (ii)], any symplectic capacity of a star-shaped domain coincides with its area. This is somehow expected, since in dimension two the symplectic and volume-preserving phenomena coincide. This is, for example, a crucial observation used in the proof of the simplicity conjecture [62], stating that the group of compactly supported area-preserving homeomorphisms of the unit disc in the plane admits many non-trivial proper normal subgroups and is therefore not simple.

Corollary 9.17 *Let $A \subseteq \mathbb{R}^2$ be a star-shaped domain and denote by $\ell(\partial A)$ the length of the boundary with respect to the standard Euclidean metric. Then*

$$\text{sys}(\partial A) \leq \frac{1}{4\pi}\ell(\partial A)^2,$$

with equality if and only if A is a disc.

Proof The statement follows from Theorem 9.14 and the celebrated isoperimetric inequality [159, Section 2.1]. See also [136, Lemma 4.4.4] for a symplectic version of the isoperimetric inequality. □

Remark 9.18 Theorem 9.14 is valid in a more general setting. Let $U \subseteq \mathbb{R}^2$ be a simply connected domain with smooth boundary ∂U. Choose $\lambda \in \Omega^1(\mathbb{R}^2)$ with $d\lambda = \omega$ and such that the vector field $X \in \mathfrak{X}(\mathbb{R}^2)$ with $i_X\omega = \lambda$ is nowhere tangent to ∂U. This guarantees that $(\partial U, \lambda|_{\partial U})$ is a contact manifold. Then by Stokes' theorem $\text{sys}(\partial U) = \text{Area}(U)$, and $\text{Area}(U) = e(U)$ by [166, Corollary C.10 (ii)]. For star-shaped hypersurfaces, $X = \frac{1}{2}(x\partial_x + y\partial_y)$ is the radial vector field, and thus the Liouville vector field X is tangent at some points on ∂U as depicted in Fig. 9.1a.

Remark 9.19 For higher dimensions, the equality $\text{sys}(\partial A) = e(A)$ in Theorem 9.14 fails. For example, consider a star-shaped domain $A \subset \mathbb{R}^{2n}, n \geq 2$, obtained by gluing a piece of a symplectic cylinder $Z(r) \subset \mathbb{R}^2 \times \{0\}$ along the boundary of the ball $B(R)$ for $r < R$ that resembles a Bordeaux bottle [107, Section 3.5] as illustrated in Fig. 9.1b. On the piece of $Z(r)$, there is a periodic orbit with period πr^2, implying that the systole of ∂A is less than or equal to πr^2. On the other hand, since $B(R) \subseteq A$, the monotonicity property of a symplectic capacity now tells us that $e(A) \geq e(B(R)) = \pi R^2$, which implies $\text{sys}(\partial A) < e(A)$.

So, if $n = 1$, we have equality in (9.1) by Theorem 9.14. The case $n = 2$ is also quite well understood, but much harder to prove. Note that by Remark 9.19, the equality in (9.1) cannot be true for arbitrary star-shaped domains. Following

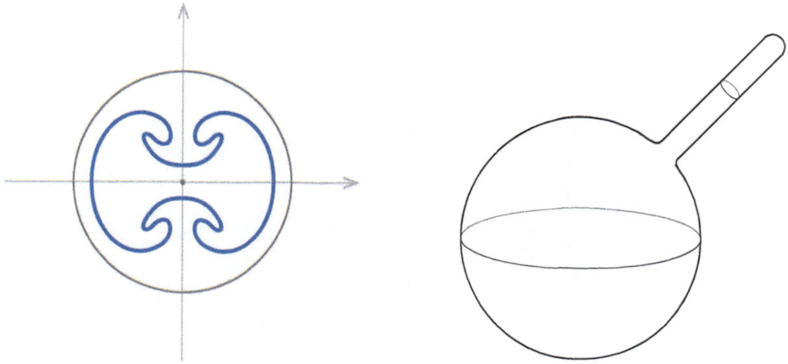

(a) (b)

Fig. 9.1 (**a**) A hypersurface in the plane $\mathbb{R}^2$ that is not star-shaped. (**b**) The sphere $\mathbb{S}^2 \subseteq \mathbb{R}^3$ with a neck resembling a bottle

[77, Definition 12.1.1], the boundary of a star-shaped domain is said to be ***strictly convex***, if the Gauss–Kronecker curvature is strictly positive at every point of ∂A. In this case, we call the convex domain a ***convex body*** by [77, Lemma 12.1.3]. Let $K \subseteq \mathbb{R}^4$ be a convex body. Then $\operatorname{sys}(\partial K) = e(K)$ holds by [8, Corollary 1]. The proof is a case where it is more convenient to use a variational principle in the Lagrangian formalism [3]. Motivated by this result, we prompt the following question.

Conjecture 9.20 For every convex body $K \subseteq \mathbb{R}^{2n}$ we have $\operatorname{sys}(\partial K) = e(K)$.

Remark 9.21 (Viterbo's Conjecture) For many symplectic capacities c it is known that $c(K) = \operatorname{sys}(\partial K)$ for all smooth convex bodies $K \subseteq \mathbb{R}^{2n}$. It was the content of Viterbo's conjecture that this equality holds for all symplectic capacities. This question has been negatively answered in [101], where convex bodies in $\mathbb{R}^{2n}$, $n \geq 2$, were constructed such that the ball capacity c_B is strictly smaller than the systole of the boundary. Conjecture 9.20 has been shown for some normalised symplectic capacities as explained in [7]. Moreover, by [7, Theorem 1] there exists a C^2-neighbourhood of the Euclidean ball in $\mathbb{R}^{2n}$ where this equality holds for all normalised symplectic capacities, implying that they are all equal there, and wrong for fibrewise star-shaped hypersurfaces in cotangent bundles [41, Corollary 1.4].

Systoles in contact geometry should be particularly useful for planning fuel-efficient space missions. A particularly interesting question was stated in [77, Section 1.3]. Using Birkhoff's shooting method in [77, Theorem 8.3.2] for the restricted three-body problem, one can show the existence of a retrograde periodic orbit, that is, an orbit in the opposite direction of the rotation of its primary. However, it is unknown whether the retrograde periodic orbit represents the systole for energy values below the first critical value. This is known as the ***Birkhoff conjecture*** [113].

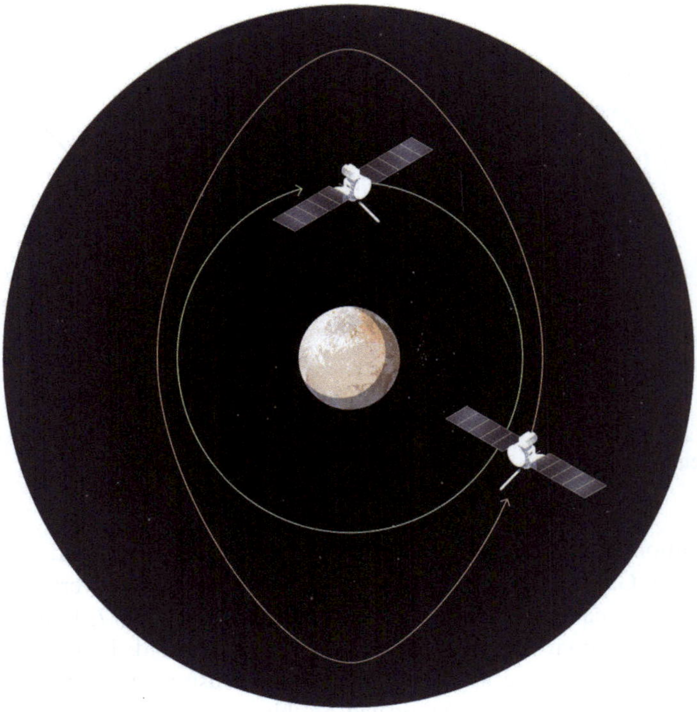

Fig. 9.2 The Jupiter–Europa–Satellite system with a distant direct and retrograde periodic orbit around Europa lying on the same regular energy hypersurface

A retrograde periodic orbit should be compared to a prograde or direct orbit, that is, a periodic orbit in the same direction as the rotation of its corresponding primary. In Fig. 9.2 the Jupiter–Europa–Satellite system is depicted with two periodic orbits lying on the same energy hypersurface. The retrograde periodic orbit depicted in green has a period of approximately ten hours, whereas the direct periodic orbit coloured orange has a period of approximately forty hours according to the NASA database of periodic orbits in different three-body problems [147]. By its construction, the retrograde periodic orbit is symmetric with respect to the involution $(z, w) \mapsto (-z, -w)$ of $\mathbb{C}^2$ and non-contractible in $\mathbb{RP}^3$. The existence of such an orbit follows from Theorem 8.17 which is similar to [18, Corollary 1.6]. By [18, Corollary 1.5], the regularisation of the energy hypersurface near one of the primaries for fixed energy values below the first critical value is contactomorphic to the tight contact structure on $\mathbb{RP}^3$, coming from the unique tight contact structure on the sphere $\mathbb{S}^3$, see Problem 9.36. Consequently, for any fixed energy value below the first critical value, the regularised planar circular restricted three-body problem of arbitrary mass ratio admits a noncontractible periodic orbit of constant energy. The period of such orbits can be quite different, as illustrated in Fig. 9.2 and might be a collision orbit. Periodic collision orbits or consecutive collision orbits are known to exist below the first critical value for any mass ratio by [80, Theorem 2.2] and [162, Theorem A]. In the following final section, we introduce symmetries to our considerations of systoles for further studying the retrograde periodic orbit! With the help of emerging numerical methods, this should be of practical relevance for space mission design, that is, travelling from A to B in the words of Hofer. Consecutive collision orbits, for example, are of use in engineering problems because a slight perturbation of such an orbit avoids collision as in the Oberth manoeuvre or gravitational slingshots. So, it is desirable to study symmetric systoles on fixed-energy hypersurfaces.

9.2 The Systolic Ratio of Convex Symmetric Hypersurfaces

Let M^d, $d \geq 2$, be an orientable compact smooth manifold without boundary, and consider the infinite-dimensional space $\mathscr{R}(M)$ of all Riemannian metrics on M. This space is infinite-dimensional even after identifying metrics that are related by isometries. Studying the space $\mathscr{R}(M)$ in general is quite difficult or impossible. The following assumption on M simplifies the study. A path-connected topological space whose fundamental group is isomorphic to a given group G and which has a contractible universal covering space is called an ***Eilenberg–MacLane space***, and is denoted by $K(G, 1)$. Following Gromov [95, Section 0], we call a compact and connected smooth manifold M without boundary ***essential***, if there exists a continuous map $f \in C^0(M, K(\pi_1(M), 1))$ such that the induced map in homology

$$\mathrm{H}_d(M; \mathbb{Z}) \to \mathrm{H}_d(K(\pi_1(M), 1); \mathbb{Z}), \qquad [\sigma] \mapsto [f \circ \sigma],$$

maps the fundamental class $[M] \in \mathrm{H}_d(M; \mathbb{Z}) \cong \mathbb{Z}$ to a nonzero element. Note that this fundamental class exists by [104, Theorem 3.26] as M is orientable.

Example 9.22 (Lens Spaces) Lens spaces are essential. In fact, following [104], consider the infinite-dimensional sphere

$$\mathbb{S}^\infty := \bigcup_{n \geq 1} \mathbb{S}^{2n-1} \subseteq \mathbb{C}^\infty.$$

Fix an integer $m \geq 1$ and let $(k_j) \subseteq \mathbb{Z}$ be a sequence such that k_j is coprime to m for all $j \geq 1$. Then there is a natural $\mathbb{Z}_m$-action on $\mathbb{S}^\infty$ generated by the rotation

$$\varphi(z_1, z_2, \dots) := \left(e^{2\pi i k_1/m} z_1, e^{2\pi i k_2/m} z_2, \dots, e^{2\pi i k_j/m} z_j, \dots\right).$$

This is a covering space action and $\mathbb{S}^\infty$ is a contractible space; consequently, the ***infinite-dimensional lens space*** $\mathbb{S}^\infty/\mathbb{Z}_m$ is a $K(\mathbb{Z}_m, 1)$. The inclusion map

$$\iota_n \colon L_m(k_1, \dots, k_n) \hookrightarrow L_m(k_1, k_2, \dots) := \mathbb{S}^\infty/\mathbb{Z}_m \qquad \forall n \geq 1,$$

gives rise to $0 \neq \mathrm{H}_{2n-1}(\iota_n)[L_m(k_1,\dots,k_n)] \in \mathrm{H}_{2n-1}(L_m(k_1,k_2,\dots);\mathbb{Z}) \cong \mathbb{Z}$. Note that any lens space is orientable since $\mathrm{H}_{2n-1}(L_m(k_1,\dots,k_n);\mathbb{Z}) \cong \mathbb{Z} \neq 0$.

Let M^d be an essential manifold. Then in particular $\pi_1(M) \neq 0$ and by a celebrated result of Gromov [95, Theorem 0.1.A], there exists a constant $C_d > 0$, only depending on the dimension of M, with

$$C_d \leq 6(d+1)d^d\sqrt{(d+1)!},$$

such that

$$\mathrm{sys}^{\pi_1(M),g}(M) \leq C_d \,\mathrm{Vol}_g(M)^{1/d} \qquad \forall g \in \mathscr{R}(M), \tag{9.2}$$

where $\mathrm{sys}^{\pi_1(M),g}(M)$ denotes the infimum over all lengths of noncontractible loops in M with respect to a fixed Riemannian metric $g \in \mathscr{R}(M)$. Note that $\mathrm{sys}^{\pi_1(M),g}(M)$ is the length of a closed geodesic. By Example 9.22 lens spaces are essential and thus satisfy (9.2). In particular, (9.2) holds for all real projective spaces $\mathbb{RP}^n$. However, even in this simple case, the precise value of the optimal constant C_d is not known for $d \geq 3$. For $d = 2$, there are several classical results relying on the uniformisation theory for surfaces. In 1952, Pu showed the inequality

$$\mathrm{sys}^{\pi_1(\mathbb{RP}^2),g}(\mathbb{RP}^2) \leq \sqrt{\frac{\pi}{2}\,\mathrm{Vol}_g(\mathbb{RP}^2)} \qquad \forall g \in \mathscr{R}(\mathbb{RP}^2).$$

This result was improved in [116]. In higher dimensions, proving normal forms of all Riemannian metrics on a given manifold is quite hard! We will not pursue this question further. What we will instead consider is the setting when an essential manifold M admits a discrete symmetry. For example, consider the torus $\mathbb{T}^2 \subseteq \mathbb{R}^3$ under rotations. More precisely, let $\varphi \in \mathrm{Diff}(M)$ be a diffeomorphism of finite order. Then we can also consider the infimum $\mathrm{sys}_\varphi^{\pi_1(M),g}(M)$ over all lengths of noncontractible loops γ in M with respect to the Riemannian metric g, which are invariant under φ, that is $\varphi(\gamma(\mathbb{T})) = \gamma(\mathbb{T})$. Of course, we have that

$$\mathrm{sys}^{\pi_1(M),g}(M) \leq \mathrm{sys}_\varphi^{\pi_1(M),g}(M) \qquad \forall g \in \mathscr{R}(M).$$

Therefore, the estimate (9.2) does not give a uniform upper bound for $\mathrm{sys}_\varphi^{\pi_1(M)}(M)$. Instead, for a fixed Riemannian metric $g \in \mathscr{R}(M)$, we are interested in studying the ***systolic ratio***

$$\mathfrak{R}_g(M,\varphi) := \frac{\mathrm{sys}_\varphi^{\pi_1(M),g}(M)}{\mathrm{sys}^{\pi_1(M),g}(M)} \in [1,+\infty].$$

Of course, we have $\mathfrak{R}_g(M,\varphi) = +\infty$ if there does not exist any φ-invariant noncontractible loop in M. As the simple example of rotations of the flat torus

$\mathbb{T}^2$ already shows, it is quite difficult to study the systolic ratio $\mathfrak{R}_g(M, \varphi)$ for all Riemannian metrics $g \in \mathscr{R}(M)$ simultaneously. However, we will see that we can also study the systolic ratio from a symplectic point of view. Due to rigidity, it is much easier to find uniform upper bounds for the systolic ratio there! In this final part of the book, we will prove such a bound for suitable Liouville automorphisms of star-shaped hypersurfaces in $\mathbb{R}^{2n}$ relying on the Forcing Theorem 8.43. First, we recall the precise setup. Let $\Sigma = \Sigma_\rho \subseteq \mathbb{R}^{2n}$ be a star-shaped hypersurface and let $\varphi \in \mathrm{Symp}(\mathbb{R}^{2n}, \omega_0)$ be a symplectomorphism of finite order, such that $\varphi(\Sigma) = \Sigma$ and $\varphi^* X = X$, for the Liouville vector field $X \in \mathfrak{X}(\mathbb{R}^{2n})$ implicitly defined by $i_X \omega_0 = \lambda_0$. If we assume additionally that $\mathrm{Fix}(\varphi|_\Sigma) \neq \varnothing$, there exists by Exercise 8.46 a nonconstant contractible parametrised periodic Reeb orbit $(v, \tau) \in \mathscr{L}\Sigma \times (0, +\infty)$ satisfying

$$0 < |\tau| = \left| \int_0^1 v^* \lambda_0 \right| = \left| \int_{\mathbb{D}} \bar{v}^* \omega_0 \right| \leq \mathrm{ord}(\varphi) e(\Sigma) < +\infty.$$

Any such periodic Reeb orbit satisfies $\varphi(v(\mathbb{T})) = v(\mathbb{T})$, since $v(\frac{t}{\mathrm{ord}(\varphi)}) \in \mathscr{L}_\varphi \Sigma$, and is therefore symmetric as in Remark 4.39. Therefore, we can define the ***symmetric systole of* Σ** to be the number $\mathrm{sys}_\varphi(\Sigma) \in (0, +\infty)$ defined as the minimum over all symmetric periodic Reeb orbits on Σ by adapting Proposition 9.11 and in analogy to the definition in [119, Equation (1.1)]. It immediately follows that

$$\mathrm{sys}(\Sigma) \leq \mathrm{sys}_\varphi(\Sigma) \qquad \text{and} \qquad \mathrm{sys}_\varphi(\Sigma) \leq \mathrm{ord}(\varphi) e(\Sigma) \tag{9.3}$$

for all φ-symmetric star-shaped hypersurfaces $\Sigma \subseteq \mathbb{R}^{2n}$.

Definition 9.23 (Symmetric Ratio) Let $\Sigma \subseteq \mathbb{R}^{2n}$ be a φ-symmetric star-shaped hypersurface for some $\varphi \in \mathrm{Symp}(\mathbb{R}^{2n}, \omega_0)$. Define the ***symmetric ratio***

$$\mathfrak{R}(\Sigma, \varphi) := \frac{\mathrm{sys}_\varphi(\Sigma)}{\mathrm{sys}(\Sigma)} \in [1, +\infty). \tag{9.4}$$

Exercise 9.24 Explain why for $n = 1$, we have $\mathfrak{R}(\Sigma, \varphi) = 1$ for all star-shaped hypersurfaces $\Sigma \subseteq \mathbb{R}^2$.

Example 9.25 (Singular Lens Spaces) Let $m \geq 1$ be an integer and let $k_1, \ldots, k_n \in \mathbb{Z}$ not necessarily be coprime to m. Consider the rotation

$$\varphi \colon \mathbb{C}^n \to \mathbb{C}^n, \qquad \varphi(z_1, \ldots, z_n) := \left(e^{2\pi i k_1/m} z_1, \ldots, e^{2\pi i k_n/m} z_n \right).$$

If $k_j = 0$ for some $j = 1, \ldots, n$, then $\mathrm{Fix}(\varphi|_\Sigma) \neq \varnothing$ for all φ-symmetric star-shaped hypersurfaces $\Sigma \subseteq \mathbb{R}^{2n}$.

In the case where $n = 2$ one can deduce a bound on $\mathfrak{R}(\Sigma, \varphi)$ if Σ is strictly convex similar to [119, Theorem 1.3]. A subset $A \subseteq \mathbb{R}^{2n}$ is ***φ-invariant***, if $\varphi(A) = A$.

Theorem 9.26 *Let $K \subseteq \mathbb{R}^4$ be a φ-invariant smooth convex body for some symplectomorphism $\varphi \in \mathrm{Symp}(\mathbb{R}^{2n}, \omega_0)$ such that the boundary ∂K is φ-symmetric. If we have $\mathrm{Fix}(\varphi|_{\partial K}) \neq \varnothing$, then $1 \leq \mathfrak{R}(\partial K, \varphi) \leq \mathrm{ord}(\varphi)$.*

Proof By [8, Corollary 1] we have $\mathrm{sys}(\partial K) = e(K)$, since $K \subseteq \mathbb{R}^4$ is convex. Note that for this equality to be true, we need to assume that $K \subseteq \mathbb{R}^4$. Thus, using the monotonicity property of the displacement energy from Exercise 2.45 and the second estimate from (9.3), we conclude

$$1 \leq \mathfrak{R}(\partial K, \varphi) = \frac{\mathrm{sys}_\varphi(\partial K)}{\mathrm{sys}(\partial K)} \leq \frac{\mathrm{ord}(\varphi) e(\partial K)}{e(K)} \leq \frac{\mathrm{ord}(\varphi) e(K)}{e(K)} = \mathrm{ord}(\varphi).$$

This shows the stated upper bound for the symmetric ratio $\mathfrak{R}(\partial K, \varphi)$. □

Remark 9.27 In view of Conjecture 9.20, we expect Theorem 9.26 to be true for all convex bodies $K \subseteq \mathbb{R}^{2n}$ for all $n \geq 1$.

What can we say in the case where $\mathrm{Fix}(\varphi|_\Sigma) = \varnothing$? Let $K \subseteq \mathbb{R}^{2n}$ be a smooth convex body invariant under a root of unity, that is, $e^{2\pi i/m} K = K$ for some $m \geq 1$ an integer. Then by [11, Theorem 2.3], we find that any closed characteristic of minimal action on the boundary ∂K is symmetric with respect to the symplectomorphism defined by $\varphi(z) := e^{2\pi i/m} z$. In particular, we conclude that $\mathrm{sys}_\varphi(\partial K) = \mathrm{sys}(\partial K)$ implies $\mathfrak{R}(\partial K, \varphi) = 1$, independent of the order m of φ, and therefore Theorem 9.26 also holds in this case. However, for a general φ-symmetric boundary ∂K, we cannot directly estimate the symmetric systole $\mathrm{sys}_\varphi(\partial K)$ in terms of the displacement energy $e(\partial K)$ in view of the Forcing Theorem 8.43. With the theory developed so far, we are not able to give general existence results of symmetric periodic Reeb orbits if $\mathrm{Fix}(\varphi|_\Sigma) = \varnothing$ for φ-symmetric star-shaped hypersurfaces by Exercise 7.24. Hence, we do not even know if $\mathrm{sys}_\varphi(\Sigma) < +\infty$ in that case. Still, there are some partial results in this direction, as we explain now.

Definition 9.28 (Translated Point, [165, Definition 2.1]) Let (Σ, α) be a compact contact manifold without boundary and let $\varphi \in \mathrm{Cont}_0(\Sigma, \alpha)$ be a contactomorphism isotopic to the identity. A point $x \in \Sigma$ is called a ***translated point for φ***, if there exists $\tau > 0$ such that $\varphi(x) = \phi_\tau^R(x)$ and $\varphi^*\alpha|_x = \alpha_x$.

Exercise 9.29 (Translated Point for Strict Contactomorphism) Show that every translated point $x \in \Sigma$ for a strict contactomorphism φ, which is isotopic to the identity, gives rise to a twisted periodic Reeb orbit in $\mathscr{P}_\varphi(\Sigma, \alpha)$.

Consider the contact manifold $(\Sigma, \alpha) := (\mathbb{S}^{2n-1}, \lambda_0|_{\mathbb{S}^{2n-1}})$. Then by [165, Theorem 1.1 (i)], there exist at least two translated points for any *generic* contactomorphism of $\mathbb{S}^{2n-1}$ that is contact isotopic to the identity. What does generic mean in this context? A translated point $x \in \Sigma$ of minimal period $\tau > 0$ is said to be nondegenerate, if for all $v \in T_x\Sigma$ the two conditions

$$D(\phi^R_{-\tau} \circ \varphi)_x(v) = v \qquad \text{and} \qquad v(f) = 0$$

do not hold simultaneously, where $f \in C^\infty(\Sigma)$ satisfies $\varphi^*\alpha = e^f\alpha$. A generic contactomorphism of $\mathbb{S}^{2n-1}$ that is contact isotopic to the identity is by definition a contactomorphism such that all its translated points are nondegenerate. Unfortunately, if we consider the rotation $\varphi(z_1, \dots, z_n) = \left(e^{2\pi i k_1/m} z_1, \dots, e^{2\pi i k_n/m} z_n\right)$ for some even $m \geq 2$ and $k_1, \dots, k_n \in \mathbb{Z}$ coprime to m, then there are no nondegenerate translated points for φ and thus Sandon's result does not imply the existence result stated in Theorem 8.17! However, by Exercise 9.29 it provides an abstract existence result of twisted Reeb orbits.

Exercise 9.30 Let (Σ, α) be a compact contact manifold and let $\varphi \in \mathrm{Cont}_0(\Sigma, \alpha)$ be a contactomorphism.

1. Show that a translated point gives rise to a leaf-wise intersection point for a suitable lift $\widehat{\varphi} \in \mathrm{Ham}_c(\mathbb{R} \times \Sigma, d(e^r\alpha))$ on the symplectisation.
2. What are the connections of the results of [165, Theorem 1.1] with the results stated in Problem 8.54 and Theorem 8.24?

There are similar existence results for translated points for more general contact manifolds. A compact contact manifold (Σ, α) is called ***hypertight***, if there do not exist any contractible periodic Reeb orbits. By [19] it is possible to construct Rabinowitz–Floer homology of such a hypertight contact manifold of its symplectisation $\mathbb{R} \times \Sigma$. Without the hypertightness assumption, compactness of negative gradient flow lines usually fails since they escape to $(-\infty, 0)$. Basically speaking, if a negative gradient flow line of the Rabinowitz action functional escapes on $(-\infty, 0)$, it converges to a contractible periodic Reeb orbit [19, Theorem 3.9]. This is not possible since, by assumption, there exist no contractible periodic Reeb orbits! Choose a contactomorphism $\varphi \in \mathrm{Cont}_0(\Sigma, \alpha)$. Then by [138, Theorem 1.3 (i)], there exists a translated point for φ and thus a twisted Reeb orbit by Exercise 9.29. Moreover, for any generic choice of φ, the number of translated points is bounded below by

$$\#\{\text{translated points for } \varphi\} \geq \sum_{j=0}^{\dim \Sigma} \dim \mathrm{H}_j(\Sigma; \mathbb{Z}_2).$$

If we assume that φ is a strict contactomorphism, that is, we have $\varphi^*\alpha = \alpha$, then either there exists a φ-symmetric Reeb orbit on the hypertight contact manifold or there exists a fixed point of φ by [138, Corollary 1.6]. In the case of the cotangent bundle T^*Q of a compact and connected smooth manifold Q without boundary and fibre-wise star-shaped hypersurfaces $\Sigma \subseteq T^*Q$ and every strict contactomorphism contact isotopic to the identity $\varphi \in \mathrm{Cont}_0(\Sigma)$ one can even say more. If the Betti numbers of the free loop space of Q are asymptotically unbounded, then there are infinitely many geometrically distinct φ-invariant Reeb orbits in Σ. Finally, we consider some existence results for perturbations of the unit sphere $\mathbb{S}^{2n-1} \subseteq \mathbb{R}^{2n}$. We start with a little warm-up exercise from linear symplectic geometry.

Exercise 9.31 For $n \geq 1$ show that $\mathrm{Sp}(2n) \cap \mathrm{O}(2n) \cong \mathrm{U}(n)$, where

$$\mathrm{U}(n) := \{A \in \mathbb{C}^{n\times n} : \bar{A}^T A = I_n\}$$

denotes the ***unitary group of degree n***.

Assume now that $\Sigma \subseteq \mathbb{R}^{2n}$ is a strictly convex hypersurface invariant under some symplectic orthogonal map $\varphi \in \mathrm{Sp}(2n) \cap \mathrm{O}(2n) = \mathrm{U}(n)$, such that there exist $r > 0$ and $R \in (r, \sqrt{2}r)$ with

$$B^{2n}(r) \subseteq \Sigma \subseteq B^{2n}(R).$$

Then by [198, Theorem 1.1], the hypersurface Σ admits at least n geometrically distinct twisted Reeb orbits. For an earlier work and a historical overview of multiplicity results of periodic Reeb orbits, see [132]. In view of the Hofer–Wysocki–Zehnder result reviewed in Problem 8.51, a more natural setting would be to consider some appropriate notion of dynamical convexity to study multiplicity questions as in [118]. Generalising Conjecture 9.20, we ask the following concluding question:

Conjecture 9.32 For every star-shaped domain $A \subseteq \mathbb{R}^{2n}$ such that the boundary ∂A is dynamically convex, we have $\mathrm{sys}(\partial A) = e(A)$.

Exercise 9.33 Let (Σ, α) be a compact contact manifold.

1. Can we always define the notion of dynamical convexity for (Σ, α)? Explain your answer!
2. Explain why we restricted ourselves to star-shaped domains in Conjecture 9.32.

Exercise 9.34 Let $U \subseteq \mathbb{R}^2$ be a simply connected open subset with a smooth boundary ∂U. For the standard volume form $\omega_0 = dx \wedge dy$ on $\mathbb{R}^2$ we choose $\lambda \in \Omega^1(\mathbb{R}^2)$ such that $d\lambda = \omega_0$. By the nondegeneracy of ω, there exists a unique vector field $X \in \mathfrak{X}(\mathbb{R}^2)$ satisfying $i_X\omega_0 = \lambda$. Moreover, we assume that X is nowhere tangent to ∂U.

1. Explain why $(\partial U, \lambda|_{\partial U})$ is a contact manifold.
2. Show that $\operatorname{sys}(\partial U) = \operatorname{Area}(U)$.
3. Show that in this setup Conjecture 9.32 holds.

Problems

9.35 (Extrinsic and Intrinsic Symplectic Capacities) The notion of a symplectic capacity introduced in Definition 9.1 is usually referred to as ***extrinsic symplectic capacity***. The notion of an intrinsic symplectic capacity is stated in Definition A.16. Both are functions that assign to each subset of $\mathbb{R}^{2n}$ a number in $[0, +\infty]$ that satisfies a conformality and a normalisation property. However, they do differ in the monotonicity axiom. An extrinsic symplectic capacity satisfies monotonicity only for *globally* defined symplectomorphisms of $\mathbb{R}^{2n}$, whereas an intrinsic symplectic capacity is also monotonic under local symplectic embeddings. In particular, every intrinsic symplectic capacity is extrinsic, but the converse need not be true, as we will see. Let $\mathbb{T}^n := \left(\mathbb{S}^1\right)^n \subseteq \mathbb{R}^{2n}$ denote the standard torus.

1. Show that $e(\mathbb{T}^n) = c_Z(\mathbb{T}^n) = \pi$.
2. Let c be any intrinsic symplectic capacity on $\mathbb{R}^{2n}$. Show that $c(\mathbb{T}^n) = 0$. Conclude that not every extrinsic symplectic capacity can be intrinsic!

Hint: See [166, Proposition C.7].

9.36 (Overtwisted Contact Structures) The standard contact structure ξ_0 on the sphere $\mathbb{S}^3$ is given by $\xi_0 := \ker \lambda_0|_{\mathbb{S}^3}$, where

$$\lambda_0 = x_1 dy_1 - y_1 dx_1 + x_2 dy_2 - y_2 dx_2.$$

It is a natural question if there exist other contact structures on $\mathbb{S}^3$ and if one can classify them. First we explain another representation of $\mathbb{S}^3 \subseteq \mathbb{R}^4$. Consider the product $\mathbb{H} := \mathbb{C} \times \mathbb{C}$ as a real vector space and define a bilinear map

$$\mathbb{H} \times \mathbb{H} \to \mathbb{H}, \qquad (a, b)(c, d) := (ac - d\bar{b}, \bar{a}d + cb).$$

With this bilinear map, $\mathbb{H}$ is a 4-dimensional algebra over $\mathbb{R}$, called the algebra of ***quaternions***. A convenient basis for $\mathbb{H}$ is provided by $\{(1, 0), (i, 0), (0, 1), (0, -i)\}$. The quaternions admit an inner product defined by

$$\langle \cdot, \cdot \rangle_{\mathbb{H}} \colon \mathbb{H} \times \mathbb{H} \to \mathbb{R}, \qquad \langle x, y \rangle_{\mathbb{H}} := \frac{1}{2}(x^* y + y^* x),$$

where $(a, b)^* := (\bar{a}, -b)$. With respect to the norm $\|\cdot\|_{\mathbb{H}}$ induced by this product, we have that $\mathbb{S}^3 \cong \{x \in \mathbb{H} : \|x\|_{\mathbb{H}} = 1\}$.

1. Explain why $\mathbb{S}^3$ is parallelizable and construct an explicit left-invariant global frame (X_1, X_2, X_3) for $\mathbb{S}^3$ with $X_1, X_2, X_3 \in \mathfrak{X}(\mathbb{S}^3)$ by representing $\mathbb{S}^3$ as unit quaternions. *Hint*: Exercise 2.62 and [129, Problem 8-6].
2. Show that ξ_0 corresponds to $0 \in \pi_3(\mathbb{S}^2)$ by using the global frame from the first part. It is now a deep result due to Eliashberg [83, Theorem 4.10.3] that up to isotopy, the class $0 \in \pi_3(\mathbb{S}^2)$ contains precisely two contact structures: The standard contact structure ξ_0, called ***tight contact structure***, and another contact structure, called ***overtwisted contact structure***.
3. Consider cylindrical coordinates (r, θ, z) on $\mathbb{R}^3$. Define $\alpha_1 \in \Omega^1(\mathbb{R}^3)$ by

$$\alpha_1 := \cos(r)dz + r\sin(r)d\theta.$$

Show that $\xi_1 := \ker\alpha_1$ is a contact structure and that $\xi_1 \neq \ker\lambda_0$. In fact, by a deep result in contact topology, the contact distributions $\ker\lambda_0$ and $\ker\alpha_1$ are not contactomorphic! In order to explain this result, consider the disc

$$D := \{z = 0, r \leq \pi\} \subseteq \mathbb{R}^3.$$

Show that the boundary ∂D belongs to ξ_1. This disc is called the ***standard overtwisted disc***. By [83, Theorem 4.6.36], the standard contact structure on $\mathbb{S}^3$ or $\mathbb{R}^3$ cannot contain such an overtwisted disc.

9.37 (Symmetric Ratio) Consider the symmetric ratio $\mathfrak{R}(\Sigma, \varphi)$ as defined in (9.4) for a φ-symmetric hypersurface $\Sigma \subseteq \mathbb{R}^{2n}$. Construct a φ-symmetric star-shaped domain $A \subseteq \mathbb{R}^{2n}$ such that the symmetric ratio $\mathfrak{R}(\partial A, \varphi)$ is arbitrarily large. *Hint:* See [119, Section 2] for inspiration.

9.38 (Lagrangian Intersections) The aim of this problem is to give an alternative proof of Proposition 9.7 for $n = 1$ based on a more elementary observation in [47] only using Floer–Gromov compactness established in Proposition 6.5.

1. First, we consider the finite-dimensional setting. Let (Q, m) be a compact Riemannian manifold, and let $f \in C^\infty(Q)$ be a Morse–Bott function. Denote by $C_1, \ldots, C_k$ the connected components of $\mathrm{Crit}(f)$ and define

$$\mathrm{gap}(f) := \min_{i \neq j} |f(C_i) - f(C_j)| \in [0, +\infty)$$

to be the minimal difference between the critical values of f. Moreover, for an arbitrary smooth function $g \in C^\infty(Q)$ consider its oscillation

$$\mathrm{osc}(g) := \max_{x\in Q} g(x) - \min_{x\in Q} g(x) \in [0, +\infty).$$

Show that if $\mathrm{osc}(g) < \mathrm{gap}(f)$ and $f + g$ is a Morse function, then we have that

$$\#\mathrm{Crit}(f+g) \geq \sum_{k=0}^{\dim Q} \dim \mathrm{H}_k(\mathrm{Crit}(f); \mathbb{Z}_2).$$

2. Consider the closed ball $B(r) \subseteq \mathbb{R}^2$ of radius $r > 0$ and let $\varphi_F \in \mathrm{Ham}_c(\mathbb{R}^2, \omega_0)$ be a Hamiltonian symplectomorphism. By [47, Main Theorem], we have that if the Hofer norm satisfies $\|F\| < \mathrm{Area}(B(r)) = \pi r^2$, then

$$\#(\partial B(r) \cap \varphi_F(\partial B(r))) \geq \dim \mathrm{H}_0(\mathbb{S}^1; \mathbb{Z}_2) + \dim \mathrm{H}_1(\mathbb{S}^1; \mathbb{Z}_2) = 2, \tag{9.5}$$

provided the intersection $\partial B(r) \cap \varphi_F(\partial B(r))$ is transverse. Show that if $c > \pi r^2$, then there does always exist a displacing Hamiltonian symplectomorphism φ_F with $\|F\| = c$. Conclude that the assumption on F for (9.5) is sharp and explicitly construct for every $c > \pi r^2$ such a time-dependent Hamiltonian function! Do the same thing for $c < \pi r^2$ such that (9.5) holds.
3. The aim of this part is to sketch a proof of (9.5) in a more general setting. A ***Lagrangian submanifold*** of a symplectic manifold (M^{2n}, ω) is defined to be a smooth embedded submanifold $L \subseteq M$ such that $\dim L = n$ and $\omega|_L = 0$. Consider the path space

$$\mathscr{P}_L M := \{\gamma \in C^\infty([0,1], M) : \gamma(0), \gamma(1) \in L\}.$$

For simplicity, we assume that $\omega = d\lambda$ is exact. For a Hamiltonian symplectomorphism $\varphi_F \in \mathrm{Ham}_c(M, \omega)$ we can consider the smooth family $(\mathscr{A}_\sigma)_{\sigma\in[0,1]}$ of symplectic action functionals

$$\mathscr{A}_\sigma : \mathscr{P}_L M \to \mathbb{R}, \qquad \mathscr{A}_\sigma(\gamma) := \int_0^1 \gamma^*\lambda - \sigma \int_0^1 F_t(\gamma(t))dt$$

Show that $\mathrm{Crit}(\mathscr{A}_\sigma) = L \cap \varphi_{\sigma F}(L)$ holds for all $\sigma \in [0,1]$ (Fig. 9.3). In particular, if L and $\varphi_F(L)$ intersect transversely, we have that both $\mathscr{A}_0$ and $\mathscr{A}_1$ are Morse–Bott functions by [129, Theorem 6.30 (b)]. The estimate (9.5) now follows from

$$\#(L \cap \varphi_F(L)) \geq \dim \mathrm{HM}(\mathscr{A}_1; \mathbb{Z}_2) = \dim \mathrm{HM}(\mathscr{A}_0; \mathbb{Z}_2) = \bigoplus_{k\geq 0} \dim \mathrm{H}_k(L; \mathbb{Z}_2),$$

Fig. 9.3 A typical element of $\mathrm{Crit}(\mathscr{A}_\sigma)$

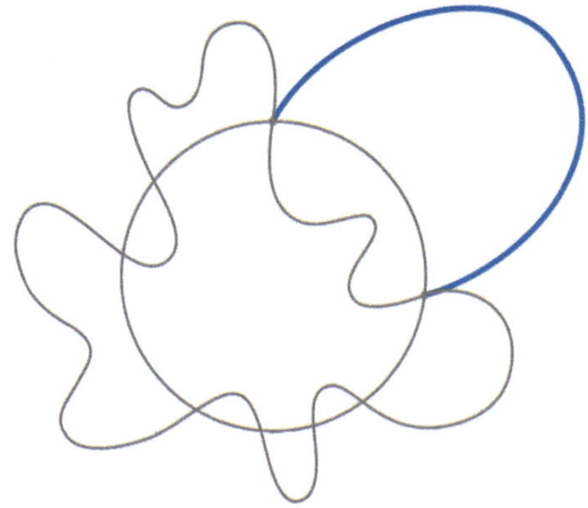

where for the invariance $\mathrm{HM}(\mathscr{A}_1;\mathbb{Z}_2) \cong \mathrm{HM}(\mathscr{A}_0;\mathbb{Z}_2)$ a suitable smallness assumption on $\|F\|$ is needed. Indeed, without any smallness assumption on the Hofer norm of F, compactness fails due to the existence of nonconstant bubbles and discs with boundary contained in the Lagrangian submanifold L by [136, Theorem 4.1.6]. Explain why we cannot use this argumentation for proving the estimate $e(B^{2n}(r)) \geq \pi r^2$ for $n \geq 2$!.

Basic Symplectic Topology

A

"The name 'complex group' formerly advocated by me in allusion to line complexes,... has become more and more embarrassing through collision with the word 'complex' in the connotation of complex number. Therefore, I propose to replace it with the Greek adjective 'symplectic' (which is also a name for a bone in the jaw of a fish)."

— *Hermann Weyl in* [196, p. 165]

The modern language of classical mechanics is provided by symplectic geometry, as discussed in [92]. For a first introduction to symplectic geometry, we recommend the book [64] and for a more extensive treatment [137]. For a historical overview, we suggest also [148] and for an introduction to Hamiltonian dynamics see [9,107,200] for a view towards global symplectic invariants. The main aim of this short appendix is to compare and contrast symplectic geometry with Riemannian geometry, arriving at the starting point of the modern era of symplectic geometry, the famous Gromov non-squeezing Theorem A.15.

A.1 Linear Symplectic Geometry

Definition A.1 (Nondegenerate Bilinear Form) Let V be a finite-dimensional real vector space and $\beta\colon V \times V \to \mathbb{R}$ a symmetric or antisymmetric bilinear map. The bilinear map β is said to be ***nondegenerate***, if the induced map

$$\beta\colon V \to V^*, \qquad \beta(x)(y) := \beta(x, y),$$

is an isomorphism, where $V^* := \{f\colon V \to \mathbb{R} \text{ linear}\}$ denotes the dual space of V.

Y. Bähni, *Lectures on Twisted Rabinowitz-Floer Homology*, Pathways in Mathematics, https://doi.org/10.1007/978-3-032-10673-5

Definition A.2 (Symplectic Vector Space) A ***symplectic vector space*** is defined to be a tuple (V, ω), where V is a finite-dimensional real vector space and $\omega\colon V \times V \to \mathbb{R}$ is a nondegenerate antisymmetric bilinear map, called a ***symplectic structure***.

Lemma A.3 *Let (V, ω) be a symplectic vector space. Then there exists an ordered basis $(x_1, \ldots, x_n, y_1, \ldots, y_n)$ of V with $2n = \dim V$, and the matrix representation of ω with respect to this ordered basis is given by*

$$\omega = \begin{pmatrix} 0 & I_n \\ -I_n & 0 \end{pmatrix} \in \mathbb{R}^{2n\times 2n}.$$

Proof For $\dim V = 0$ there is nothing to show, so assume that $\dim V \geq 1$. Thus, there exists a nonzero vector $x \neq 0$ in V. By the non-degeneracy of ω, the nonzero vector x admits a "friend" $y \neq 0$ in V such that $\omega(x, y) \neq 0$. Otherwise, the induced map $V \to V^*$ would not be an isomorphism, since the kernel is not trivial in that case. By setting $x_1 := x$ and $y_1 := \frac{y}{\omega(x,y)}$ we even get $\omega(x_1, y_1) = 1$. Set $U := \operatorname{span}(x_1, y_1)$. We claim $\dim U = 2$. In fact, if $y_1 = \lambda x_1$ for some $\lambda \in \mathbb{R}$, then $\omega(x_1, y_1) = 0$ by antisymmetry of ω contradicts $\omega(x_1, y_1) = 1$. Moreover, we claim that we have the decomposition

$$V = U \oplus U^{\perp}, \tag{A.1}$$

where

$$U^{\perp} := \{v \in V : \omega(v, u) = 0\ \forall u \in U\}$$

is the symplectic complement of U in V.

- $\underline{U \cap U^{\perp} = \{0\}}$. If $v \in U \cap U^{\perp}$, then we can write

  $$v = ax_1 + by_1 \in U$$

  for a suitable choice of $a, b \in \mathbb{R}$. As $v \in U^{\perp}$ we compute

  $$0 = \omega(v, x_1) = -b \qquad \text{and} \qquad 0 = \omega(v, y_1) = a,$$

  implying $v = 0$.
- $\underline{U + U^{\perp} = V}$. Every $v \in V$ can be decomposed as

  $$v = \underbrace{\omega(v, y_1)x_1 - \omega(v, x_1)y_1}_{=:u} + \underbrace{v - \omega(v, y_1)x_1 + \omega(v, x_1)y_1}_{=:u^{\perp}}.$$

  It is clear that $u \in U$ and $u^{\perp} \in U^{\perp}$ by

  $$\omega(u^{\perp}, x_1) = 0 \qquad \text{and} \qquad \omega(u^{\perp}, y_1) = 0.$$

Next, we observe that ω restricted to the symplectic complement $U^{\perp}$ is again nondegenerate. In fact, suppose that $v \in U^{\perp} \setminus \{0\}$ is such that there does not exist any $u \in U^{\perp}$ with $\omega(v, u) \neq 0$. This would imply $v \in (U^{\perp})^{\perp} = U$ by Exercise A.4 below. But $U \cap U^{\perp} = \{0\}$ and so $v = 0$ contradict our assumption that v is nonzero. So we can inductively apply the direct sum decomposition to achieve

$$V = \bigoplus_{i=1}^{n} U_i, \qquad U_i = \operatorname{span}(x_i, y_i).$$

By construction it holds that

$$\omega(x_i, x_j) = 0 = \omega(y_i, y_j) \qquad \text{and} \qquad \omega(x_i, y_i) = \delta_{ij},$$

giving rise to our desired basis, where δ_{ij} denotes the Kronecker-delta. □

Exercise A.4 Let (V, ω) be a symplectic vector space and $U \subseteq V$ a linear subspace.

1. Show the dimension formula $\dim U + \dim U^{\perp} = \dim V$.
2. Show that $(U^{\perp})^{\perp} = U$.

Exercise A.5 (Gram-Schmidt Algorithm) Let V be a finite-dimensional real vector space and $\langle\cdot,\cdot\rangle\colon V \times V \to \mathbb{R}$ an inner product on V, that is, a nondegenerate symmetric bilinear map.

1. By imitating the proof of Lemma A.3, show that V with $\dim V = n$ admits an ordered basis $(x_1, \ldots, x_n)$ such that the matrix representing the inner product $\langle\cdot,\cdot\rangle$ is given by $\langle\cdot,\cdot\rangle = I_n \in \mathbb{R}^{n\times n}$.
2. Explain why every finite-dimensional real vector space admits an inner product but not a symplectic structure, in general!

What the orthogonal group

$$\mathrm{O}(n) := \{A \in \mathbb{R}^{n\times n} : A^T A = I_n\},$$

means for an inner product space, does the symplectic group

$$\mathrm{Sp}(2n) := \{A \in \mathbb{R}^{2n\times 2n} : A^T J_0 A = J_0\},$$

mean for a symplectic vector space, where

$$J_0 := \begin{pmatrix} 0 & I_n \\ -I_n & 0 \end{pmatrix} \in \mathbb{R}^{2n\times 2n}.$$

Indeed, elements of the orthogonal group preserve the standard inner product on $\mathbb{R}^n$ since

$$\langle Ax, Ay\rangle = (Ax)^T Ay = x^T A^T Ay = x^T I_n y = x^T y = \langle x, y\rangle$$

for all $x, y \in \mathbb{R}^n$ and $A \in \mathrm{O}(n)$. Similarly, elements of the symplectic group preserve the standard symplectic structure

$$\omega_0(x, y) := x^T J_0 y \qquad \forall x, y \in \mathbb{R}^{2n}.$$

Indeed, we compute

$$(A^*\omega_0)(x, y) = \omega_0(Ax, Ay) = (Ax)^T J_0(Ay) = x^T A^T J_0 Ay = x^T J_0 y = \omega_0(x, y)$$

for all $x, y \in \mathbb{R}^{2n}$ and $A \in \mathrm{Sp}(2n)$. Of course, this symplectic structure comes from the standard symplectic form

$$\omega_0 := \sum_{i=1}^{n} dx_i \wedge dy_i \in \Omega^2(\mathbb{R}^{2n}). \tag{A.2}$$

Exercise A.6 Show $\det(A) = \pm 1$ for all $A \in \mathrm{Sp}(2n)$.

Exercise A.7 Show that $\mathrm{Sp}(2n)$ is a group with respect to matrix multiplication.

In fact, it does hold that $\mathrm{Sp}(2n) \subseteq \mathrm{SL}(2n)$, where

$$\mathrm{SL}(2n) := \{A \in \mathbb{R}^{2n\times 2n} : \det(A) = 1\}$$

denotes the special linear group.

Lemma A.8 (Liouville) *It holds that* $\det(A) = 1$ *for all* $A \in \mathrm{Sp}(2n)$.

Proof Let $A \in \mathrm{Sp}(2n)$. We observe that

$$\Omega_0 := \frac{\omega_0^n}{n!} = x_1 \wedge y_1 \wedge \cdots \wedge x_n \wedge y_n \in \Lambda^{2n}(\mathbb{R}^{2n})$$

is a volume form. Since $A^*\omega_0 = \omega_0$, we also have $A^*\Omega_0 = \Omega_0$. Hence

$$\Omega_0(Av_1, \dots, Av_{2n}) = \Omega_0(v_1, \dots, v_{2n}) \qquad \forall v_1, \dots, v_{2n} \in \mathbb{R}^{2n}.$$

But by [129, Proposition 14.9], we also have that

$$\Omega_0(Av_1, \dots, Av_{2n}) = \det(A)\Omega_0(v_1, \dots, v_{2n}) \qquad \forall v_1, \dots, v_{2n} \in \mathbb{R}^{2n},$$

implying $\det(A) = 1$. □

Exercise A.9 Define

$$\mathfrak{sp}(2n) := \{A \in \mathbb{R}^{2n\times 2n} : A^T J_0 + J_0 A = 0\},$$

and denote by

$$e^A := \sum_{k=0}^{\infty} \frac{A^k}{k!} \in \mathbb{R}^{2n\times 2n}$$

the usual matrix exponential.

1. Show that $\mathfrak{sp}(2n)$ is a real vector space and compute its dimension.
2. Show that $\det(e^A) = 1$ holds for all $A \in \mathfrak{sp}(2n)$.
3. Show that $e^A \in \mathrm{Sp}(2n)$ for all $A \in \mathfrak{sp}(2n)$.

Exercise A.10 Let $A \in \mathrm{Sp}(2n)$ be a symplectic matrix.

1. Show that if $\lambda \in \mathbb{C}$ is an eigenvalue of A, then so are λ^{-1}, $\bar{\lambda}$ and $\bar{\lambda}^{-1}$. Moreover, show that all these eigenvalues have the same algebraic multiplicity.
2. If -1 or 1 is an eigenvalue of A, then these eigenvalues have even multiplicity.

A.2 Symplectic Embeddings

Consider the symplectic vector space $(\mathbb{R}^{2n}, \omega_0)$ with the corresponding volume form

$$\Omega_0 = \frac{\omega_0^n}{n!} = dx_1 \wedge dy_1 \wedge \cdots \wedge dx_n \wedge dy_n \in \Omega^{2n}(\mathbb{R}^{2n}).$$

As the symplectic group $\mathrm{Sp}(2n)$ is contained in the special linear group, we have that

$$\Omega_0(Av_1, \ldots, Av_{2n}) = \Omega_0(v_1, \ldots, v_{2n}) \qquad \forall v_1, \ldots, v_{2n} \in \mathbb{R}^{2n}$$

for every matrix $A \in \mathrm{Sp}(2n)$. This observation motivates the following question.

Question A.11 Is every volume-preserving smooth map also symplectic?

For $n = 1$, we have $\Omega_0 = \omega_0$. Thus, in the plane $\mathbb{R}^2$ the notion of symplectic and volume-preserving linear maps coincides. However, for $n \geq 2$ the difference is huge and is the foundation of modern symplectic geometry!

Exercise A.12 Show that $\mathrm{SL}(2n)$ is a Lie group of dimension $4n^2 - 1$ with corresponding Lie algebra

$$\mathfrak{sl}(2n) := \{A \in \mathbb{R}^{2n\times 2n} : \operatorname{tr}(A) = 0\}.$$

Exercise A.13 Show that $\mathrm{Sp}(2n)$ is a Lie group of dimension $2n^2 + n$ with the corresponding Lie algebra $\mathfrak{sp}(2n)$.

Exercise A.14 Explain why $\mathrm{Sp}(2) = \mathrm{SL}(2)$ holds and that for $n \geq 2$ there are strictly more volume-preserving linear maps than symplectic ones.

Thus, we have answered Question A.11 for linear symplectic and volume-preserving maps. For nonlinear maps, the story is much more difficult. As a starting point, we consider symplectic embeddings as in [166]. A ***symplectic embedding*** is an injective smooth map $\varphi\colon U \to \mathbb{R}^{2n}$ defined in an open subset $U \subseteq \mathbb{R}^{2n}$ satisfying $\varphi^*\omega_0 = \omega_0$ for the standard symplectic form on $\mathbb{R}^{2n}$. For any open and connected subset $V \subseteq \mathbb{R}^{2n}$, one may ask if there exists a symplectic embedding $\varphi\colon U \to V$. If we only assume the embedding to preserve the volume form Ω_0, then the condition

$$\mathrm{Vol}(U) = \int_U \Omega_0 \leq \int_V \Omega_0 = \mathrm{Vol}(V) \tag{A.3}$$

is both necessary and sufficient by [166, Theorem B.1]. So in dimension 2, there is always a symplectic embedding from U into V if and only if the volume constraint (A.3) is satisfied. In higher dimensions, much stronger obstructions for the existence of symplectic embeddings occur. The easiest case is that of a closed ball

$$B^{2n}(r) := \left\{(x, y) \in \mathbb{R}^{2n} : \|x\|^2 + \|y\|^2 \leq r^2\right\}$$

of radius $r > 0$ and a closed cylinder

$$Z^{2n}(R) := \left\{(x, y) \in \mathbb{R}^{2n} : x_1^2 + y_1^2 \leq R^2\right\}$$

of radius $R > 0$. As the volume of the cylinder is infinite, there is always a volume-preserving embedding from $B^{2n}(r)$ to $Z^{2n}(R)$ irrespective of the radii by (A.3). So, you can always squeeze an arbitrarily large ball into a cylinder. In the symplectic case, this is no longer true and is considered the most geometric expression of the phenomenon of symplectic rigidity.

Theorem A.15 (Gromov's Nonsqueezing Theorem, [33, Theorem 1.1]) *For $n \geq 2$, there exists a symplectic embedding of the ball $B^{2n}(r)$ into the cylinder $Z^{2n}(R)$ if and only if $r \leq R$.*

Proof A proof via polyfold theory can be found in [33] and a reproduction of the original argument is provided in [97, pp. 175–176] or [84]. □

Gromov's nonsqueezing theorem A.15 is intimately connected to the notion of symplectic capacities. Generally speaking, these invariants measure the 2-dimensional size of a symplectic manifold and can be thought of as a symplectic area or a symplectic width, a terminology suggested by Gromov.

Definition A.16 (Intrinsic Symplectic Capacity, [137, p. 460], [166, Definition C.1]) An ***intrinsic symplectic capacity on*** $\mathbb{R}^{2n}$ is a map c which assigns to each subset $A \subseteq \mathbb{R}^{2n}$ a number $c(A) \in [0, +\infty]$ such that the following three properties hold:

1. (***Monotonicity***) If there exists a symplectic embedding φ of an open neighbourhood of A in $\mathbb{R}^{2n}$ with $\varphi(A) \subseteq B$, then $c(A) \leq c(B)$.
2. (***Conformality***) $c(\lambda A) = \lambda^2 c(A)$ for all $\lambda \in \mathbb{R}$.
3. (***Normalisation***) It holds that

$$c(B^{2n}(r)) = c(Z^{2n}(r)) = \pi r^2 \qquad \forall r > 0.$$

Remark A.17 For all measureable $A \subseteq \mathbb{R}^{2n}$ define $c(A) := \operatorname{Vol}(A)^{1/n} \in [0, +\infty]$. Then c satisfies the first and second condition defining an intrinsic symplectic capacity. Moreover, $c(B^{2n}(r))$ is always finite, but $c(Z^{2n}(r)) = +\infty$ for all $r > 0$ if $n \geq 2$. Thus, volume is never an intrinsic symplectic capacity, and the requirement $c(Z^{2n}(r)) < +\infty$ allows for the detection of symplectic phenomena not connected to volume!

Lemma A.18 *The existence of a symplectic capacity on* $\mathbb{R}^{2n}$ *is equivalent to Gromov's nonsqueezing theorem A.15.*

Proof Assume that Gromov's Nonsqueezing Theorem A.15 holds and define

$$\bar{c}_B(A) := \sup\{\pi r^2 > 0 : \exists \text{ symplectic embedding } B^{2n}(r) \hookrightarrow A\},$$

and

$$\bar{c}_Z(A) := \inf\{\pi r^2 > 0 : \exists \text{ symplectic embedding } \varphi(A) \hookrightarrow Z^{2n}(r)\},$$

for any subset $A \subseteq \mathbb{R}^{2n}$. Then we check that both $\bar{c}_B$ and $\bar{c}_Z$ are intrinsic symplectic capacities. Conversely, assume that c is any intrinsic symplectic capacity and that there exists a symplectic embedding $B^{2n}(r) \hookrightarrow Z^{2n}(r)$ for some $r, R > 0$. Then by normalisation and monotonicity we compute

$$\pi r^2 = c(B^{2n}(r)) \leq c(Z^{2n}(R)) = \pi R^2 \qquad \Rightarrow \qquad r \leq R.$$

This finishes the proof. □

Remark A.19 Gromov's nonsqueezing theorem A.15 has many important applications [97, 165–181]. For example, a C^0-rigidity result can be proved for symplectomorphisms on $\mathbb{R}^{2n}$. Consider a sequence (φ_j) in $\mathrm{Symp}(\mathbb{R}^{2n}, \omega_0)$ such that

$$\varphi_j \xrightarrow{C^0_{\mathrm{loc}}} \varphi \in C^0(\mathbb{R}^{2n}, \mathbb{R}^{2n}), \qquad j \to \infty.$$

If φ is smooth, then it also belongs to $\mathrm{Symp}(\mathbb{R}^{2n}, \omega_0)$. In particular, the group of symplectic diffeomorphisms of a compact symplectic manifold is C^0-closed in the group of all diffeomorphisms of the manifold. Hence, volume-preserving diffeomorphisms cannot be approximated by symplectomorphisms! If we assume that

$$\varphi_j \xrightarrow{C^1_{\mathrm{loc}}} \varphi \in C^1(\mathbb{R}^{2n}, \mathbb{R}^{2n}), \qquad j \to \infty,$$

the result would be obvious. Indeed, for every j we have that

$$D\varphi_j(x)^T J_0 D\varphi_j(x) = J_0 \qquad \forall x \in \mathbb{R}^{2n},$$

as $D\varphi_j(x) \in \mathrm{Sp}(2n)$ by assumption. Consequently

$$D\varphi(x)^T J_0 D\varphi(x) = \lim_{j\to\infty} D\varphi_j(x)^T J_0 D\varphi_j(x) = J_0,$$

for all $x \in \mathbb{R}^{2n}$ and thus φ is symplectic. That φ is still symplectic if it is differentiable and the limit is only with respect to the C^0_{loc}-topology is by no means obvious and a remarkable rigidity result of symplectic maps [166, p. 5]! An elegant proof based on the existence of a symplectic capacity can be found in [107, Section 2.2].

Solutions

In this chapter, we give solutions or references to the relevant literature to all exercises found in this book and some of the problems. You should only look at the solutions if you have put in some significant effort to solve the exercises and problems yourself or if you want to check your answer!

Exercises and Problems of Chap. 1

1.1 One possible route is depicted in Fig. 1.3.

1.2

1. Following [141, Section 2.7], the map

$$S^*\mathbb{S}^2 \to \mathrm{SO}(3), \qquad (q, p) \mapsto (q, p, q \times p)$$

   is a diffeomorphism.
2. See [103, Proposition 1.17] or [103, Proposition 1.19], and show that the homeomorphisms are actually diffeomorphisms.

1.4 We compute

$$\begin{aligned} i_{X_{H_\Sigma}} d\lambda_0 &= \sum_{j=1}^{n} \left(i_{X_{H_\Sigma}} dx_j \wedge dy_j - dx_j \wedge i_{X_{H_\Sigma}} dy_j \right) \\ &= \sum_{j=1}^{n} \left(\frac{\partial H_\Sigma}{\partial x_j} dx_j + \frac{\partial H_\Sigma}{\partial y_j} dy_j \right) \\ &= dH_\Sigma. \end{aligned}$$

From this, we conclude that

$$dH_\Sigma(X_{H_\Sigma}) = d\lambda_0(X_{H_\Sigma}, X_{H_\Sigma}) = 0$$

Y. Bähni, *Lectures on Twisted Rabinowitz-Floer Homology*, Pathways in Mathematics, https://doi.org/10.1007/978-3-032-10673-5

by antisymmetry of the symplectic form $d\lambda_0$. Thus, we conclude

$$X_{H_\Sigma}(x) \in T_x\Sigma = \ker dH_\Sigma(x) \qquad \forall x \in \Sigma.$$

1.5 Let $\Sigma = H_\Sigma^{-1}(0) \subseteq \mathbb{R}^{2n}$ be a star-shaped hypersurface. Then the sublevel set $W_\Sigma := H_\Sigma^{-1}((-\infty, 0])$ is a properly embedded submanifold of dimension $2n$ with boundary $\partial W = \Sigma$ by [129, Proposition 5.47 (a)]. Let $z \in \mathbb{C}^n \setminus \{0\}$ satisfy

$$0 < \|z\| \leq \rho\left(\frac{z}{\|z\|}\right).$$

Then $z \in W_\Sigma$ and also the ray σz, $\sigma \in [0, 1]$, is contained in W_Σ since

$$H_\Sigma(\sigma z) = \begin{cases} h\left(\sigma^2\|z\|^2\rho(\frac{z}{\|z\|})^{-2} - 1\right), & z \neq 0, \\ -\frac{1}{2}, & z = 0, \end{cases}$$

is a monotonically increasing function for all $\sigma > 0$ and $H_\Sigma(0) = -\frac{1}{2}$.

1.6 Writing the vector field in the complex coordinates $z_j = x_j + iy_j$ yields

$$R = 2i\sum_{j=1}^{n}\left(\bar{z}_j\frac{\partial}{\partial\bar{z}_j} - z_j\frac{\partial}{\partial z_j}\right)$$

with

$$\frac{\partial}{\partial z_j} := \frac{1}{2}\left(\frac{\partial}{\partial x_j} - i\frac{\partial}{\partial y_j}\right) \qquad \text{and} \qquad \frac{\partial}{\partial \bar{z}_j} := \frac{1}{2}\left(\frac{\partial}{\partial x_j} + i\frac{\partial}{\partial y_j}\right).$$

Thus the periodic flow is given by

$$\mathbb{R} \times \mathbb{S}^{2n-1} \to \mathbb{S}^{2n-1},\ (t, z) \mapsto e^{-2it}z.$$

1.7 Similar to Exercise 1.6, in complex coordinates $z_j = x_j + iy_j$ the vector field is given by

$$2i\sum_{j=1}^{n}\frac{\pi}{a_j}\left(\bar{z}_j\frac{\partial}{\partial\bar{z}_j} - z_j\frac{\partial}{\partial z_j}\right).$$

Thus the flow is

$$\mathbb{R} \times E(a_1, \ldots, a_n) \to E(a_1, \ldots, a_n), \quad (t, z) \mapsto \left(e^{-2i\pi t/a_1}z_1, \ldots, e^{-2\pi it/a_n}z_n\right).$$

Now by [166, p. 26], the periodic orbits depend on the numbers a_j. If the numbers a_j are linearly independent over $\mathbb{Z}$, the only periodic orbits are of the form

$$\mathbb{R} \to E(a_1, \ldots, a_n), \qquad t \mapsto \left(0, \ldots, e^{-2\pi i t/a_j} z_j, \ldots, 0\right).$$

The general case is slightly more difficult.

Exercises and Problems of Chap. 2

2.3 Let M be a smooth manifold. Which statements are correct?

☒ $L_X(\omega\wedge\eta) = L_X\omega\wedge\eta+\omega\wedge L_X\eta$ for all $\omega \in \Omega^k(M)$, $\eta \in \Omega^l(M)$ and $X \in \mathfrak{X}(M)$.
☐ $L_X(\omega \wedge \eta) = L_X\omega \wedge \eta + (-1)^{kl}\omega \wedge L_X\eta$ for all $\omega \in \Omega^k(M)$, $\eta \in \Omega^l(M)$ and all vector fields $X \in \mathfrak{X}(M)$.
☐ $d(\omega \wedge \eta) = d\omega \wedge \eta + \omega \wedge d\eta$ for all $\omega \in \Omega^k(M)$ and $\eta \in \Omega^l(M)$.
☒ $d(\omega \wedge \eta) = d\omega \wedge \eta + (-1)^{kl}\omega \wedge d\eta$ for all $\omega \in \Omega^k(M)$ and $\eta \in \Omega^l(M)$.

2.4 Let M be a smooth manifold. Which statements are correct?

☐ $i_{[X,Y]} = L_Y i_X - i_X L_Y$ for all $X, Y \in \mathfrak{X}(M)$.
☒ $i_{[X,Y]} = L_X i_Y - i_Y L_X$ for all $X, Y \in \mathfrak{X}(M)$.
☐ $i_X(\omega\wedge\eta) = i_X\omega\wedge\eta+\omega\wedge i_X\eta$ for all $\omega \in \Omega^k(M)$, $\eta \in \Omega^l(M)$ and $X \in \mathfrak{X}(M)$.
☒ $i_X(\omega \wedge \eta) = i_X\omega \wedge \eta + (-1)^{kl}\omega \wedge i_X\eta$ for all $\omega \in \Omega^k(M)$, $\eta \in \Omega^l(M)$ and all vector fields $X \in \mathfrak{X}(M)$.

2.6 Let (M^{2n}, ω) be a symplectic manifold. Which statements are correct?

☐ For the second de Rham cohomology group, we have $H^2_{\mathrm{dR}}(M) \neq 0$.
☒ If M is compact, then $[\omega] \neq 0$ in $H^2_{\mathrm{dR}}(M)$, which means that ω is not exact.
☐ Every sphere $\mathbb{S}^{2n}$ admits a symplectic form for $n \geq 1$.
☒ The cotangent bundle $M = T^*Q$ of any smooth manifold Q (orientable or not!) admits a symplectic form and is therefore orientable.

2.9 Let ϕ denote the global flow of the Liouville vector field $X = X_\lambda \in \mathfrak{X}(M)$ associated with a compact exact symplectic manifold (M, λ). Then we compute

$$\begin{aligned}
\frac{d}{dt}\phi_t^*\lambda &= \phi_t^* L_X\lambda && \text{(Fisherman's Formula in Lemma 2.2)}\\
&= \phi_t^*(i_X d\lambda + d i_X\lambda) && \text{(Cartan's Magic Formula in Lemma 2.1)}\\
&= \phi_t^*(\lambda + d i_X i_X d\lambda) && \text{(Definition of the Liouville vector field)}\\
&= \phi_t^*\lambda && (i_X \circ i_X = 0 \text{ by antisymmetry}).
\end{aligned}$$

Thus, we get the differential equation

$$\frac{d}{dt}\phi_t^*\lambda = \phi_t^*\lambda \qquad \forall t \in \mathbb{R}.$$

This is only satisfied if

$$\phi_t^*\lambda = e^t\lambda \qquad \forall t \in \mathbb{R}.$$

2.13 This immediately follows from the nondegeneracy of the symplectic form, as this induces a tangent-cotangent-bundle isomorphism

$$\mathfrak{X}(M) \to \Omega^1(M), \qquad X \mapsto i_X\omega.$$

Taking the inverse yields for every differential form in $\Omega^1(M)$ a unique vector field. In particular, we can do that for $dH \in \Omega^1(M)$ for any smooth function $H \in C^\infty(M)$.

2.15 In Darboux coordinates (q_j, p_j) we have

$$\omega = \sum_{j=1}^{n} dp_j \wedge dq_j.$$

Thus, the Hamiltonian vector field X_H is locally given by

$$X_H = \sum_{j=1}^{n} \left(\frac{\partial H}{\partial p_j}\frac{\partial}{\partial q_j} - \frac{\partial H}{\partial q_j}\frac{\partial}{\partial p_j} \right).$$

Consequently, integral curves of X_H satisfy ***Hamilton's equations***

$$\dot{q} = \frac{\partial H}{\partial p} \qquad \text{and} \qquad \dot{p} = -\frac{\partial H}{\partial q}.$$

Suppose that H is a mechanical Hamiltonian function. Then the Hamiltonian equations are given by

$$\ddot{q} = -\frac{\partial V}{\partial q},$$

which is precisely ***Newton's second law***. It is interesting to note that the form of the Hamiltonian vector field in coordinates strongly depends on the existence of Darboux coordinates in Lemma 2.5, which depends on the assumption that any symplectic form is closed. Thus, we may think of $d\omega = 0$ as an analytical condition, whereas the nondegeneracy of a symplectic form ω is an algebraic condition. Finally, we show that the flow of a Hamiltonian vector field preserves volume. It is enough to consider the case $M = \mathbb{R}^{2n}$. For every bounded open subset $U \subseteq \mathbb{R}^{2n}$, we compute using [178, Lemma 8.8]

$$\begin{aligned}\frac{d}{dt}\,\mathrm{Vol}(\phi_t^H(U)) &= \int_{\phi_t^H(U)} \mathrm{div}\, X_H(x, y)dxdy \\ &= \sum_{j=1}^{n} \int_{\phi_t^H(U)} \left(\frac{\partial}{\partial x_j}\frac{\partial H}{\partial y_j} - \frac{\partial}{\partial y_j}\frac{\partial H}{\partial x_j} \right) dxdy \\ &= 0\end{aligned}$$

by the interchangeability of second derivatives of a smooth function. Consequently, the volume $\mathrm{Vol}(\phi_t^H(U))$ is constant and coincides with $\mathrm{Vol}(\phi_0^H(U)) = \mathrm{Vol}(U)$.

2.16 Let ϕ denote the local flow of a Hamiltonian system (M, ω, H). Then we compute

$$\begin{aligned}\frac{d}{dt}\phi_t^*\omega &= \phi_t^* L_{X_H}\omega && \text{(Fisherman's Formula in Lemma 2.2)} \\ &= \phi_t^*(i_{X_H}d\omega + di_{X_H}\omega) && \text{(Cartan's Magic Formula in Lemma 2.1)} \\ &= \phi_t^*(di_{X_H}\omega) && (\omega \text{ is closed by definition}) \\ &= -\phi_t^*(ddH) && \text{(Definition of the Hamiltonian vector field} X_H) \\ &= 0 && (d \circ d = 0).\end{aligned}$$

Hence, $\phi_t^*\omega$ is constant for any time t in a neighbourhood of 0. But $\phi_0^*\omega = \mathrm{id}_M^*\,\omega = \omega$, and so the statement follows. We again see why a symplectic form is assumed to be closed and nondegenerate. First, nondegeneracy implies that there always exists a unique Hamiltonian vector field associated with any Hamiltonian function, and second, that the flow of any such Hamiltonian vector field is a local symplectomorphism. Using the computation in the proof of [129, Proposition 16.33], we calculate for any compact regular domain $U \subseteq M^n$

$$\frac{d}{dt}\,\mathrm{Vol}(\phi_t(U)) = \int_{\phi_t(U)} \omega^n = \int_U (\phi_t)^*\,\omega^n = \frac{d}{dt}\int_U \left(\phi_t^*\omega\right)^n = \frac{d}{dt}\int_U \omega^n = 0.$$

2.23

1. For all $f, g \in C^\infty(M)$ we compute

$$\begin{aligned}i_{[X_f, X_g]}\omega &= L_{X_f} i_{X_g}\omega - i_{X_g} L_{X_f}\omega \\ &= -L_{X_f}dg - i_{X_g}(i_{X_f}d\omega + di_{X_f}\omega) \\ &= -(i_{X_f}ddg + di_{X_f}dg) + i_{X_g}(ddf)\end{aligned}$$

$$
\begin{aligned}
&= -di_{X_f}dg \\
&= di_{X_f}i_{X_g}\omega \\
&= -d\{f, g\}.
\end{aligned}
$$

2. The bilinearity and antisymmetry of the Poisson bracket is immediate from the definition. Moreover, the Leibniz rule follows from the computation

$$
\begin{aligned}
\{f, gh\} &= \omega(X_f, X_{gh}) \\
&= \omega(X_f, hX_g + gX_h) \\
&= h\omega(X_f, X_g) + g\omega(X_f, X_h) \\
&= h\{f, g\} + g\{f, h\}
\end{aligned}
$$

for all $f, g, h \in C^\infty(M)$. For proving the Jacobi identity, we compute

$$
\begin{aligned}
\{f, \{g, h\}\} &= \omega(X_f, X_{\{g,h\}}) \\
&= \omega(X_f, [X_g, X_h]) \\
&= i_{X_f}\omega[X_g, X_h] \\
&= -df[X_g, X_h] \\
&= -[X_g, X_h]f \\
&= -X_gX_hf + X_hX_gf \\
&= -X_g\{h, f\} + X_h\{g, f\} \\
&= -\{g, \{h, f\}\} + \{h, \{g, f\}\} \\
&= -\{g, \{h, f\}\} - \{h, \{f, g\}\}.
\end{aligned}
$$

2.26 Lemma 2.24 immediately follows from formula (2.5) and Lemma 2.25 follows from Lemma 2.17.

2.29 Let $H \in C^\infty(M \times [0, 1])$ be a time-dependent Hamiltonian function. There are two physical interpretations of H that are important to us:

- Time-dependent Hamiltonian functions are perturbations of Hamiltonian systems, since the total energy changes. For example, in the restricted three-body problem, a perturbation could be a comet passing by and interfere with the orbit of the satellite. This physical phenomenon leads to the notion of leaf-wise intersection points, as in [16] considered later in this book.
- Time-dependent Hamiltonian functions describe smooth perturbations of energy surfaces. For example, think of the sphere $\mathbb{S}^{2n-1}$ deformed into a star-shaped

hypersurface in $\mathbb{C}^n$. This leads to the notion of homotopies of regular energy hypersurfaces as in Definition 4.26.

2.31 See [102, Theorem 3.3] and [102, Theorem 3.4].

2.32 By the Morse lemma, we get

$$H(q,p) = f\left((2k+1)\pi, 0\right) + \frac{1}{2}\operatorname{Hess} H_{((2k+1)\pi,0)}\left((q,p),(q,p)\right) = \frac{1}{2}\left(p^2 - q^2\right).$$

Thus, the level sets are hyperbolic paraboloids.

2.33

1. Using Hamilton's equations in coordinates from Exercise 2.15, we get

$$\dot{q} = p \qquad \text{and} \qquad \dot{p} = -\sin(q).$$

   Combining these equations yields

$$\ddot{q} = -\sin(q).$$

2. The resulting force is given as the tangential component of the gravitational force, and thus $F = -\sin(q)$. Using Newton's second law yields

$$\ddot{q} = F = -\sin(q).$$

In this case, both methods are relatively easy to handle. However, computing the resulting force in a general physical system is usually hard, and in some cases, such as the brachistochrone problem, practically impossible. The drawback of the Hamiltonian equations is that there is no way to model, for example, a mathematical pendulum with friction, and deriving the equations from a variational principle is conceptually challenging.

2.37 The equation of motion modelling linear friction is given by

$$\ddot{q} = -\mu\dot{q} - \sin q, \qquad \mu > 0.$$

This equation is equivalent to the first order system

$$\begin{pmatrix} \dot{q} \\ \dot{p} \end{pmatrix} = \begin{pmatrix} p \\ -\mu p - \sin(q) \end{pmatrix}.$$

This system yields the Hamiltonian function

$$H\colon \mathbb{R}^2 \to \mathbb{R}, \qquad H(q,p) = \frac{1}{2}p^2 + \mu q p - \cos(q),$$

whose corresponding Hamiltonian equations do not give rise to the right equation for the acceleration $\ddot{q}$. Even by introducing a time-dependent Hamiltonian function

$$H\colon \mathbb{R}^2 \times \mathbb{R} \to \mathbb{R}, \qquad H(q,p) := e^{-\mu t}\left(\frac{1}{2}p^2 + (1-\cos(q))\right)$$

does not give the right equation of motion as

$$\dot{q} = \frac{\partial H}{\partial p} = e^{-\mu t} p \qquad \text{and} \qquad \dot{p} = -\frac{\partial H}{\partial q} = -e^{-\mu t}\sin(q)$$

implies

$$\ddot{q} = -\mu e^{-\mu t} p + e^{-\mu t}\dot{p} = -\mu \dot{q} - e^{-2\mu t}\sin(q).$$

Therefore, Hamilton's formalism is of limited use for studying systems with friction, but fortunately we are interested in celestial mechanics!

2.45 The inequality $e(A) \leq e(B)$ for subsets $A \subseteq B \subseteq M$ immediately follows from the definition of the displacement energy as an infimum. Let $\varphi \in \mathrm{Symp}(M,\omega)$ be a symplectomorphism. To show that $e(A) = e(\varphi(A))$ for all subsets $A \subseteq M$, we first note that if $\varphi_F \in \mathrm{Ham}_c(M,\omega)$ is a displacing Hamiltonian sympletomorphism for A, then it is also a displacing Hamiltonian symplectomorphism for $\varphi(A)$. In fact, we have $\varphi_F \circ \varphi = \varphi \circ \varphi_{F\circ\varphi}$. Consequently, we compute

$$\varnothing = \varphi(\varphi_F(A)) \cap \varphi(A) = \varphi_{F\circ\varphi}(\varphi(A)) \cap \varphi(A).$$

Moreover, one readily sees from the definition of the Hofer norm that $\|F \circ \varphi\| = \|F\|$. This implies that $e(\varphi(A)) = e(A)$ for all subsets $A \subseteq M$.

2.54 In the spatial case $n = 3$ we have that

$$e_1 := \begin{pmatrix} 0 & 0 & 0 \\ 0 & 0 & 1 \\ 0 & -1 & 0 \end{pmatrix}, \qquad e_2 := \begin{pmatrix} 0 & 0 & 1 \\ 0 & 0 & 0 \\ -1 & 0 & 0 \end{pmatrix} \qquad \text{and} \qquad e_3 := \begin{pmatrix} 0 & 1 & 0 \\ -1 & 0 & 0 \\ 0 & 0 & 0 \end{pmatrix}$$

is a basis for $\mathfrak{so}(3)$. The integrals of motion $\mu(e_j)$ are precisely the components L_j of the ***angular momentum***

$$L\colon T^*(\mathbb{R}^3 \setminus \{0\}) \to \mathbb{R}^3, \qquad L(q,p) := q \times p.$$

2.58 The crucial portion of the Eilenberg–Chevalley cochain complex is given by

$$\ldots \longrightarrow \Lambda^2\mathfrak{sl}_2(\mathbb{R}) \xrightarrow{\ d\ } \Lambda^3\mathfrak{sl}_2(\mathbb{R}) \longrightarrow 0.$$

We compute

$$\begin{aligned} df(e_1, e_2, e_3) &= e_1 f(e_2, e_3) - e_2 f(e_1, e_3) + e_3 f(e_1, e_2) \\ &\quad - f([e_1, e_2], e_3) + f([e_1, e_3], e_2) - f([e_2, e_3], e_1) \\ &= -2f(e_2, e_3) - 2f(e_3, e_2) - f(e_1, e_1) \\ &= -2f(e_2, e_3) + 2f(e_2, e_3) \\ &= 0. \end{aligned}$$

It follows that

$$\mathrm{H}^3(\mathfrak{sl}_2(\mathbb{R}); \mathbb{R}) \cong \Lambda^3 \mathfrak{sl}_2(\mathbb{R}) \cong \mathbb{R}.$$

2.62

1. Assume that $\mathbb{S}^n$ is a Lie group for $n \geq 2$ even. By [129, Corollary 8.39], every Lie group is parallelisable and thus admits a smooth global frame, that is, an ordered n-tuple of nowhere-vanishing linearly independent vector fields $(X_1, \ldots, X_n)$ on the sphere $\mathbb{S}^n$ that spans $T\mathbb{S}^n$. But by the Hairy Ball Theorem [129, Problem 16-6], there exists a vector field vanishing nowhere on $\mathbb{S}^n$ if and only if n is odd. Hence, the sphere $\mathbb{S}^n$ cannot be a Lie group for $n \geq 2$ even.
2. First, both $\mathbb{S}^0 \cong \mathbb{Z}_2$ and $\mathbb{S}^1 \subseteq \mathbb{C}$ admit a Lie group structure. Assume that the sphere $\mathbb{S}^n$ is a Lie group for $n \geq 3$ odd. As $\mathbb{S}^n$ is compact and simply connected, it follows $\mathrm{H}^*(\mathfrak{g}; \mathbb{R}) \cong \mathrm{H}^*_{\mathrm{dR}}(\mathbb{S}^n)$ for the Lie algebra $\mathfrak{g} = T_1\mathbb{S}^n$. This Lie algebra is semi-simple since $\mathbb{S}^n$, $n \geq 3$, is simply-connected and thus has trivial fundamental group. But by (2.9), we must have that

$$0 \neq \mathrm{H}^3(\mathfrak{g}; \mathbb{R}) \cong \mathrm{H}^3_{\mathrm{dR}}(\mathbb{S}^n) \cong \begin{cases} \mathbb{R}, & n = 3, \\ 0, & n \neq 3, \end{cases}$$

by [129, Theorem 17.21]. This implies that only $\mathbb{S}^3$ admits a Lie group structure for $n \geq 3$ odd.
3. The antipodal action $x \mapsto -x$ gives a free $\mathbb{Z}_2$-action on $\mathbb{S}^n$ for all $n \geq 0$. That there are no others if n is even is proven in [104, Proposition 2.29] based on degree considerations.

2.63

- A Hamiltonian system is a triple (M, ω, H), where (M, ω) is a symplectic manifold and $H \in C^\infty(M)$ is a classical observable.
- The dynamics of a Hamiltonian system is governed by the evolution equation

$$\frac{d}{dt} f \circ \phi_t = \{H, f\} \circ \phi_t,$$

where ϕ denotes the flow of the Hamiltonian vector field X_H and $\{\cdot, \cdot\}$ is the Poisson bracket.

- In Darboux coordinates, the integral curves of the Hamiltonian vector field satisfy the Hamiltonian equations

$$\dot{q} = \frac{\partial H}{\partial p} \qquad \text{and} \qquad \dot{p} = -\frac{\partial H}{\partial q}.$$

- If G is a symmetry group, then the integrals of motions are given by $\mu(\xi)$ for every $\xi \in \mathfrak{g} = T_eG$ by Noether's theorem, where $\mu\colon \mathfrak{g} \to C^\infty(M)$ is the momentum map.
- The Hamiltonian formalism is unable to model dissipative systems as the mathematical pendulum with friction. However, the Hamiltonian formalism can be used to analyse problems in celestial mechanics qualitatively, as the Kepler problem or the restricted three body problem. Moreover, the abstract formulation of classical mechanics using the Hamiltonian formalism can also be used in modern physics in the formulation of quantum mechanics.

2.64 We show only the second part. Denote the induced action by

$$\Psi\colon G \times T^*M \to T^*M, \qquad \Psi_g := D_*\psi_g.$$

By the first part and the Momentum Lemma 2.57 we conclude that Ψ is Hamiltonian and Poisson with momentum map

$$\begin{aligned}
\mu(\xi)(q, p) &= \lambda_{(q,p)}\left(\widehat{\xi}(q, p)\right) \\
&= p\left(D\pi\left(\widehat{\xi}(q, p)\right)\right) \\
&= p\left(\left.\frac{d}{dt}\right|_{t=0} \pi\left(\Psi_{\exp(-t\xi)}(q, p)\right)\right) \\
&= p\left(\left.\frac{d}{dt}\right|_{t=0} \psi_{\exp(-t\xi)}(\pi(q, p))\right) \\
&= p\left(\left.\frac{d}{dt}\right|_{t=0} \psi_{\exp(-t\xi)}(q)\right) \\
&= p(\widehat{\xi})
\end{aligned}$$

for all $(q, p) \in T^*M$ and $\xi \in \mathfrak{g}$.

2.66 We will only elaborate on the second part. Symmetries of Hamiltonian systems partially allow us to solve the corresponding Hamiltonian equations. If there are many, as in the case of integrable Hamiltonian systems, one is also able to completely solve them quantitatively. This is mainly due to the fact that in the completely integrable case there exist so-called action-angle coordinates trivialising the flow of the Hamiltonian vector field giving rise to Liouville tori. See, for example, [63, Chapter IX] or [77, Theorem 3.4.1]. The idea of perturbation then leads to KAM theory as in [22, Section 6.3]. However, checking integrability in a general Hamiltonian system is no easy matter and is the focus of active research.

Exercises and Problems of Chap. 3

3.9 If m is a Riemannian metric, we compute using (3.1)

$$\begin{aligned}\tau_L(\dot{q})(v) &= \frac{1}{2}\frac{d}{d\varepsilon}\bigg|_{\varepsilon=0} \|\dot{q}+\varepsilon v\|^2 + \frac{d}{d\varepsilon}\bigg|_{\varepsilon=0} \theta_q(\dot{q}+\varepsilon v) \\ &= \frac{1}{2}\frac{d}{d\varepsilon}\bigg|_{\varepsilon=0} m_q(\dot{q}+\varepsilon v, \dot{q}+\varepsilon v) + \theta_q(v) \\ &= m_q(\dot{q}, v) + \theta_q(v).\end{aligned}$$

We claim that τ_L is surjective. As $\dim TQ = \dim T^*Q$ it then follows that τ_L is a vector bundle isomorphism. Let $(q, p) \in T^*Q$. Then there exists $(q, \dot{q}) \in TQ$ such that $m_q(\dot{q}, \cdot) = p(\cdot) - \theta_q(\cdot)$ since $(q, \dot{q}) \mapsto m_q(\dot{q}, \cdot)$ is a vector bundle isomorphism according to the definition of a Riemannian metric. Hence we compute

$$\tau_L(\dot{q}) = m_q(\dot{q}, \cdot) + \theta_q(\cdot) = p(\cdot).$$

The physical interpretation of a mechanical Lagrangian function L is kinetic *minus* potential energy.

3.12 Let $(q, p) \in T^*Q$ and pick $(q, \dot{q}) \in TQ$ such that

$$m_q(\dot{q}, \cdot) = p(\cdot) - \theta_q(\cdot) \qquad \Leftrightarrow \qquad \tau_L^{-1}(q, p) = (q, \dot{q})$$

as in Exercise 3.9. Then we compute

$$\begin{aligned}H_L(q, p) &= E_L\left(\tau_L^{-1}(q, p)\right) \\ &= E_L(q, \dot{q}) \\ &= \tau_L(\dot{q})(\dot{q}) - L(q, \dot{q}) \\ &= m_q(\dot{q}, \dot{q}) + \theta_q(\dot{q}) - \frac{1}{2}\|\dot{q}\|_m^2 - \theta_q(\dot{q}) + V(q) \\ &= \frac{1}{2}\|\dot{q}\|_m^2 + V(q) \\ &= \frac{1}{2}\|p - \theta_q\|_{m^*}^2 + V(q).\end{aligned}$$

The physical interpretation of H_L is kinetic *plus* potential energy, so H_L is just the total energy of the system, with the coordinate q being the position and p being the momentum.

3.14 Locally we compute

$$L(q,\dot{q}) = \frac{1}{2}\sum_{i,j=1}^{n} m_{ij}(q)\dot{q}_i\dot{q}_j + \sum_{i=1}^{n}\theta_i\dot{q}_i - V(q).$$

Therefore, we get

$$\sum_{i,j=1}^{n} \frac{\partial^2 L}{\partial\dot{q}_i\partial\dot{q}_j}(q,\dot{q})v_iv_j = \frac{1}{2}\sum_{i,j=1}^{n} m_{ij}(q)v_iv_j > 0$$

for all $(q,\dot{q}) \in TM$ and $v := \sum_{i=1}^{n} v_i\partial_i \in T_qQ$ such that $v \neq 0$, as m is positive-definite by definition of a Riemannian metric. A magnetic Lagrangian function L is fibrewise supercoersive since for all $(q,\dot{q}) \in TQ$ we compute

$$\lim_{\|\dot{q}\|_m\to\infty} \frac{L(q,\dot{q})}{\|\dot{q}\|_m} = \lim_{\|\dot{q}\|_m\to\infty}\left(\frac{1}{2}\|\dot{q}\|_m + \theta_q\left(\frac{\dot{q}}{\|\dot{q}\|_m}\right) - \frac{V(q)}{\|\dot{q}\|_m}\right) = +\infty.$$

3.17

1. By Theorem 3.16, the momentum map $\mu\colon \mathbb{R} \to C^\infty(T^*Q)$ is given by

$$\mu(s)(q,p) = \tau_L(\dot{q})(sq_0) = s\sum_{j=1}^{n}\frac{\partial L}{\partial\dot{q}}(\dot{q})(q_0)_j = s\sum_{j=1}^{n} m_j\dot{q}_j(q_0)_j,$$

and $\mu(1)$ is an integral of motion. If $Q = \left(\mathbb{R}^3\right)^N$ is a Lagrangian system of N interacting particles with masses $m_1,\dots,m_N > 0$, then the total momentum

$$P := \sum_{j=1}^{N} m_j\dot{q}_j, \qquad \dot{q}_j \in \mathbb{R}^3,$$

is conserved, if $q_0 := (1,\dots,1) \in \mathbb{R}^{3N}$. Conservation of momentum is a fundamental physical law that reflects *homogeneity of space*. Moreover, we compute using Hamilton's equations

$$0 = \dot{P} = -\frac{\partial H_L}{\partial q} = -\frac{\partial V}{\partial q} = \sum_{j=1}^{N} F_j,$$

for the forces F_j. In particular, for a system of two interacting particles, we conclude $F_1 = -F_2$, which is precisely ***Newton's third law***.

2. By Theorem 3.16, the momentum map $\mu\colon \mathbb{R} \to C^\infty(T^*Q)$ is given by

$$\mu(s)(q, p) = \tau_L(\dot{q})(sq_0) = \frac{\partial L}{\partial \dot{q}}(sq_0) = s\sum_{j=1}^{n} m_j\dot{q}_j(Aq)_j.$$

and $\mu(1)$ is an integral of motion. If $Q = \left(\mathbb{R}^3\right)^N$, then the total angular momentum

$$M := \sum_{j=1}^{N} q_j \times m_j\dot{q}_j, \qquad q_j, \dot{q}_j \in \mathbb{R}^3,$$

is conserved, similar to Exercise 2.54. Conservation of angular momentum is a fundamental physical law that reflects *isotropy of space*.

3. By Theorem 3.16, the momentum map $\mu\colon \mathbb{R} \to C^\infty(T^*(Q \times \mathbb{R}))$ is given by

$$\mu(s)(q, \tau, p, \dot{\tau}) = \tau_{\widehat{L}}(\dot{q}, \dot{\tau})(0, s) = s\frac{\partial \widehat{L}}{\partial \dot{\tau}}(q, \tau, \dot{q}, \dot{\tau}),$$

Moreover, we compute

$$\begin{aligned}
\mu(1)(q, \tau, p, 1) &= \left.\frac{\partial \widehat{L}}{\partial \dot{\tau}}\right|_{\dot{\tau}=1}(q, \tau, \dot{q}, \dot{\tau}) \\
&= -\dot{\tau}\left.\frac{\partial}{\partial \dot{\tau}}\right|_{\dot{\tau}=1} L\left(q, \frac{\dot{q}}{\dot{\tau}}\right) - L(q, \dot{q}) \\
&= \frac{\partial L}{\partial \dot{q}}\left(q, \frac{\dot{q}}{\dot{\tau}}\right)\left.\frac{\dot{q}}{\dot{\tau}}\right|_{\dot{\tau}=1} - L(q, \dot{q}) \\
&= \frac{\partial L}{\partial \dot{q}}(q, \dot{q})\dot{q} - L(q, \dot{q}) \\
&= E_L(q, \dot{q}) \\
&= H_L(q, p),
\end{aligned}$$

since $p = \tau_L(\dot{q})$. Conservation of energy is a fundamental physical law that reflects *homogeneity of time* in an autonomous Lagrangian system. To summarise:

- Space translational symmetries lead to conservation of momentum.
- Space rotational symmetries lead to conservation of angular momentum.
- Time translational symmetry leads to conservation of total energy.

3.19 For all $q \in \mathscr{L}Q$ we compute

$$\begin{aligned}\mathscr{E}_L(q) &= \int_0^1 L(\dot{q}(t))dt \\ &= \int_0^1 \tau_L(\dot{q})(\dot{q}) - \int_0^1 E_L(\dot{q}(t))dt \\ &= \int_0^1 \tau_L(\dot{q})(\dot{q}) - \int_0^1 H_L(\tau_L(\dot{q}(t)))dt \\ &= \int_0^1 \dot{q}^*\lambda_L - \int_0^1 H_L(\tau_L(\dot{q}(t)))dt \\ &= \int_0^1 (\tau_L \circ \dot{q})^*(\tau_L)_*\lambda_L - \int_0^1 H_L(\tau_L(\dot{q}(t)))dt \\ &= \int_0^1 (\tau_L \circ \dot{q})^*\lambda - \int_0^1 H_L(\tau_L(\dot{q}(t)))dt.\end{aligned}$$

3.25 Assume that $f \neq 0$. Then there exists $x_0 \in Q$ such that $f(x_0) \neq 0$. Without loss of generality, we may assume that $f(x_0) > 0$. By the continuity of f, there exists a constant $\delta > 0$ with $f(x) \in B_{f(x_0)/2}(f(x_0))$ for all $x \in B_\delta(x_0)$. Choose a smooth bump function $\varphi \in C_c^\infty(M)$ supported in $B_\delta(x_0)$ and with $\varphi = 1$ on $\bar{B}_{\delta/2}(x_0)$. Then

$$0 = \int_Q f\varphi = \int_{B_\delta(x_0)} f\varphi \geq \int_{B_{\delta/2}(x_0)} f\varphi > \frac{1}{2} f(x_0)\,\mathrm{Vol}(B_{\delta/2}(x_0)) > 0.$$

A contradiction and so $f = 0$ follows.

3.26 For $f \in C^\infty(Q)$ we compute

$$\mathscr{E}_{L+df}(q) = \mathscr{E}_L(q) + \int_0^1 df(\dot{q}(t))dt = \mathscr{E}_L(q) + f(q(1)) - f(q(0)) = \mathscr{E}_L(q)$$

for all $q \in \mathscr{L}Q$.

3.28 Both proofs make use of the Gâteaux derivative of a functional defined on an infinite-dimensional space. There is no need to consider the tangent space of the infinite-dimensional manifold $\mathscr{L}M$ in both cases, as one can elegantly define the tangent space at a loop γ as vector fields along γ. As $\mathscr{L}M$ is only a Fréchet manifold and not even a Banach manifold, this would introduce unnecessary difficulties. To avoid this issue, one usually works with paths belonging to a suitable Sobolev space. The proof of Lemma 3.23 is local in nature, appealing to coordinates, whereas the proof of Lemma 3.27 is global. Both perspectives are important in mathematics and physics, as one should be able to grasp the global concept as well as being able to perform dirty computations in coordinates!

3.29 In coordinates we have

$$H_L(q,p) = \left(\frac{\partial L}{\partial \dot{q}}\dot{q} - L(q,\dot{q})\right)\Big|_{p=\frac{\partial L}{\partial \dot{q}}}.$$

Therefore

$$\frac{\partial H_L}{\partial p} = \frac{\partial}{\partial p}\left(p\dot{q} - L(q,\dot{q})\right)\big|_{p=\frac{\partial L}{\partial \dot{q}}} = \dot{q}$$

and

$$\frac{\partial H_L}{\partial q} = \frac{\partial}{\partial q}\left(p\dot{q} - L(q,\dot{q})\right)\Big|_{p=\frac{\partial L}{\partial \dot{q}}} = -\frac{\partial L}{\partial q}\Big|_{p=\frac{\partial L}{\partial \dot{q}}} = -\frac{d}{dt}\frac{\partial L}{\partial \dot{q}}\Big|_{p=\frac{\partial L}{\partial \dot{q}}} = -\dot{p}$$

using the Euler–Lagrange equations (3.5). Those are precisely Hamilton's equations in the standard coordinates (q, p) on the phase space T^*Q from Exercise 2.15.

3.31

	Hamiltonian formalism	Lagrangian formalism
Configuration Space	Smooth manifold Q	Smooth manifold Q
Energy Function	Hamiltonian function $H \in C^\infty(T^*Q)$	Lagrangian function $L \in C^\infty(TQ)$
Domain	Phase space cotangent bundle T^*Q	State space tangent bundle TQ
Coordinates	Position q and momentum p	Position q and velocity $\dot{q}$
Dimension	$2\dim(Q)$	$2\dim(Q)$
Functional	Hamiltonian action functional $\mathscr{A}_H$	Lagrangian action functional $\mathscr{E}_L$
Hamilton's Principle	$\frac{d}{d\varepsilon}\Big\vert_{\varepsilon=0} \mathscr{A}_H(q_\varepsilon, p_\varepsilon) = 0$ for all variations $(q_\varepsilon, p_\varepsilon)$	$\frac{d}{d\varepsilon}\Big\vert_{\varepsilon=0} \mathscr{E}_L(q_\varepsilon) = 0$ for all variations q_ε
Equations of Motion	Hamilton's equations $\dot{q} = \frac{\partial H}{\partial p}$ and $\dot{p} = -\frac{\partial H}{\partial q}$	Euler–Lagrange equations $\frac{d}{dt}\frac{\partial L}{\partial \dot{q}} = \frac{\partial L}{\partial q}$
Transformation	Inverse Legendre transformation $\tau_L^{-1}\colon T^*Q \to TQ$	Legendre transformation $\tau_L\colon TQ \to T^*Q$
Symmetries	Symmetry group G with $\forall g \in G : H \circ \psi_g = H$	Symmetry group G with $\forall g \in G : L \circ D\psi_g = L$
Momentum Map	$\mu\colon \mathfrak{g} \to C^\infty(T^*Q)$	$\tau_L^* \circ \mu\colon \mathfrak{g} \to C^\infty(TQ)$, $\tau_L^*\colon C^\infty(T^*Q) \to C^\infty(TQ)$

3.33 We prove only the second part. By assumption, L is fibrewise convex and a diffeomorphism, thus L is Tonelli by Remark 3.13. In particular Hess L_x is positive definite for all $x \in \mathbb{R}^n$. Using part (b) of [150, Theorem 6.1.8 (b)] we compute

$$\begin{aligned} H_L(x,y) &= \left(E_L \circ \tau_L^{-1}\right)(x,y) \\ &= E_L\left(x, (\nabla L_x)^{-1}(y)\right) \\ &= \tau_L|_{(x,(\nabla L_x)^{-1}(y))}\left((\nabla L_x)^{-1}(y)\right) - L_x\left((\nabla L_x)^{-1}(y)\right) \\ &= \left\langle \nabla L_x\left((\nabla L_x)^{-1}(y)\right), (\nabla L_x)^{-1}(y)\right\rangle_{\mathbb{R}^n} - L_x\left(\nabla L_x)^{-1}(y)\right) \\ &= \langle y, (\nabla L_x)^{-1}(y)\rangle_{\mathbb{R}^n} - L_x\left((\nabla L_x)^{-1}(y)\right) \\ &= (L_x)^*(y) \end{aligned}$$

for all $x, y \in \mathbb{R}^n$.

3.34 The geodesic vector field V gives rise to a vector field $V^* \in \mathfrak{X}(T^*Q)$, called the ***cogeodesic vector field***, via the commutative diagram

$$\begin{array}{ccc} TQ & \xrightarrow{\;V\;} & TTQ \\ {\scriptstyle \widehat{m}}\downarrow & & \downarrow{\scriptstyle D\widehat{m}} \\ T^*Q & \xrightarrow[V^*]{} & TT^*Q, \end{array}$$

where $\widehat{m}\colon TQ \to T^*Q$ denotes the tangent-cotangent bundle isomorphism. Let ϕ^V denote the flow of V. Then we have $\phi_t^{V^*} = \widehat{m} \circ \phi_t^V \circ \widehat{m}^{-1}$. Indeed, we compute

$$\begin{aligned} \frac{d}{dt}(\widehat{m} \circ \phi_t^V \circ \widehat{m}^{-1}) &= D\widehat{m} \circ \frac{d}{dt}\phi_t^V \circ \widehat{m}^{-1} \\ &= D\widehat{m} \circ V \circ \phi_t^V \circ \widehat{m}^{-1} \\ &= D\widehat{m} \circ V \circ \widehat{m}^{-1} \circ \widehat{m} \circ \phi_t^V \circ \widehat{m}^{-1} \\ &= V^* \circ \widehat{m} \circ \phi_t^V \circ \widehat{m}^{-1}. \end{aligned}$$

Now the statement follows from [77, Theorem 2.3.1]. Note that in this situation, you cannot avoid performing a computation in coordinates!

Exercises and Problems of Chap. 4

4.2 The hypersurface $\Sigma^{2n-1} \subseteq M^{2n}$ is clearly of odd dimension and $\omega|_\Sigma$ is closed. It remains to show that $\ker \omega|_\Sigma \to \Sigma$ is a line bundle spanned by $X_H|_\Sigma$. Using the dimension formula from Exercise A.4

$$\dim T_x\Sigma + \dim(T_x\Sigma)^{\perp} = \dim T_xM = 2n$$

yields $\dim(T_x\Sigma)^{\perp} = 1$ for the symplectic orthogonal complement

$$(T_x\Sigma)^{\perp} = \left\{v \in T_xM : i_v\omega|_{T_x\Sigma} = 0\right\}.$$

Now we have that $X_H(x) \in (T_x\Sigma)^{\perp}$ for every $x \in \Sigma$ as

$$\omega(X_H(x), v) = -dH_x(v) = 0 \qquad \forall v \in T_x\Sigma = \ker dH_x.$$

Also $X_H(x) \in T_x\Sigma$ as

$$dH_x(X_H(x)) = \omega(X_H(x), X_H(x)) = 0.$$

Consequently $\ker\omega|_{\Sigma} = (T\Sigma)^{\perp} = \langle X_H|_{\Sigma}\rangle$ as claimed.

4.4 Let $x \in \Sigma^{2n-1}$ and write $T_x\Sigma = \ker\omega_x \oplus S_x$ for some subspace $S_x \subseteq T_x\Sigma$. By Lemma A.3, there exists a basis (u, a_i, b_i) of $T_x\Sigma$ such that u spans $\ker\omega_x$ and (a_i, b_i) is a symplectic basis for S_x. Then the statement follows immediately from

$$\alpha_x \wedge \omega_x^{n-1}(u, a_i, b_i) = \alpha_x(u)\omega_x^{n-1}(a_i, b_i),$$

and the fact that ω_x is a symplectic tensor if and only if $\omega_x^{n-1} \neq 0$.

4.5 From Exercise 4.2 it follows that $(\Sigma, \omega|_{\Sigma})$ is a Hamiltonian manifold. The condition that $\alpha \in \Omega^1(\Sigma^{2n-1})$ is a stabilising form is by Exercise 4.4 equivalent to the condition that $\alpha \wedge \omega^{n-1} \in \Omega^{2n-1}(\Sigma)$ is a volume form and thus nowhere vanishing on Σ. For $\alpha := i_X\omega|_{\Sigma}$ we compute

$$i_X\omega|_{\Sigma} \wedge \omega|_{\Sigma}^{n-1} = \frac{1}{n} i_X\omega|_{\Sigma}^{n}.$$

As X is nowhere tangent to Σ, the form $i_X\omega|_{\Sigma}^{n} \in \Omega^{2n-1}(\Sigma)$ is a volume form by [129, Proposition 15.21]. Finally, using Cartan's magic formula (Lemma 2.1), we compute

$$\ker\omega|_{\Sigma} \subseteq \ker L_X\omega|_{\Sigma} = \ker(i_Xd\omega + di_X\omega)|_{\Sigma} = \ker d\alpha.$$

4.12

1. By the definition of a stable Hamiltonian manifold, $\ker\omega$ is a line distribution. Thus, there exists a smooth nowhere-vanishing section $X \in \mathfrak{X}(\Sigma)$. Set $R = \alpha(X)^{-1}X$. This is well-defined, as α is nowhere-vanishing on the characteristic line bundle $\ker\omega$ by the definition of a stable Hamiltonian structure.

2. Using Fisherman's formula (Lemma 2.2) and Cartan's magic formula (Lemma 2.1) we compute

$$\frac{d}{dt}\phi_t^*\alpha = \phi_t^* L_R\alpha = \phi_t^*(i_R d\alpha + d i_R\alpha) = 0$$

as $R \in \ker\omega \subseteq \ker d\alpha$ and $i_R\alpha = 1$. Thus, $\phi_t^*\alpha$ is constant and so $\phi_t^*\alpha = \phi_0^*\alpha = \alpha$.
3. In the definition of the Reeb vector field we see that we need the condition of α being nowhere-vanishing on the characteristic distribution $\ker\omega$ to have existence and uniqueness. The condition $\ker\omega \subseteq \ker d\alpha$ ensures that the first part is true, which has important dynamical consequences, as we shall see later.
4. By the first part, we have $\alpha = \phi_t^*\alpha = \alpha \circ D\phi_t$. As $D\phi_t$ is an isomorphism, we conclude $\ker\alpha = \ker\alpha \circ D\phi_t$.

4.18

1. Similarly to Lemma 3.27 we compute

$$d\mathscr{A}_0|_\gamma(X) = \int_0^1 d\alpha(X(t), \dot\gamma(t))dt$$

for all $X \in \Gamma(\gamma^*T\Sigma)$. Thus, if $(\gamma, \tau) \in \mathscr{P}(\Sigma, \omega, \alpha)$, we get

$$d\mathscr{A}_0|_\gamma(X) = \tau\int_0^1 d\alpha(X(t), R(\gamma(t)))dt = 0$$

since R belongs to $\ker\omega \subseteq \ker(d\alpha)$. As $X \in \Gamma(\gamma^*T\Sigma)$ was arbitrary, we conclude that $(\gamma, \tau) \in \mathrm{Crit}(\mathscr{A}_0)$.
2. We already know from the first part that $\mathscr{P}(\Sigma, \alpha) \subseteq \mathrm{Crit}(\mathscr{A}_0)$. Thus we only have to show that also $\mathrm{Crit}(\mathscr{A}_0) \subseteq \mathscr{P}(\Sigma, \alpha)$ holds. Assume that $\gamma \in \mathscr{L}\Sigma$ satisfies

$$d\mathscr{A}_0|_\gamma(X) = 0 \qquad \forall X \in \Gamma(\gamma^*T\Sigma).$$

It follows that $\dot\gamma$ belongs to $\ker d\alpha$. From Exercise 4.12, we have the splitting

$$T\Sigma = \ker\alpha \oplus \langle R\rangle.$$

Moreover, $\langle R\rangle = \ker d\alpha$ since $\ker d\alpha$ is a line bundle. It follows, that $\gamma \in \mathrm{Crit}(\mathscr{A}_0)$ is a parametrised periodic Reeb orbit of period $\int_\gamma \alpha$, when we reparametrise time such that $\dot\gamma = R \circ \gamma$ holds. This observation is the starting point for contact homology theory surveyed in [5, Chapter 8].
3. By considering the mapping torus from Example 4.7, we know that there exist exact stable Hamiltonian forms. But then $\mathrm{Crit}(\mathscr{A}_0) = \mathscr{L}\Sigma$.

4.20

1. It is convenient to use the identity

$$\lambda_0 = \frac{i}{4}\sum_{j=1}^{n}(z_j d\bar{z}_j - \bar{z}_j dz_j) \in \Omega^1(\mathbb{R}^{2n}).$$

Then we compute

$$\begin{aligned} f^*\lambda_0 &= \frac{i}{4}\sum_{j=1}^{n}(f^*z_j df^*\bar{z}_j - f^*\bar{z}_j df^*z_j) \\ &= \frac{i}{4}\sum_{j=1}^{n}(\rho(z)z_j d\left(\rho(z)\bar{z}_j\right) - \rho(z)\bar{z}_j d\left(\rho(z)z_j\right)) \\ &= \frac{i}{4}\sum_{j=1}^{n}(\rho(z)z_j\bar{z}_j d\rho(z) + \rho(z)^2 z_j d\bar{z}_j - \rho(z)\bar{z}_j z_j d\rho(z) - \rho(z)^2\bar{z}_j dz_j) \\ &= \rho(z)^2\frac{i}{4}\sum_{j=1}^{n}(z_j d\bar{z}_j - \bar{z}_j dz_j) \\ &= \rho(z)^2\lambda_0. \end{aligned}$$

The equality $\ker(f^*\lambda_0|_{\Sigma_\rho}) = \ker(\lambda_0|_{\mathbb{S}^{2n-1}})$ follows immediately from $\rho > 0$.

2. Let $R \in \mathfrak{X}(\mathbb{S}^{2n-1})$ denote the Reeb vector field of the contact form $\lambda_0|_{\mathbb{S}^{2n-1}}$. Then the Reeb vector field of the contact form $\rho^2\lambda_0|_{\mathbb{S}^{2n-1}}$ is given by $\rho^{-2}R$. For $\rho \neq 1$, the flows of R and $\rho^{-2}R$ can be very different as the case of an ellipsoid with rationally independent weights shows.

4.24

1. Both functionals are defined on infinite-dimensional spaces and differ only by the appearance of the Lagrange multiplier τ in the Rabinowitz action functional. Adapting the proof of Lemma 3.27, one sees that

$$(\gamma, \tau) \in \mathrm{Crit}(\mathscr{A}^H) \qquad \Leftrightarrow \qquad \begin{cases} \dot{\gamma}(t) = \tau X_H(\gamma(t)) & \forall t \in \mathbb{T}, \\ \int_0^1 H(\gamma(t))dt = 0. \end{cases}$$

By conservation of energy (Corollary 2.28), the second equation combined with the first implies that $\gamma \in C^\infty(\mathbb{T}, \Sigma)$. Therefore, the critical points of $\mathscr{A}^H$ are τ-periodic loops of constant energy of the Hamiltonian vector field X_H, while the critical points of the Hamiltonian action functional $\mathscr{A}_H$ are 1-periodic loops of arbitrary energy since

$$\gamma \in \mathrm{Crit}(\mathscr{A}_H) \qquad \Leftrightarrow \qquad \dot{\gamma}(t) = X_H(\gamma(t)) \quad \forall t \in \mathbb{T},$$

Hence the functional $\mathscr{A}_H$ is better suited for problems involving fixed period but arbitrary energy, and the Rabinowitz action functional is better suited for problems involving fixed energy but arbitrary period. In particular, the Rabinowitz action functional is suited for studying periodic orbits on regular energy hypersurfaces. Below, a scenario is sketched where on the left a compact symplectic manifold with boundary $\partial W = \Sigma$ is depicted and on the right a non-compact symplectic manifold $\Sigma \times [0, +\infty)$ with boundary also being Σ.

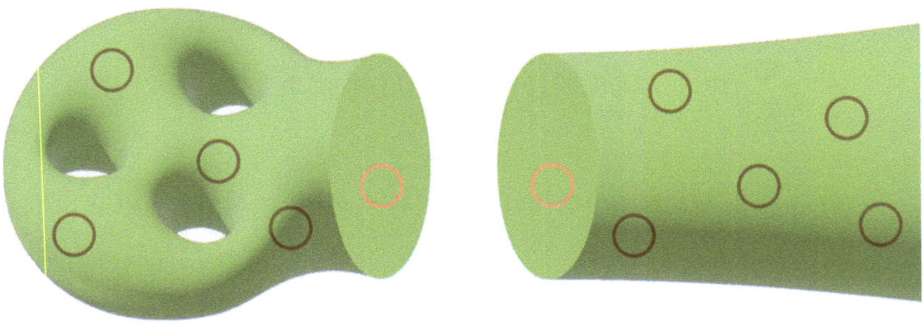

(a) Illustration of Crit $\mathscr{A}_H$

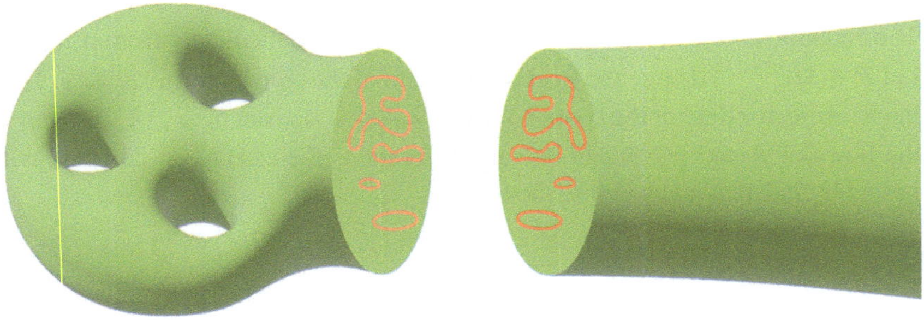

(b) Illustration of Crit $\mathscr{A}^H$

4.25 Particularly interesting in the context of this book are the constrained Lagrangian systems in [50, Example 4.1] and the discussion of symplectic field theory in [50, Example 4.6].

4.32 We only need to show that $\varphi^*\lambda_0 = \lambda_0$, which follows from the computation

$$\varphi^*\lambda_0 = \frac{i}{4}\sum_{j=1}^{n}\left(e^{2\pi i k_j/m} z_j d\left(e^{-2\pi i k_j/m}\bar{z}_j\right) - e^{-2\pi i k_j/m}\bar{z}_j d\left(e^{2\pi i k_j/m} z_j\right)\right) = \lambda_0.$$

4.36 See [104, pp. 137–141].

4.37 We prove a more general statement for covering spaces. Suppose that Σ is a simply connected topological manifold and $\varphi\colon \Sigma \to \Sigma$ a homeomorphism such that $\varphi \neq \mathrm{id}_\Sigma$ and $m := \operatorname{ord}\varphi < \infty$. If the induced discrete action of φ on Σ

$$\mathbb{Z}_m \times \Sigma \to \Sigma, \qquad [k]\cdot x := \varphi^k(x)$$

is free, then $\pi\colon \Sigma \to \Sigma/\mathbb{Z}_m$ is a normal covering map by [128, Theorem 12.26] and consequently, Σ is the universal covering manifold of the topological manifold $\Sigma/\mathbb{Z}_m$ by [128, Corollary 12.27]. Denote by $\mathrm{Aut}_\pi(\Sigma)$ the group of deck transformations. Then, by construction, we have $\varphi \in \mathrm{Aut}_\pi(\Sigma)\setminus\{\mathrm{id}_\Sigma\}$. Using the long exact sequence of homotopy groups of a fibration [104, Theorem 4.41], there is a short exact sequence

$$0 \longrightarrow \pi_1(\Sigma) \longrightarrow \pi_1(\Sigma/\mathbb{Z}_m) \longrightarrow \pi_0(\mathbb{Z}_m) \longrightarrow 0.$$

In particular, $\pi_1(\Sigma/\mathbb{Z}_m) \cong \mathbb{Z}_m \cong \langle\varphi\rangle$. Suppose that $\pi\circ\gamma$ is contractible. Then we have $\pi\circ\gamma \simeq c_{\pi(x)}$, where $c_{\pi(x)}$ denotes the constant loop based at $\pi(x)$. The Monodromy Theorem [128, Theorem 11.15 (b)] implies $x \in \mathrm{Fix}(\varphi)$. But the only deck transformation of π that fixes any point of Σ is id_Σ by [128, Proposition 12.1 (a)]. Conversely, assume that $\gamma \in \mathscr{L}(\Sigma/\mathbb{Z}_m, \pi(x))$ is not contractible. Again, the Monodromy Theorem [128, Theorem 11.15 (b)] yields $\widetilde{\gamma}_x(1) \neq x$. In fact, if $\widetilde{\gamma}_x(1) = x$, then we conclude that $\gamma \simeq c_{\pi(x)}$ and, consequently, γ would be contractible. For this, note that part (b) of the Monodromy Theorem is actually an if and only if statement if the covering space is simply connected. As normal covering maps have transitive automorphism groups by [128, Corollary 12.5], there exists $\psi \in \mathrm{Aut}_\pi(\Sigma)\setminus\{\mathrm{id}_\Sigma\}$ such that $\widetilde{\gamma}_x(1) = \psi(x)$. But $\mathrm{Aut}_\pi(\Sigma) \cong \pi_1(\Sigma/\mathbb{Z}_m) \cong \langle\varphi\rangle$ by [128, Corollary 12.9] and the statement follows.

4.42 Write $f := (\mathrm{ev}_0, \mathrm{ev}_1)$. Then $\mathscr{L}_\varphi M = f^{-1}(\Gamma_\varphi(M))$. Thus, in order to show that the free twisted loop space $\mathscr{L}_\varphi M$ is a smooth Banach manifold, it is enough to show that f is transverse to the properly embedded smooth submanifold $\Gamma_\varphi(M) \subseteq M \times M$. By [125, Proposition 2.4] we need to show that the composition

$$\Phi_\gamma\colon T_\gamma\mathscr{P}M \xrightarrow{Df_\gamma} T_{(x,\varphi(x))}(M\times M) \to T_{(x,\varphi(x))}(M\times M)/T_{(x,\varphi(x))}\Gamma_\varphi(M)$$

is surjective and $\ker\Phi_\gamma$ is complemented for all $\gamma \in f^{-1}(\Gamma_\varphi(M))$, where we abbreviate $x := \gamma(0)$. Note that we have a canonical isomorphism

$$T_{(x,\varphi(x))}(M\times M)/T_{(x,\varphi(x))}\Gamma_\varphi(M) \to T_{\varphi(x)}M, \quad [(v,u)] := u - D\varphi(v).$$

Under this canonical isomorphism, the linear map Φ_γ is given by

$$\Phi_\gamma(X) = X(1) - D\varphi(X(0)), \qquad \forall X \in \Gamma^0(\gamma^*TM).$$

Let $X_v \in \Gamma(\gamma^*TM)$ be the unique parallel vector field with $X_v(1) = v \in T_{\varphi(x)}M$ for some Riemannian metric. Fix a cut-off function $\beta \in C^\infty([0,1])$ with $\operatorname{supp}\beta \subseteq [\frac{1}{2}, 1]$ and $\beta = 1$ in a neighbourhood of 1. Then $\Phi_\gamma(\beta X_v) = v$ and Φ_γ are surjective. Moreover,

$$\ker \Phi_\gamma = \{X \in \Gamma^0(\gamma^*TM) : X(1) = D\varphi(X(0))\}$$

is complemented by the finite-dimensional vector space

$$V := \{\beta X_v \in \Gamma(\gamma^*TM) : v \in T_{\varphi(x)}M\}.$$

Indeed, any $X \in \Gamma^0(\gamma^*TM)$ can be decomposed uniquely as

$$X = X - \beta X_v + \beta X_v, \qquad v := X(1) - D\varphi(X(0)).$$

Abbreviating $Y := X - \beta X_v \in \Gamma^0(\gamma^*TM)$, we have that

$$Y(1) = D\varphi(X(0)) = D\varphi(Y(0)),$$

implying $Y \in \ker \Phi_\gamma$. Hence, $\mathscr{L}_\varphi M$ is a smooth Banach manifold. Now note that $\mathscr{L}_\varphi M$ can be identified with the pullback

$$f^*\mathscr{P}M = \{(x,\gamma) \in M \times \mathscr{P}M : (\gamma(0), \gamma(1)) = (x, \varphi(x))\},$$

making the diagram

$$\begin{array}{ccc} f^*\mathscr{P}M & \xrightarrow{\ \mathrm{pr}_2\ } & \mathscr{P}M \\ {\scriptstyle \mathrm{pr}_1}\downarrow & & \downarrow{\scriptstyle f} \\ M & \xrightarrow[\ \Gamma_\varphi\ ]{} & M \times M \end{array}$$

commute, via the homeomorphism

$$\mathscr{L}_\varphi M \to f^*\mathscr{P}M, \qquad \gamma \mapsto (\gamma(0), \gamma).$$

Finally, one computes

$$T_{(x,\gamma)} f^*\mathscr{P}M = \{(v, X) \in T_xM \times T_\gamma \mathscr{P}M : Df_\gamma X = D\Gamma_\varphi|_x(v)\}$$

for all $(x,\gamma) \in f^*\mathscr{P}M$, implying the stated formula.

4.43

1. For every $X \in T_\gamma \mathscr{L}_\varphi M$ we compute

$$\begin{aligned} d\mathscr{A}^0_\varphi|_\gamma(X) &= \frac{d}{d\varepsilon}\bigg|_{\varepsilon=0} \mathscr{A}^0_\varphi(\gamma_\varepsilon) \\ &= \int_0^1 \gamma^* L_X \lambda \\ &= \int_0^1 \gamma^* i_X d\lambda + \int_0^1 \gamma^* d i_X \lambda \\ &= \int_0^1 d\lambda(X(t), \dot\gamma(t)) + \int_0^1 d\gamma^* i_X \lambda \\ &= \int_0^1 d\lambda(X(t), \dot\gamma(t)) + \lambda(X(1)) - \lambda(X(0)) \\ &= \int_0^1 d\lambda(X(t), \dot\gamma(t)) + \lambda(D\varphi(X(0))) - \lambda(X(0)) \\ &= \int_0^1 d\lambda(X(t), \dot\gamma(t)) + (\varphi^*\lambda - \lambda)(X(0)) \\ &= \int_0^1 d\lambda(X(t), \dot\gamma(t)). \end{aligned}$$

   It follows that $\gamma \in \operatorname{Crit}(\mathscr{A}^0_\varphi)$ if and only if $\dot\gamma(t) = 0$ for all $t \in [0, 1]$. Thus, we see that $\gamma \in \mathscr{L}_\varphi M$ is constant. But by the definition of the free twisted loop space, we have $\gamma(0) = \gamma(1) = \varphi(\gamma(0))$ implying $\gamma(0) \in \operatorname{Fix}(\varphi)$.
2. By [137, Lemma 5.5.7], the fixed point set $\operatorname{Fix}(\varphi)$ is a symplectic submanifold of M. Hence, $\operatorname{Fix}(\varphi)$ always forms a whole critical submanifold of M. In particular, if $\varphi = \mathrm{id}_M$, we have that $\operatorname{Fix}(\varphi) = M$ and the critical points are not isolated! Consequently, the twisted symplectic area functional is not a Morse function in that case.
3. Similar to the definition of the Rabinowitz action functional in Definition 4.23, one could introduce a Lagrange multiplier. Indeed, one could define

$$\mathscr{A}^{H_\Sigma}_\varphi \colon \mathscr{L}_\varphi \mathbb{R}^{2n} \times (0, +\infty) \to \mathbb{R}, \quad \mathscr{A}^{H_\Sigma}_\varphi(\gamma, \tau) := \mathscr{A}^0_\varphi(\gamma) - \tau \int_0^1 H_\Sigma(\gamma(t))dt.$$

   Then the critical points project down to noncontractible periodic Reeb orbits by Exercise 4.37.

Exercises and Problems of Chap. 5

5.1

1. Let $x^{\pm} \in \mathrm{Crit}(h)$ with $x^- \neq x^+$. Then the reparametrisation action of $\mathbb{R}$ on negative gradient flow lines is free. In particular, there would be uncountably many parametrised negative gradient flow lines for which it does not make sense to count! Dividing the reparametrisation action out is also important to get compactness modulo breaking. Indeed, consider a sequence of negative gradient flow lines with the same asymptotic ends.

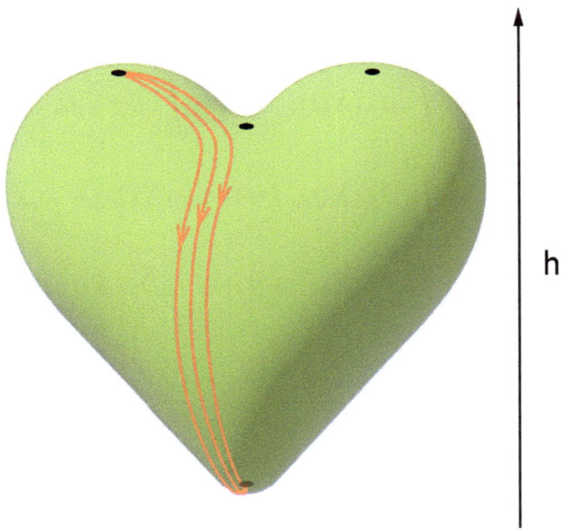

By using different time-reparametrisations of the negative gradient flow lines, the sequence can converge to two different negative gradient flow lines.

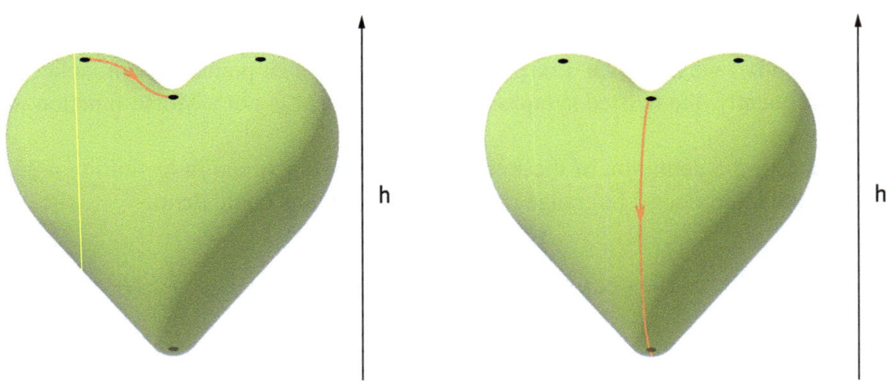

So if we instead consider a sequence of unparametrised negative gradient flow lines, we get a broken negative gradient flow line as the limit. This is usually referred to as ***Floer–Gromov compactness*** in the infinite-dimensional setting.

2. Suppose that ∇ and $\tilde{\nabla}$ are two connections on M. For $X, Y \in \mathfrak{X}(M)$ we compute

$$\begin{aligned}\mathrm{Hess}^{\nabla}(h)(X,Y) &= \nabla_X(dh)(Y)\\ &= X(Y(h)) - dh(\nabla_X Y)\\ &= \mathrm{Hess}^{\tilde{\nabla}}(h)(X,Y) + dh(\tilde{\nabla}_X Y - \nabla_X Y).\end{aligned}$$

Thus if $v, w \in T_xM$ and $V, W \in \mathfrak{X}(M)$ denote any extensions of v and w, respectively, we have that

$$\begin{aligned}\mathrm{Hess}^{\nabla}(h)|_x(v,w) &= \mathrm{Hess}^{\tilde{\nabla}}(h)|_x(v,w) + dh_x(\tilde{\nabla}_V W - \nabla_V W)|_x\\ &= \mathrm{Hess}^{\tilde{\nabla}}(h)|_x(v,w)\end{aligned}$$

since $x \in \mathrm{Crit}(h)$ by assumption, implying $\mathrm{Hess}^{\nabla}(h)|_x = \mathrm{Hess}^{\tilde{\nabla}}(h)|_x$. Thus, the definition of the pointwise Hessian at a critical point is independent of the choice of connection and is therefore well-defined. Assume that ∇ is torsion free, then

$$\mathrm{Hess}^{\nabla}(X,Y) = \mathrm{Hess}^{\nabla}(Y,X)$$

for all $X, Y \in \mathfrak{X}(M)$. This immediately follows from the computation

$$\mathrm{Hess}^{\nabla}(h)(X,Y) = \mathrm{Hess}^{\nabla}(h)(Y,X) - dh(T^{\nabla}(X,Y)) = \mathrm{Hess}^{\nabla}(h)(Y,X),$$

where

$$T^{\nabla}(X,Y) := \nabla_X Y - \nabla_Y X - [X,Y]$$

denotes the torsion tensor with respect to the connection ∇.

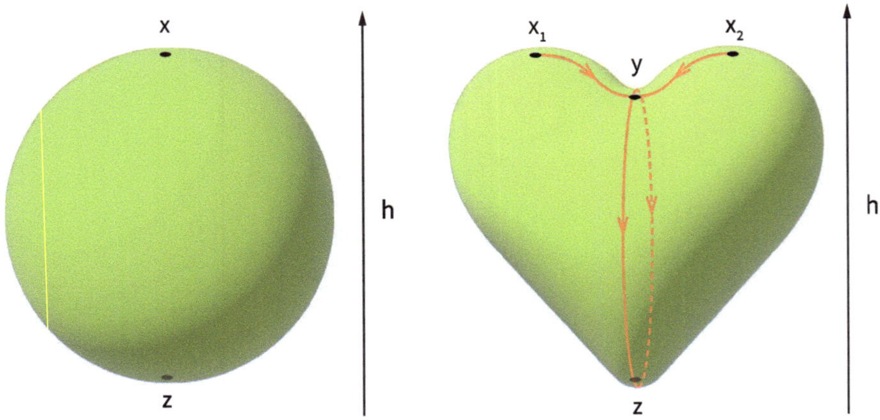

3. The height function h on the sphere $\mathbb{S}^2$ has precisely two critical points: A maximum x and a minimum z. Moreover, there are infinitely many negative gradient flow lines from x to z. But as $\operatorname{ind}_h(x) = 2$ and $\operatorname{ind}_h(z) = 0$, there are no negative gradient flow lines between critical points of index difference one contributing to the boundary operators. Thus the associated chain complex is given by

$$0 \xrightarrow{0} \mathbb{Z}_2 \xrightarrow{0} 0 \xrightarrow{0} \mathbb{Z}_2 \xrightarrow{0} 0.$$

We conclude that

$$\mathrm{HM}_k(h) = \begin{cases} \mathbb{Z}_2 & k = 0, 2, \\ 0 & \text{else.} \end{cases}$$

This is precisely the singular homology $\mathrm{H}_k(\mathbb{S}^2; \mathbb{Z}_2)$. The height function h on the deformed sphere resembling a heart has four critical points: Two maxima denoted by x_1 and x_2, a saddle point y, as well as one minimum z. The corresponding boundary maps are given by

$$\partial x_1 = y = \partial x_2 \qquad \text{and} \qquad \partial y = z + z = 0.$$

Thus the resulting chain complex is

$$0 \xrightarrow{0} \mathbb{Z}_2 \oplus \mathbb{Z}_2 \xrightarrow{(1,1)} \mathbb{Z}_2 \xrightarrow{0} \mathbb{Z}_2 \xrightarrow{0} 0.$$

We again conclude that

$$\mathrm{HM}_k(h) = \begin{cases} \mathbb{Z}_2 & k = 0, 2, \\ 0 & \text{else.} \end{cases}$$

4. The upright torus has precisely four critical points: One maximum x, two saddle points y_1 and y_2, as well as one minimum z. We note that $W^u(y_1)$ and $W^s(y_2)$ intersect in two open intervals, with coinciding tangent spaces. In particular, they do not intersect transversally. In order to achieve the Morse–Smale condition, we have to slightly tilt the torus. Then one checks that the Morse–Smale condition holds and we can compute the associated Morse homology. The tilting of the torus can also be interpreted as a change of the Riemannian metric on the torus. For a generic choice of the Riemannian metric on any compact manifold we get a well-defined count of gradient flow lines leading to a boundary operator. However, there can be some degenerate metrics with degenerate gradient flow lines which one first has to perturb a little. The analytical problem is the well-definedness of the boundary map and that $\partial \circ \partial = 0$ holds. For that one needs that the moduli space $\mathscr{M}(x^-, x^+)$ is a compact smooth manifold of dimension zero. In this case, the boundary maps are given by

$$\partial x = 2y_1 + 2y_2 = 0 \qquad \text{and} \qquad \partial y_1 = 2z = 0 = \partial y_2.$$

Thus the resulting chain complex is

$$0 \xrightarrow{0} \mathbb{Z}_2 \xrightarrow{0} \mathbb{Z}_2 \oplus \mathbb{Z}_2 \xrightarrow{0} \mathbb{Z}_2 \xrightarrow{0} 0,$$

and we conclude that

$$\mathrm{HM}_k(h) = \begin{cases} \mathbb{Z}_2 & k = 0, 2, \\ \mathbb{Z}_2 \oplus \mathbb{Z}_2 & k = 1, \\ 0 & \text{else.} \end{cases}$$

Again, the Morse homology $\mathrm{HM}_*(h)$ of the height function h coincides with the singular homology $\mathrm{H}_*(\mathbb{T}^2; \mathbb{Z}_2)$ of the torus $\mathbb{T}^2 \subseteq \mathbb{R}^3$.

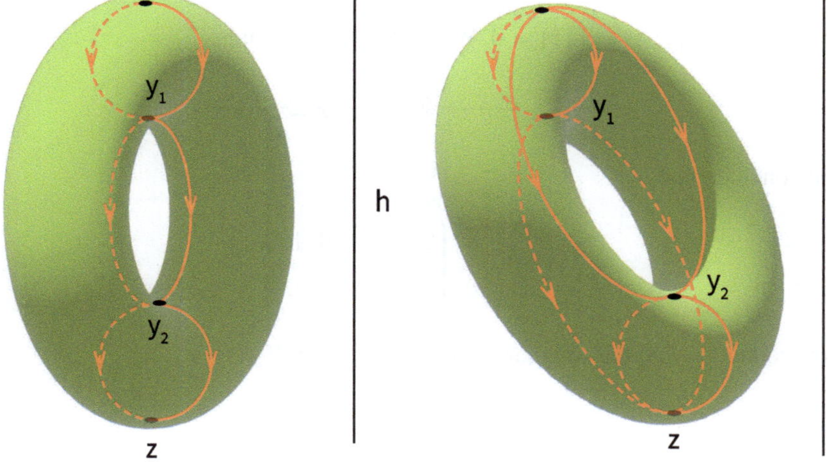

5. A suitable $\mathbb{Z}_2$-invariant Morse function $h \in C^\infty(\mathbb{S}^2)$ is given by

$$h\colon \mathbb{S}^2 \to \mathbb{R}, \qquad h(x,y,z) := ax^2 + by^2 + cz^2$$

for some real numbers $0 < c < b < a$. This is a Morse function on $\mathbb{S}^2$ with two maxima $x^\pm := (\pm 1, 0, 0)$, two saddle points $y^\pm := (0, \pm 1, 0)$ and two minima $z^\pm := (0, 0, \pm 1)$. The corresponding boundary maps are

$$\partial x^\pm = y^- + y^+ \qquad \text{and} \qquad \partial y^\pm = z^- + z^+.$$

Thus, we get the chain complex

$$0 \xrightarrow{0} \mathbb{Z}_2 \oplus \mathbb{Z}_2 \xrightarrow{\begin{pmatrix}1 & 1\\ 1 & 1\end{pmatrix}} \mathbb{Z}_2 \oplus \mathbb{Z}_2 \xrightarrow{\begin{pmatrix}1 & 1\\ 1 & 1\end{pmatrix}} \mathbb{Z}_2 \oplus \mathbb{Z}_2 \xrightarrow{0} 0$$

with resulting homology

$$\mathrm{HM}_k(h) = \begin{cases} \mathbb{Z}_2 & k = 0, 2, \\ 0 & \text{else,} \end{cases}$$

coinciding with the singular homology of the sphere $\mathbb{S}^2$ with coefficients in $\mathbb{Z}_2$ as expected. As h is invariant under the antipodal $\mathbb{Z}_2$-action $z \mapsto -z$, we can consider the induced map $\bar{h}\colon \mathbb{RP}^2 \to \mathbb{R}$ on the real projective plane $\mathbb{RP}^2 = \mathbb{S}^2/\mathbb{Z}_2$. This map is again a Morse function with precisely three critical points: A maximum $[x^\pm]$, a saddle point $[y^\pm]$ and a minimum $[z^\pm]$. We see that we can define a boundary map on the quotient by $\partial[a] := [\partial a]$ for all $a \in \mathrm{Crit}(\bar{h})$. Thus we get

$$\partial[x^\pm] = [\partial x^\pm] = [y^- + y^+] = 2[y^\pm] = 0,$$

and similarly $\partial[y^\pm] = 2[z^\pm] = 0$. We get the chain complex

$$0 \xrightarrow{0} \mathbb{Z}_2 \xrightarrow{0} \mathbb{Z}_2 \xrightarrow{0} \mathbb{Z}_2 \xrightarrow{0} 0$$

and conclude that

$$\mathrm{HM}_k(\bar{h}) = \begin{cases} \mathbb{Z}_2 & k = 0, 1, 2, \\ 0 & \text{else.} \end{cases}$$

This coincides with the singular homology $\mathrm{H}_*(\mathbb{RP}^2; \mathbb{Z}_2)$.

6. If one wants to proceed as in the example of the real projective space $\mathbb{RP}^2$, we first have to find a $\mathbb{Z}_m$-invariant Morse function. One could for example try

$$f : \mathbb{S}^{2n-1} \to \mathbb{R}, \qquad f(z) := \sum_{j=1}^{n} j|z_j|^2.$$

However, this is not a Morse function as we will see later. Of course, the space of all Morse function on a compact manifold is dense in the space of all smooth functions $C^\infty(M)$ endowed with the C^∞-topology, but computing the homology associated with such a perturbation of f is not very illuminating. Probably the best option for now is sticking to Morse theory and the cellular decomposition of lens spaces. This cellular decomposition yields the chain complex

$$0 \longrightarrow \mathbb{Z} \xrightarrow{0} \mathbb{Z} \xrightarrow{m} \mathbb{Z} \xrightarrow{0} \dots \xrightarrow{0} \mathbb{Z} \xrightarrow{m} \mathbb{Z} \xrightarrow{0} \mathbb{Z} \longrightarrow 0,$$

and thus the resulting homology is given by

$$\mathrm{HM}_k(L_m(k_1, \dots, k_n)) \cong \begin{cases} \mathbb{Z}_2 & 0 \le k \le 2n-1, \\ 0 & \text{else,} \end{cases}$$

if m is even, and $\mathrm{HM}_k(L_m(k_1, \dots, k_n)) = 0$ for all k if m is odd.

5.2

1. Assume that (M, ω) is a connected symplectic manifold. Then we have the pairing

$$[\omega]|_{\pi_2(M)} := \int_{\mathbb{S}^2} f^*\omega \qquad \forall [f] \in \pi_2(M),$$

where $f \in C^\infty(\mathbb{S}^2, M)$ is a smooth representative of the homotopy class $[f]$. Note that every homotopy class in $\pi_2(M)$ admits a smooth representative by the Whitney approximation theorem [129, Theorem 6.26]. This map is well-defined, since if $F \in C^\infty(\mathbb{S}^2 \times [0, 1]], M)$ is a homotopy between f_0 and f_1 in $C^\infty(\mathbb{S}^2, M)$, we compute

$$0 = \int_{\mathbb{S}^2 \times [0,1]} F^* d\omega = \int_{\mathbb{S}^2 \times [0,1]} dF^*\omega = \int_{\mathbb{S}^2} f_1^*\omega - \int_{\mathbb{S}^2} f_0^*\omega$$

by Stoke's theorem since ω is closed. From this, the statement follows immediately.

2. Let (M, ω) be a symplectically aspherical symplectic manifold. Consider two filling discs $\bar{\gamma}_1, \bar{\gamma}_2 \in C^\infty(\mathbb{D}, M)$ for a contractible loop $\gamma \in \Lambda M$. Then we

obtain a sphere $f \in C^\infty(\mathbb{S}^2, M)$ by gluing the two discs together, where one of the fillings has the opposite orientation. By symplectic asphericity we compute

$$0 = \int_{\mathbb{S}^2} f^*\omega = \int_{\mathbb{D}} \bar{\gamma}_2^*\omega - \int_{\mathbb{D}} \bar{\gamma}_1^*\omega,$$

implying that the symplectic action functional $\mathscr{A}_H$ is well-defined.

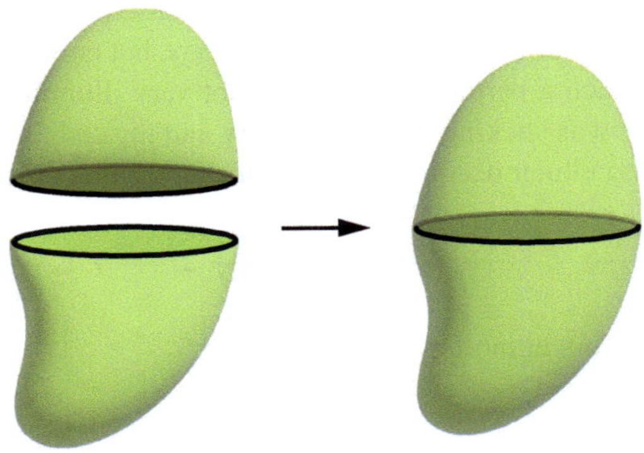

5.4

1. The calculation is similar to the one in the proof of Lemma 3.27. Let $\gamma \in \Lambda M$ be a contractible loop, and let $X \in \Gamma(\gamma^*TM)$ be a tangent vector in $T_\gamma \mathscr{L}M$. As the loop γ is contractible by assumption, there exists a filling disc $\bar{\gamma} \in C^\infty(\mathbb{D}, M)$ that gives rise to an extension $\bar{X} \in \Gamma(\bar{\gamma}^*TM)$ by [129, Lemma 8.6]. Thus, we compute

$$\begin{aligned} d\mathscr{A}_H|_\gamma(X) &= \frac{d}{d\varepsilon}\bigg|_{\varepsilon=0} \mathscr{A}_H(\gamma_\varepsilon) \\ &= \int_{\mathbb{D}} \bar{\gamma}^* L_{\bar{X}}\omega - \int_0^1 dH_t(X(t))dt \\ &= \int_{\mathbb{D}} \bar{\gamma}^* i_{\bar{X}} d\omega + \int_{\mathbb{D}} \bar{\gamma}^* d i_{\bar{X}}\omega - \int_0^1 \omega(X(t), X_{H_t}(\gamma(t)))dt \\ &= \int_{\mathbb{D}} d\bar{\gamma}^* i_{\bar{X}}\omega - \int_0^1 \omega(X(t), X_{H_t}(\gamma(t)))dt \\ &= \int_{\mathbb{S}^1} \bar{\gamma}^* i_{\bar{X}}\omega - \int_0^1 \omega(X(t), X_{H_t}(\gamma(t)))dt \\ &= \int_0^1 \omega(X(t), \dot{\gamma}(t) - X_{H_t}(\gamma(t)))dt, \end{aligned}$$

where we tacitly identify $\mathbb{S}^1$ and $[0, 1]$ without denoting this identification. The statement now follows from this computation and the nondegeneracy of ω.

2. Let $H \in C^\infty(M)$ be a smooth function and let $\gamma \in \mathrm{Crit}(\mathscr{A}_H)$ be a critical point. By the first part, the contractible loop γ solves the Hamiltonian equation

$$\dot{\gamma}(t) = X_H(\gamma(t)) \qquad \forall t \in \mathbb{T}.$$

If γ is assumed to be nonconstant, there does exist a whole $\mathbb{S}^1$-family of critical points defined by reparametrisation $\gamma_s(t) := \gamma(s+t)$ for all $s, t \in \mathbb{T}$. Indeed, we compute

$$\frac{d}{dt}\gamma_s(t) = \frac{d}{dt}\gamma(s+t) = \dot{\gamma}(s+t) = X_H(\gamma(s+t)) = X_H(\gamma_s(t)),$$

since H does not depend on time. Thus, non-constant critical points of $\mathscr{A}_H$ can never be isolated and thus $\mathscr{A}_H$ cannot be a Morse function in this case.

5.7 Let $h \in C^\infty(M)$ be a Morse function on a compact smooth manifold M. Then by the Morse Lemma, it holds that $\mathrm{Crit}(h) = \{x_1, \ldots, x_k\}$ is a finite number of points. In particular $\mathrm{Crit}(h)$ is a zero-dimensional smooth manifold with tangent space

$$T_{x_j}\mathrm{Crit}(h) = \{0\} \qquad \forall j = 1, \ldots, k.$$

But by the definition of a Morse function, every critical point is nondegenerate, meaning that also

$$\ker \mathrm{Hess}_h(x_j) = \{0\}.$$

5.9 Consider the Lagrange multiplier functional

$$\Lambda\colon \mathbb{R}^{2n} \times \mathbb{R} \to \mathbb{R}, \qquad \Lambda(x, y, \lambda) := \sum_{j=1}^n j(x_j^2 + y_j^2) + \lambda \sum_{j=1}^n (x_j^2 + y_j^2),$$

where $z_j = x_j + iy_j$. If we want to compute $\mathrm{Crit}(f)$, we can consider the equations

$$\frac{\partial \Lambda}{\partial x} = 0, \qquad \frac{\partial \Lambda}{\partial y} = 0, \qquad \text{and} \qquad \frac{\partial \Lambda}{\partial \lambda} = 0.$$

The latter equation just means $(x, y) \in \mathbb{S}^{2n-1}$. The first two equations yield

$$j(x_j + y_j) + \lambda(x_j + y_j) = 0 \qquad \Leftrightarrow \qquad jz_j + \lambda z_j = 0 \qquad \forall j = 1, \ldots, n.$$

This is only satisfied if $x_k = y_k = 0$ for all $k \neq j$, so precisely if

$$z_j \in \mathbb{S}^1_j := \left\{(z_1, \dots, z_n) \in \mathbb{S}^{2n-1} : z_k = 0 \, \forall k \neq j\right\}.$$

Hence $\mathrm{Crit}(f)$ is a disjoint union of n circles $\mathbb{S}^1_1 \cup \dots \cup \mathbb{S}^1_n$. Consider the bordered Hessian

$$\mathrm{Hess}_\Lambda := \begin{pmatrix} \frac{\partial^2 \Lambda}{\partial \lambda^2} & \frac{\partial^2 \Lambda}{\partial \lambda \partial x} & \frac{\partial^2 \Lambda}{\partial \lambda \partial y} \\ \left(\frac{\partial^2 \Lambda}{\partial \lambda \partial x}\right)^T & \frac{\partial^2 \Lambda}{\partial x^2} & \frac{\partial^2 \Lambda}{\partial x \partial y} \\ \left(\frac{\partial^2 \Lambda}{\partial \lambda \partial y}\right)^T & \frac{\partial^2 \Lambda}{\partial y \partial x} & \frac{\partial^2 \Lambda}{\partial y^2} \end{pmatrix}$$

$$= \begin{pmatrix} 0 & 2x_1 & \dots & 2x_n & 2y_1 & \dots & 2y_n \\ 2x_1 & 2+2\lambda & & & & & \\ \vdots & & \ddots & & & & \\ 2x_n & & & 2n+2\lambda & & & \\ 2y_1 & & & & 2+2\lambda & & \\ \vdots & & & & & \ddots & \\ 2y_n & & & & & & 2n+2\lambda \end{pmatrix}.$$

Consequently, we have that

$$\ker \mathrm{Hess}_f(z_j) = \ker \mathrm{Hess}_\Lambda(z_j, -j) = \{ae_j + be_{j+n} : ax_j + by_j = 0\} = T_{z_j}\mathbb{S}^1_j$$

for all $j = 1, \dots, n$ and f is a Morse–Bott function on $\mathbb{S}^{2n-1}$.

5.10 We have already seen in Exercise 5.7, that every Morse function on a compact smooth manifold is a Morse–Bott function. The only Morse function on a point is the constant one and thus has precisely one critical point. Moreover, there are no cascades in that case and thus the Morse–Bott homology just coincides with ordinary Morse homology. Another extremely important example of a Morse–Bott function on a smooth manifold M is a constant function f. In that case, $\mathrm{Crit}(f) = M$ and choosing any Morse function h on M we just get the ordinary Morse homology since there are no cascades.

5.11 Consider the real line $\mathbb{R}$ with the Morse function $h(x) := -x^2$. Then h admits only a nondegenerate maximum at $x = 0$ and thus the resulting homology is

$$\mathrm{HM}_k(h) = \begin{cases} \mathbb{Z}_2 & k = 1, \\ 0 & \text{else}, \end{cases}$$

not coincident with the singular homology of the real line with coefficients in $\mathbb{Z}_2$! Another extreme example would be the Morse function $h(x) := x$ with no critical points at all. Consequently, the homology strongly depends on the choice of a particular Morse function and is not independent of this choice. There are other reasons why the construction of Morse homology fails for noncompact manifolds. If we remove a point of the heart where one gradient flow line connects the saddle point to the minimum, then the resulting boundary operators no longer satisfy $\partial \circ \partial = 0$. Thus, the Morse homology of the punctured heart is not even defined in that case.

5.13 The resulting homology is $\mathrm{H}_*(\mathbb{S}^2; \mathbb{Z}_2)$ by considering the chain complex

$$0 \longrightarrow \mathbb{Z}_2 \oplus \mathbb{Z}_2 \xrightarrow{\begin{pmatrix} 1 & 1 \\ 0 & 0 \end{pmatrix}} \mathbb{Z}_2 \oplus \mathbb{Z}_2 \xrightarrow{\begin{pmatrix} 0 & 1 \\ 0 & 1 \end{pmatrix}} \mathbb{Z}_2 \oplus \mathbb{Z}_2 \longrightarrow 0.$$

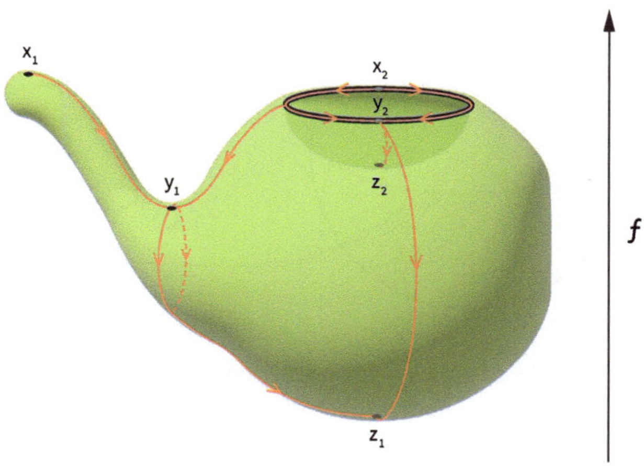

5.14

1. The indices of f directly follow from the computation in Exercise 5.9

$$\mathrm{ind}_f(z_j) = 2j - 2 \qquad \forall z_j \in \mathbb{S}^1_j.$$

On every circle $\mathbb{S}^1_j$, $j = 1, \ldots, n$, the standard height function h has precisely two critical points: one maximum x_j^+ and one minimum x_j^-. Thus, we compute for the index

$$\mathrm{ind}_f(x_j^+) + \mathrm{ind}_h(x_j^+) = 2j - 1 \qquad \text{and} \qquad \mathrm{ind}_f(x_j^-) + \mathrm{ind}_h(x_j^-) = 2j - 2.$$

2. The associated chain complex is given by

$$\mathrm{CM}_k(f) \cong \begin{cases} \mathbb{Z}_2 & 0 \leq k \leq 2n - 1, \\ 0 & \text{else.} \end{cases}$$

with boundary operator

$$\partial_k \colon \mathrm{CM}_k(f) \to \mathrm{CM}_{k-1}(f), \qquad \partial_k = \begin{cases} 1 & k \text{ even and } 1 \leq k \leq 2n-1, \\ 0 & \text{else.} \end{cases}$$

as depicted below. Thus the resulting homology is

$$\mathrm{HM}_k(f) \cong \begin{cases} \mathbb{Z}_2 & k = 0, 2n-1, \\ 0 & \text{else.} \end{cases}$$

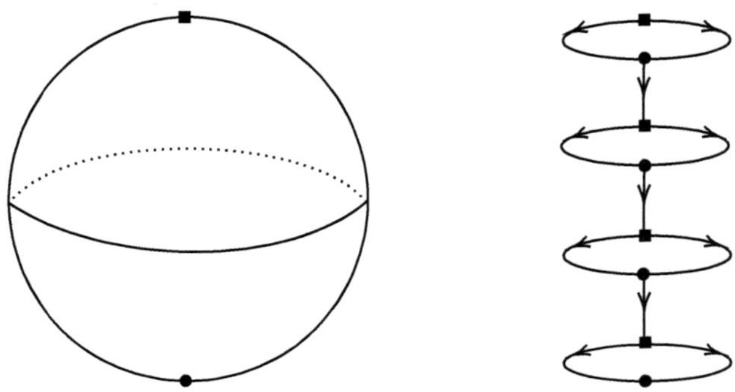

3. The homology of lens spaces with coefficients in $\mathbb{Z}$ is given in (4.6). From that we can compute the homology with coefficients in $\mathbb{Z}_2$ by means of the universal coefficient theorem for homology [104, Theorem 3A.3] by the short exact sequence

$$0 \longrightarrow \mathrm{H}_k(L_m;\mathbb{Z}) \otimes_{\mathbb{Z}} \mathbb{Z}_2 \longrightarrow \mathrm{H}_k(L_m;\mathbb{Z}_2) \longrightarrow \mathrm{Tor}(\mathrm{H}_{k-1}(L_m,\mathbb{Z}_2),\mathbb{Z}_2) \longrightarrow 0.$$

We compute

$$\mathbb{Z} \otimes_{\mathbb{Z}} \mathbb{Z}_2 \cong \mathbb{Z}_2 \qquad \text{and} \qquad \mathbb{Z}_m \otimes_{\mathbb{Z}} \mathbb{Z}_2 \cong \mathbb{Z}_2/m\mathbb{Z}_2 \cong \begin{cases} \mathbb{Z}_2 & m \text{ even,} \\ 0 & m \text{ odd,} \end{cases}$$

and

$$\mathrm{Tor}(\mathbb{Z},\mathbb{Z}_2) \cong 0 \qquad \text{and} \qquad \mathrm{Tor}(\mathbb{Z}_m,\mathbb{Z}_2) \cong \ker(\mathbb{Z}_2 \xrightarrow{m} \mathbb{Z}_2) \cong \begin{cases} \mathbb{Z}_2 & m \text{ even,} \\ 0 & m \text{ odd.} \end{cases}$$

Thus the resulting homology is

$$\mathrm{H}_k(L_m; \mathbb{Z}_2) \cong \begin{cases} \mathbb{Z}_2 & k = 0, \ldots, 2n-1, \\ 0 & \text{else}, \end{cases}$$

if m is even and $\mathrm{H}_k(L_m; \mathbb{Z}_2) = 0$ for all k if m is odd.

4. We would like to proceed as in the case of a $\mathbb{Z}_2$-invariant Morse function on the sphere $\mathbb{S}^2$ for the real projective plane $\mathbb{RP}^2$. On each critical component $\mathbb{S}^1_j$ pick the $\mathbb{Z}_m$-invariant Morse function $h(e^{i\theta}) := \cos(2\pi m\theta)$. By connecting every maximum to a minimum of index difference one as illustrated below, we get the associated chain complex

$$\mathrm{CM}_k(f) \cong \begin{cases} \mathbb{Z}_2^m & 0 \leq k \leq 2n-1, \\ 0 & \text{else}, \end{cases}$$

with boundary maps rather ugly to write down. Do this as an exercise, and check that the resulting homology is still that of the odd-dimensional sphere $\mathbb{S}^{2n-1}$! However, if we use the identification via the $\mathbb{Z}_m$-action, we get

$$\mathrm{CM}_k(f/\mathbb{Z}_m) \cong \begin{cases} \mathbb{Z}_2 & 0 \leq k \leq 2n-1, \\ 0 & \text{else}, \end{cases}$$

with boundary maps $\partial_k \colon \mathrm{CM}_k(f/\mathbb{Z}_m) \to \mathrm{CM}_{k-1}(f/\mathbb{Z}_m)$ given by

$$\partial_k = \begin{cases} 1 & k \text{ even and } 1 \leq k \leq 2n-1, m \text{ odd} \\ 0 & \text{else}. \end{cases}$$

Thus the resulting homology is

$$\mathrm{HM}_k(f/\mathbb{Z}_m) \cong \begin{cases} \mathbb{Z}_2 & k = 0, \ldots, 2n-1, \\ 0 & \text{else}, \end{cases}$$

if m is even and $\mathrm{HM}_k(f/\mathbb{Z}_m) = 0$ for all k if m is odd.

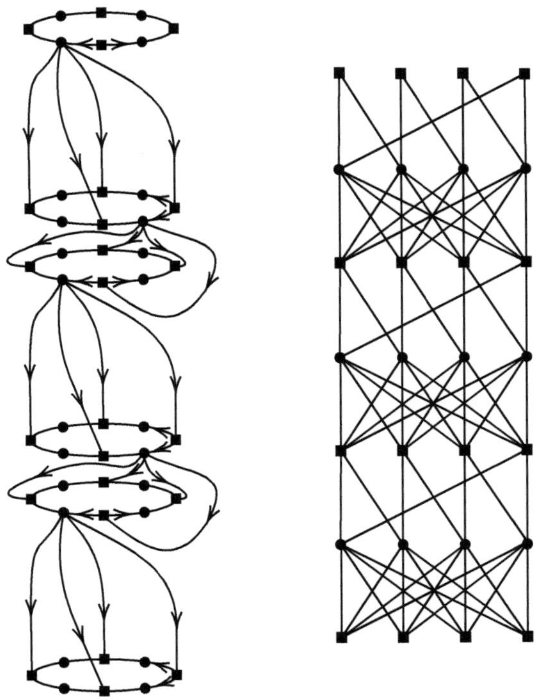

5.18 Let $R = \mathbb{Z}_2$ or $R = \mathbb{Z}$ and let $h \in C^\infty(M)$ be a Morse function on a compact smooth manifold M^n. Then there exists a polynomial Q with nonnegative integer coefficients satisfying

$$\sum_{k=0}^{n} \#\operatorname{Crit}_k(h) z^k = \sum_{k=0}^{n} \operatorname{rank} \mathrm{H}_k(M; R) z^k + (1+z) Q(z).$$

In particular, we deduce that

$$\#\operatorname{Crit}_k(h) \geq \operatorname{rank} \mathrm{H}_k(M; R) \qquad \forall 0 \leq k \leq n,$$

implying that

$$\#\operatorname{Crit}(h) \geq \sum_{k=0}^{n} \operatorname{rank} \mathrm{H}_k(M; R).$$

Without the assumption that h is a Morse function, this inequality is not true. In fact, we have $\#\operatorname{Crit}(h) \geq 4$ for any Morse function $h \in C^\infty(\mathbb{T}^2)$ and the function

$$\mathbb{T}^2 \to \mathbb{R}, \qquad (x, y) \mapsto \sin(\pi x) \sin(\pi y) \sin(\pi(x+y))$$

by the identification $\mathbb{T}^2 = \mathbb{R}^2/\mathbb{Z}^2$ has only three critical points: a maximum, a minimum, and a degenerate saddle point also called *monkey saddle*.

5.19 One immediately sees that

$$\chi(M) = \sum_{k=0}^{\dim M} (-1)^k \operatorname{rank} \mathrm{H}_k(M; R) = P_M(-1; R),$$

and similarly

$$\chi(C) = \sum_{k=0}^{\dim C} (-1)^k \operatorname{rank} \mathrm{H}_k(C; R) = P_C(-1; R)$$

for any critical submanifold $C \subseteq M$. This implies

$$P_f(-1; R) = \sum_{C \subseteq M \text{ critical submanifold}} (-1)^{\operatorname{ind}_f(C)} \chi(C)$$

for the Morse–Bott polynomial, and thus we conclude from Proposition 5.16

$$\chi(M) = P_M(-1; R) = P_f(-1; R) = \sum_{C \subseteq M \text{ critical submanifold}} (-1)^{\operatorname{ind}_f(C)} \chi(C).$$

5.20

1. The negative normal bundles $E^-(\mathbb{S}^1_j) = \mathbb{R}^{2j-2}$ for all $j = 1, \dots, n$ are $\mathbb{Z}$-orientable and the corresponding polynomials are given by

$$P_{\mathbb{S}^{2n-1}}(z; R) = \sum_{j=0}^{2n-1} \operatorname{rank} \mathrm{H}_j(\mathbb{S}^{2n-1}; R) z^j = 1 + z^{2n-1} = (1+z) \sum_{j=0}^{2n-2} (-1)^j z^j,$$

and

$$P_f(z; R) = \sum_{j=1}^{n} P_{\mathbb{S}^1_j}(z; R) z^{\operatorname{ind}_f(\mathbb{S}^1_j)} = (1+z) \sum_{j=1}^{n} z^{2j-2}.$$

Thus an explicit polynomial Q with non-negative integer coefficients is

$$Q := \sum_{j=1}^{n-1} z^{2j-1}.$$

2. The negative normal bundles

$$E^-(\mathbb{S}^1_j/\mathbb{Z}_m) = \mathbb{R}^{2j-2} \qquad \forall j = 1, \ldots, n$$

are $\mathbb{Z}$-orientable. First, consider the case where <u>m is odd</u> and let $R = \mathbb{Z}_2$. Then we compute

$$P_{\mathbb{S}^{2n-1}/\mathbb{Z}_m}(z; \mathbb{Z}_2) = \sum_{j=0}^{2n-1} \operatorname{rank} \mathrm{H}_j(\mathbb{S}^{2n-1}/\mathbb{Z}_m; \mathbb{Z}_2) z^j = 0,$$

and

$$P_f(z; \mathbb{Z}_2) = \sum_{j=1}^{n} P_{\mathbb{S}^1_j/\mathbb{Z}_m}(z; \mathbb{Z}_2) z^{\operatorname{ind}_f(\mathbb{S}^1_j/\mathbb{Z}_m)} = (1+z) \sum_{j=1}^{n} z^{2j-2} = (1+z)Q(z).$$

Consider the case $R = \mathbb{Z}$ now. Then we compute

$$P_{\mathbb{S}^{2n-1}/\mathbb{Z}_m}(z; \mathbb{Z}) = \sum_{j=0}^{2n-1} \operatorname{rank} \mathrm{H}_j(\mathbb{S}^{2n-1}/\mathbb{Z}_m; \mathbb{Z}) z^j = 1 + z^{2n-1}.$$

Thus, we are in the setting of the first part. Assume that <u>m is even</u>. For $R = \mathbb{Z}$ nothing changes and Q can be computed as in the previous case. However, for $R = \mathbb{Z}_2$, we compute

$$P_{\mathbb{S}^{2n-1}/\mathbb{Z}_m}(z; \mathbb{Z}_2) = \sum_{j=0}^{2n-1} \operatorname{rank} \mathrm{H}_j(\mathbb{S}^{2n-1}/\mathbb{Z}_m; \mathbb{Z}_2) z^j = \sum_{j=0}^{2n-1} z^j,$$

and

$$P_f(z; \mathbb{Z}_2) = \sum_{j=1}^{n} P_{\mathbb{S}^1_j/\mathbb{Z}_m}(z; \mathbb{Z}_2) z^{\operatorname{ind}_f(\mathbb{S}^1_j/\mathbb{Z}_m)} = (1+z) \sum_{j=1}^{n} z^{2j-2}.$$

Thus we get

$$P_f(z; \mathbb{Z}_2) - P_{\mathbb{S}^{2n-1}/\mathbb{Z}_m}(z; \mathbb{Z}_2) = 0 = (1+z)Q$$

for $Q := 0$.

3. On the circles $\mathbb{S}^1_j$, the negative gradient flow lines flow into opposite directions, and thus the respective boundary map is still zero. The negative gradient flow lines of the Morse–Bott function $f/\mathbb{Z}_m$ are all oriented the same and thus the associated boundary map is just multiplication by m. This gives the right singular homology with integer coefficients.

5.21 By differentiating the shift map with respect to the $\mathbb{R}$-variable, one loses a derivative of the loop. Moreover, the shift map is merely continuous with respect to the compact open topology [78, Lemma 2.1], but not in the norm topology [78, Lemma 2.2]. For more details, see [69, Example 2.4].

5.23 For a proof of the claimed properties of weighted Sobolev spaces, see [69, Lemma 4.10]. Suppose that $u \in W_\delta^{k,p}(\mathbb{R}, \mathbb{R})$ for $k \geq 1$. Then, by definition of weighted Sobolev spaces, we have $\gamma_\delta u \in W_\delta^{k,p}(\mathbb{R}, \mathbb{R})$. By [43, Corollary 8.9] it follows that

$$\lim_{|s|\to\infty} \gamma_\delta(s)u(s) = 0.$$

This can only happen if u decays "exponentially fast" at its ends.

5.34 In the case of a Morse function, exponential decay guarantees that any negative gradient flow line asymptotically converges to a unique critical point. Indeed, for functions which are not Morse, it might happen that a negative gradient flow line spirals into a set of critical points as described in [152, Example 3]. In the Morse–Bott case the interpretation is similar: The exponential decay guarantees that a negative gradient flow line of a Morse–Bott function asymptotically converges to a unique critical submanifold at both ends.

5.36

1. Let $u \in C^\infty(\mathbb{R}, M)$ be a negative gradient flow line of a Morse-Bott function f. As $\lim_{s\to\pm\infty} u(s) = x^\pm \in \operatorname{Crit}(f)$ by assumption, there exists some $T \in \mathbb{R}$ such that $u(s)$ belongs to some Morse–Bott chart about x^- for all $s \leq T$ and to some Morse–Bott chart about x^+ for all $s \geq T$. Locally we can therefore write

$$f(x) = \frac{1}{2}\sum_{j=1}^{n} a_j^\pm x_j^2$$

where $a_j^\pm$ are the eigenvalues of $\operatorname{Hess}_f(x^\pm)$, respectively. The negative gradient flow line u thus locally satisfies

$$\begin{aligned}
\dot{u}(s) &= -\operatorname{grad}_f(u(s)) \\
&= -\sum_{i,j=1}^{n} m_{ij}\frac{\partial f}{\partial x_i}(u(s))\frac{\partial}{\partial x_j} \\
&= -\sum_{i,j=1}^{n} \delta_{ij}\frac{\partial f}{\partial x_i}(u(s))\frac{\partial}{\partial x_j} \\
&= -\sum_{j=1}^{n} a_j^\pm u_j(s)\frac{\partial}{\partial x_j}
\end{aligned}$$

for all $|s| \geq T$ as we are working with the standard Euclidean metric. Explicitly solving this ordinary differential equation yields $u_j(s) = u_j(\pm T)e^{-a_j^{\pm}s}$ for all $|s| \geq T$. Moreover, we see that

$$\partial_s^k u_j(s) = (-a_j^{\pm})^k u_j(\pm T)e^{-a_j^{\pm}s} \qquad \forall k \in \mathbb{N}_0.$$

2. If one uses a general metric instead of the standard Euclidean one, then the negative gradient flow equation cannot easily be solved. This was crucial for concluding exponential decay in the first part.

5.37

1. The reason for this terminology comes from the fact that for a negative gradient flow line $u \in C^\infty(\mathbb{R}, M)$ we compute

$$|f(u(s))| \leq \kappa \| \operatorname{grad}_m f(u(s))\|_m^2 = \kappa \|\partial_s u(s)\|_m^2$$

for all $s \in \mathbb{R}$, when we assume that an action-energy inequality (5.7) for a Morse–Bott function f holds. Thus the term $\| \operatorname{grad}_m f(u(s))\|_m^2$ can be interpreted as an energy density by considering the energy $E_m(u)$.
2. As in Exercise 5.36, we write

$$f(x) = \frac{1}{2}\sum_{j=1}^{n} a_j x_j^2, \qquad n := \dim M.$$

We claim the action-energy inequality (5.7) then holds with

$$\kappa := \frac{1}{2\min\{|a_j| : a_j \neq 0, 1 \leq j \leq n\}} > 0.$$

Indeed, we estimate

$$|f(x)| \leq \frac{1}{2}\sum_{j=1}^{n} |a_j| x_j^2 = \frac{1}{2}\kappa \sum_{j=1}^{n} \frac{|a_j|}{\kappa} x_j^2 \leq \kappa \sum_{j=1}^{n} a_j^2 x_j^2 = \kappa \|\nabla f(x)\|^2,$$

where ∇f denotes the local gradient of f with respect to the Euclidean metric.
3. We estimate

$$\begin{aligned} 0 &\leq m_1\left(\kappa^2 \operatorname{grad}_{m_1} f - \operatorname{grad}_{m_2} f, \kappa^2 \operatorname{grad}_{m_1} f - \operatorname{grad}_{m_2} f\right) \\ &= \kappa^4 \| \operatorname{grad}_{m_1} f\|_{m_1}^2 - 2\kappa^2 m_1\left(\operatorname{grad}_{m_1} f, \operatorname{grad}_{m_2} f\right) + \| \operatorname{grad}_{m_2} f\|_{m_1}^2. \end{aligned}$$

The second term we rewrite using the definition of the gradient two times for both metrics m_1 and m_2

$$m_1\left(\operatorname{grad}_{m_1} f, \operatorname{grad}_{m_2} f\right) = df\left(\operatorname{grad}_{m_2} f\right) = \| \operatorname{grad}_{m_2} f \|_{m_2}^2.$$

The third term can be estimated using the assumption of the exercise

$$\| \operatorname{grad}_{m_2} f \|_{m_1}^2 \leq \kappa^2 \| \operatorname{grad}_{m_2} f \|_{m_2}^2.$$

Combining these three facts we obtain the estimate

$$\begin{aligned} 0 &\leq \kappa^4 \| \operatorname{grad}_{m_1} f \|_{m_1}^2 - 2\kappa^2 \| \operatorname{grad}_{m_2} f \|_{m_2}^2 + \kappa^2 \| \operatorname{grad}_{m_2} f \|_{m_2}^2 \\ &= \kappa^4 \| \operatorname{grad}_{m_1} f \|_{m_1}^2 - \kappa^2 \| \operatorname{grad}_{m_2} f \|_{m_2}^2 \end{aligned}$$

implying

$$\| \operatorname{grad}_{m_2} f \|_{m_2}^2 \leq \kappa^2 \| \operatorname{grad}_{m_1} f \|_{m_1}^2.$$

5.40 We explain the main idea for a negative gradient flow line with precisely one cascade. For $x^\pm \in \operatorname{Crit}(h)$, a negative gradient flow line with one cascade is a negative gradient flow line $u \in C^\infty(\mathbb{R}, M)$ of the Morse–Bott function f, such that there exist $y^\pm \in W^\pm(x^\pm)$ satisfying $\lim_{s\to\pm\infty} u(s) = y^\pm$. In other words, y^- lies in the unstable manifold of x^-, and y^+lies in the stable manifold of x^+. Moreover, this implies the existence of negative gradient flow lines $u^\pm \in C^\infty(\mathbb{R}, \operatorname{Crit}(f))$ of the Morse function h with

$$\lim_{s\to\pm\infty} u^\pm(s) = x^\pm \qquad \text{and} \qquad u^\pm(t^\pm) = y^\pm$$

for some times $t^\pm \in \mathbb{R}$. Consequently, the negative gradient flow lines $u^\pm$ are partial negative gradient flow lines of the Morse function h and it does not make sense to reparametrise them via a $\mathbb{R}$-time shift. However, reparametrisation makes sense for the negative gradient flow line u of f. This implies that $\mathbb{R}^m$ acts only on the m negative gradient flow lines of the Morse–Bott function for a negative gradient flow line with m cascades.

5.43 The main difference between the polyfold approach and the classical approach is that the classical approach perturbs the Riemannian metric, whereas the polyfold approach perturbs a section of suitable bundles. In Morse–Bott homology, this might be a matter of taste, but since we are ultimately interested in constructing a Morse–Bott homology associated with the symplectic action functional or the Rabinowitz action functional we roughly explain an advantage of the polyfold approach there. As explained in [76, pp. 296–297], the set of almost complex structures for defining Floer homology depends on the method that is used to establish transversality. Indeed, if one uses the language of polyfolds, fixing a suitable almost complex structure works. However, for the classical approach, one needs to consider a smooth family $(J_t)_{t\in\mathbb{T}}$ in the case of Floer homology, and even a smooth family $(J_{t,\tau})_{(t,\tau)\in\mathbb{T}\times\mathbb{R}}$ for Rabinowitz–Floer homology by [4, Section 4.3]. This introduces some cumbersome notation in the proofs, and especially in defining the L^2-metric. However, the language of polyfold theory is cumbersome as well, and often technicalities and issues are hidden behind the abstract framework. So in both cases, one has to be very careful if one wants to establish new results!

Exercises and Problems of Chap. 6

6.1

1. By Exercise 5.4, we have the formula

$$\begin{aligned} d\mathscr{A}_H|_\gamma(X) &= \int_0^1 \omega(X(t), \dot{\gamma}(t) - X_{H_t}(\gamma(t)))dt \\ &= \int_0^1 \omega(JX(t), J(\dot{\gamma}(t) - X_{H_t}(\gamma(t))))dt, \end{aligned}$$

since J is ω-compatible and thus satisfies $\omega(J\cdot, J\cdot) = \omega(\cdot, \cdot)$. By the defining equation of the gradient $\operatorname{grad}_J \mathscr{A}_H \in \mathfrak{X}(\Lambda M)$ we get

$$\begin{aligned} d\mathscr{A}_H|_\gamma(X) &= \int_0^1 \omega(J(\operatorname{grad}_J \mathscr{A}_H|_\gamma(t)), X(t))dt \\ &= \int_0^1 \omega(JX(t), \operatorname{grad}_J \mathscr{A}_H|_\gamma(t))dt \end{aligned}$$

by the symmetry of the Riemannian metric $\omega(J\cdot, \cdot)$. From these two formulas we get the desired gradient by using the nondegeneracy of ω.

2. Consider the Floer equation

$$\partial_s u(s,t) + J(\partial_t u(s,t) - X_{H_t}(u(s,t))) = 0 \qquad \forall (s,t) \in \mathbb{R}\times\mathbb{T},$$

for the Floer cylinder $u \in C^\infty(\mathbb{R}\times\mathbb{T}, M)$ as in (6.1). If $H = 0$, then the Floer equation reduces to

$$\partial_s u(s,t) + J\partial_t u(s,t) = 0 \qquad \forall (s,t) \in \mathbb{R}\times\mathbb{T}.$$

This is the well-know equation of a J-holomorphic map as introduced by Gromov in 1985 in his seminal paper [96]. If we assume that the Floer cylinder u does not depend on time $t \in \mathbb{T}$, then the Floer equation reduces to

$$\partial_s u(s) - JX_H(u(s)) = 0 \qquad \forall s \in \mathbb{R}.$$

If we denote by $m_J(\cdot,\cdot) := \omega(J\cdot,\cdot)$ the induced Riemannian metric, one can rewrite the above equation as

$$\partial_s u(s) - \operatorname{grad}_{m_J} H(u(s)) = 0 \qquad \forall s \in \mathbb{R}.$$

Consequently, $u \in C^\infty(\mathbb{R}, M)$ is a positive gradient flow line of H giving rise to Morse cohomology. Finally, if we assume that $u \in C^\infty(\mathbb{R}\times\mathbb{T}, M)$ is s-independent, the Floer equation reduces to

$$\partial_t u(t) - X_{H_t}(u(t))) = 0 \qquad \forall t \in \mathbb{T}.$$

This is a time-dependent 1-periodic Hamiltonian equation from Hamiltonian dynamics from Exercise 5.4 or Lemma 3.27. Hence, studying the Floer equation combines two theories we have already discussed in the book!

6.2 As $\mathscr{A}_H$ is a zero-order perturbation of the symplectic area functional $\mathscr{A}_0$, it is enough to compute Morse indices in that case. Moreover, using a Darboux chart, it is actually enough to consider a loop $z \in C^\infty(\mathbb{T}, \mathbb{C}^n)$. This loop can be represented by its Fourier series

$$z = \sum_{k=-\infty}^{+\infty} z_k e^{2\pi i k t}, \qquad z_k \in \mathbb{C}^n.$$

Then

$$\mathscr{A}_0(z) = -\pi \sum_{k=-\infty}^{+\infty} k\|z_k\|^2 =: \mathscr{A}(z).$$

In fact, we have $\mathscr{A}_0(0) = 0 = \mathscr{A}(0)$ and as $C^\infty(\mathbb{T}, \mathbb{C}^n)$ is connected, it suffices to show that the differential on both sides coincides. Represent the tangent vector $v \in T_z C^\infty(\mathbb{T}, \mathbb{C}^n) \cong C^\infty(\mathbb{T}, \mathbb{C}^n)$ by its Fourier series as

$$v = \sum_{k=-\infty}^{+\infty} v_k e^{2\pi i k t}, \qquad v_k \in \mathbb{C}^n.$$

Then we compute

$$\begin{aligned} d\mathscr{A}_0|_z(v) &= \int_0^1 \omega(v(t), \dot{z}(t))dt \\ &= \int_0^1 \operatorname{Re}(i\overline{v}(t)\dot{z}(t))dt \end{aligned}$$

$$= -2\pi \operatorname{Re}\left(\sum_{k=-\infty}^{+\infty} k\overline{v}_k z_k\right)$$

$$= -\pi \sum_{k=-\infty}^{+\infty} k(\overline{v}_k z_k + v_k \overline{z}_k)$$

$$= d\mathscr{A}|_z(v).$$

Thus the gradient $\operatorname{grad}\mathscr{A}_0$ with respect to the standard real inner product on $\mathbb{C}^n$ is given by

$$\operatorname{grad}\mathscr{A}_0|_z^k = -2\pi k z_k \qquad \forall k \in \mathbb{Z}.$$

Consequently, the spectrum of the Hessian of $\mathscr{A}_0$ is $2\pi\mathbb{Z}$ and we see that $\mathscr{A}_0$ has infinite Morse index and coindex.

6.12 There are several traditional results in complex analysis [158, Kapitel 8] that admit a suitable analogue in the pseudoholomorphic setting. The general theme is as follows. In the integrable case the proofs are obvious consequences of the fact that every differentiable function of one complex variable admits a power series expansion. The proof in the almost complex case is usually considerably harder and works only under some suitable additional assumptions.

- The ***identity theorem*** for complex functions states that for two complex differentiable functions $f, g\colon U \to \mathbb{C}$, where $U \subseteq \mathbb{C}$ is an open subset, the following conditions are equivalent:

  1. $f = g$
  2. The set $\{z \in U\colon f(z) = g(z)\}$ admits an accumulation point in U.
  3. There exists $z_0 \in U$ with $f^{(k)}(z_0) = g^{(k)}(z_0)$ for all $k \geq 0$.

  In the pseudoholomorphic case, there is a similar statement in [136, Theorem 2.3.2].
- The ***mean value inequality*** for a compact symplectic manifold (M, ω) with ω-compatible almost complex structure J states that for every J-holomorphic curve

$$u\colon \{w \in \mathbb{C} : |z - w| \leq r\} \to M$$

  for $r > 0$, there exist constants $c, \delta > 0$ such that

$$\int_{B_r} |du|^2 < \delta \qquad \Rightarrow \qquad |du(0)|^2 \leq \frac{c}{r^2}\int_{B_r} |du|^2.$$

  The standard mean value equality in complex analysis [158, Satz 7.2.3] states that

$$f'(z_0) = \frac{1}{2\pi}\int_0^{2\pi} f'(z_0 + re^{i\theta})d\theta$$

for all complex differentiable functions $f\colon \{w \in \mathbb{C} : |z - w| \leq r\} \to \mathbb{C}$.

- The ***maximum principle*** [158, Theorem 8.5.6] states that if $U \subseteq \mathbb{C}$ is a bounded open subset and $f\colon U \to \mathbb{C}$ is a complex differentiable function that continuously extends to the closure $\bar{U}$, then we have the inequality $|f(z)| \leq \|f|_{\partial U}\|_\infty$ for all $z \in \bar{U}$. In particular, every such function $f\colon U \to \mathbb{C}$ admits its maximum on the boundary ∂U. The classical strong maximum principle for subharmonic functions [87, Theorem 2.2], states that if $u \in C^2(U)$ for a bounded open subset $U \subseteq \mathbb{R}^2$ satisfies $\Delta u \geq 0$ and there exists a point $z_0 \in U$ satisfying $f(z_0) = \sup_{z\in U} u(z)$, then u is constant. In particular, every harmonic function, that is, $\Delta u = 0$, cannot assume an interior maximum unless it is constant. An even more advanced maximum principle is presented in Problem 7.62, following [136, Corollary 9.2.11]. Maximum principles are crucial for developing Floer homology for non-compact manifolds as they provide uniform bounds for sequences of negative gradient flow lines.

6.15 First of all, it is clear by Exercise 5.2, that without further assumptions on the symplectic manifold (M, ω) the symplectic action functional is not well-defined in general. The next inaccuracy is the word "simply". The construction of Morse–Bott homology in Part 5 considered Morse–Bott functions $f \in C^\infty(M)$, where M is a finite-dimensional compact smooth manifold. However, in Floer homology we consider the symplectic action functional $\mathscr{A}_H$ which is defined on the infinite-dimensional contractible loop space ΛM. Even worse, the space of smooth loops $\mathscr{L}M$ is not a nicely behaved smooth infinite-dimensional manifold like the Banach manifolds in [125, p. viii]. This is only true for regularities $C^k(\mathbb{T}, M)$ for $k < \infty$, and the loop space is a Fréchet manifold by [197]. On Fréchet manifolds, doing analysis is usually quite hard, as, for example, the implicit function theorem can be wrong in general. For a general introduction to loop spaces from the point of view of topology, see [126, pp. 33–40]. So also the statement "negative gradient flow lines are maps $u \in C^\infty(\mathbb{R}, \Lambda M)$" is problematic, as this would require defining a suitable smooth structure on the infinite-dimensional space ΛM. One should rather use Floer cylinders as used in the second model, as smoothness is easy to define there, and this observation also leads to the application of holomorphic curve techniques. This was also particularly important in proving Proposition 6.5. Also, the notation $\operatorname{grad}_J \mathscr{A}_H \in \mathfrak{X}(\Lambda M)$ is slightly imprecise. Such notation might be convenient and also very didactically appealing, as we know exactly where things "live", but might introduce misconceptions, as one has to be very careful about well-definedness. It is also assumed in the first model of Hamiltonian Floer homology that there is a well-defined grading by the Conley–Zehnder index in general. This is only true under the additional assumption that $c_1(TM)|_{\pi_2(M)} = 0$. In both models, the conclusion "Hence the Floer homology of a time-dependent Hamiltonian function does not give an interesting global invariant of (M, ω)" is wrong, since, for example, Hamiltonian

Floer homology gives a tool for bounding the number of nondegenerate periodic solutions of a time-dependent Hamiltonian function from below by a constant only depending on the topology of the manifold. In particular, this gives a lower bound for the nondegenerate fixed points of a Hamiltonian symplectomorphism. This is definitely a nontrivial statement!

Exercises and Problems of Chap. 7

7.2 Let (W^n, λ) be a Liouville domain and assume that $\partial W = \varnothing$. For the symplectic form $\omega := d\lambda$ we compute

$$0 \neq \int_W \omega^n = \int_W d(\lambda \wedge d\lambda^{n-1}) = \int_{\partial W} \lambda \wedge d\lambda^{n-1} = 0.$$

Consequently, the boundary ∂W is not empty and we can consider the smooth manifold $(\partial W, \lambda|_{\partial W})$ of dimension $2n - 1$. Let $X = X_\lambda \in \mathfrak{X}(W)$ be the Liouville vector field. Then we compute

$$\lambda \wedge \omega^{n-1} = i_X\omega \wedge \omega^{n-1} = \frac{1}{n} i_X \omega^n.$$

By [129, Proposition 15.12], the differential form $i_X\omega^n|_{\partial W} \in \Omega^{2n-1}(\partial W)$ is a volume form since X is outward-pointing along ∂W and thus, in particular, nowhere tangent to ∂W. Consequently, $\lambda|_{\partial W}$ is a contact form by Exercise 4.4.

7.7

1. Suppose that $f_\varphi, f'_\varphi \in C^\infty_c(\operatorname{Int} W)$ satisfy

   $$df_\varphi = \varphi^*\lambda - \lambda = df'_\varphi \qquad \Rightarrow \qquad d(f_\varphi - f'_\varphi) = 0.$$

   Thus by [129, Proposition 11.22], the function $f_\varphi - f'_\varphi$ is constant on each connected component of $\operatorname{Int} W$. But by assumption, a Liouville domain (W, λ) is connected and so $f'_\varphi = f_\varphi + C$ for some $C \in \mathbb{R}$. But $f_\varphi(x) = f'_\varphi(x) = 0$ for all

   $$x \in \operatorname{Int} W \setminus (\operatorname{supp}(f_\varphi) \cup \operatorname{supp}(f'_\varphi)),$$

   and so $C = 0$ implying $f_\varphi = f'_\varphi$.
2. Let $\varphi, \psi \in \operatorname{Aut}(W, \lambda)$ be two Liouville automorphisms. The composition $\varphi \circ \psi$ is not necessarily a Liouville automorphism, since $\operatorname{ord}(\varphi \circ \psi)$ is not finite unless φ and ψ commute. Moreover, if we consider two Liouville automorphisms as in Example 7.6 for $m \neq m'$, then the composition thereof is in general not of the desired form.

7.8 That ψ is an exact symplectic embedding follows from the computation

$$\begin{aligned}\frac{d}{dr}\psi_r^*\lambda &= (\phi_r^X)^* L_X\lambda \\ &= \left(\phi_r^X\right)^* (di_X\lambda + i_X d\lambda) \\ &= \left(\phi_r^X\right)^* (di_X i_X d\lambda + \lambda) \\ &= \left(\phi_r^X\right)^* \lambda \\ &= \psi_r^*\lambda\end{aligned}$$

similar to Exercise 7.8, implying

$$\psi_r^*\lambda = e^r \lambda|_\Sigma \qquad \forall r \in (-\infty, 0],$$

by $\psi_0 = \iota_{\partial W}$, where $\iota_{\partial W}\colon \partial W \hookrightarrow W$ denotes inclusion.

7.10 This essentially follows from Exercise 7.8 and [129, Theorem 9.29].

7.24

1. Assume that there does not exist a nonconstant periodic Reeb orbit on Σ. Then by the second property of Rabinowitz–Floer homology in Theorem 7.22, we conclude that

$$\mathrm{RFH}_*(\Sigma, \mathbb{R}^{2n}) \cong \mathrm{H}_*(\Sigma; \mathbb{Z}_2) \neq 0.$$

But every star-shaped hypersurface Σ is displaceable in $\mathbb{R}^{2n}$ by considering a translation, so by the third property in Theorem 7.22 also

$$\mathrm{RFH}_*(\Sigma, \mathbb{R}^{2n}) \cong 0.$$

It follows that there must exist a (non-constant) periodic Reeb orbit on Σ!
2. If a Liouville automorphism $\varphi \in \mathrm{Aut}(W, \lambda)$ satisfies $\varphi \neq \mathrm{id}_{W_\Sigma}$, there are two cases to distinguish. If $\mathrm{Fix}(\varphi|_\Sigma) \neq \varnothing$, then there exists a non-constant twisted periodic Reeb orbit by arguing as in the first part. However, if $\mathrm{Fix}(\varphi|_\Sigma) = \varnothing$, this argument no longer works!

7.28 There is a natural $\mathbb{R}$-action on the twisted loop space $\mathscr{L}_\varphi M$ given by

$$(s \cdot \gamma)(t) := \gamma(t+s) \qquad \forall t \in \mathbb{R}.$$

If (γ, τ) is a critical point of $\mathscr{A}_\varphi^H$ such that γ is not constant, then so is every reparametrisation. This is a Morse–Bott scenario. If we allow $\mathrm{ord}(\varphi|_{\partial W}) = +\infty$, this reparametrisation action would be an action of the noncompact group $\mathbb{R}$ on $\mathrm{Crit}(\mathscr{A}_\varphi^H)$ as opposed to a compact group action of the circle $\mathbb{T}$ in the case of finite-order. Non-compact actions usually induce analytical difficulties, so it is better to exclude them similar to how we counted only unparametrised negative gradient flow lines in Morse–Bott homology.

7.39 We first prove $\varphi_* R = R$. Observe that

$$\begin{aligned} i_{\varphi_* R} d\alpha &= \varphi^* d\alpha(R \circ \varphi^{-1}, D\varphi^{-1} \cdot) \\ &= d\varphi^* \alpha(R \circ \varphi^{-1}, D\varphi^{-1} \cdot) \\ &= d\alpha(R \circ \varphi^{-1}, D\varphi^{-1} \cdot) \\ &= \varphi_*(i_R d\alpha) \\ &= 0, \end{aligned}$$

and

$$\alpha(\varphi_* R) = \alpha(D\varphi \circ R \circ \varphi^{-1}) = \varphi^* \alpha(R \circ \varphi^{-1}) = \alpha(R \circ \varphi^{-1}) = 1.$$

Hence the statement follows from the uniqueness of the Reeb vector field $R = R_\alpha$. Moreover, we compute

$$\frac{d}{dt} \varphi \circ \phi_t^R = D\varphi \circ R \circ \phi_t^R = \varphi_* R \circ \varphi \circ \phi_t^R = R \circ \varphi \circ \phi_t^R,$$

for all $t \in \mathbb{R}$.

7.48 We proceed as in Exercise 6.1. By (7.4) we have the formula

$$\begin{aligned} d\mathscr{A}_\varphi^H|_{(\gamma,\tau)}(X, \eta) &= \int_0^1 d\lambda(X(t), \dot{\gamma}(t) - \tau X_H(\gamma(t))) - \eta \int_0^1 H(\gamma(t)) \\ &= \int_0^1 d\lambda(JX(t), J(\dot{\gamma}(t) - \tau X_H(\gamma(t)))) - \eta \int_0^1 H(\gamma(t)). \end{aligned}$$

By the defining equation of the gradient $\mathrm{grad}_J \mathscr{A}_\varphi^H \in \mathfrak{X}(\mathscr{L}_\varphi \widehat{W} \times \mathbb{R})$ we get

$$\begin{aligned} d\mathscr{A}_\varphi^H|_{(\gamma,\tau)}(X, \eta) &= \int_0^1 d\lambda(J\,\mathrm{grad}_J \mathscr{A}_\varphi^H|^1_{(\gamma,\tau)}, X(t))dt + \eta\, \mathrm{grad}_J \mathscr{A}_\varphi^H|^2_{(\gamma,\tau)} \\ &= \int_0^1 d\lambda(JX(t), \mathrm{grad}_J \mathscr{A}_\varphi^H|^1_{(\gamma,\tau)}dt + \eta\, \mathrm{grad}_J \mathscr{A}_\varphi^H|^2_{(\gamma,\tau)}, \end{aligned}$$

where $\operatorname{grad}_J \mathscr{A}_\varphi^H|^1_{(\gamma,\tau)}$ stands for the $T_\gamma \mathscr{L}_\varphi \widehat{W}$-component of the gradient and $\operatorname{grad}_J \mathscr{A}_\varphi^H|^2_{(\gamma,\tau)}$ for the $\mathbb{R}$-component. This yields the desired formula.

7.53 The generators of twisted Rabinowitz–Floer homology are possibly infinite sums $\zeta = \sum_{(\gamma,\tau)\in \mathrm{Crit}(h)} \zeta_{(\gamma,\tau)}(\gamma,\tau)$ with $\zeta_{(\gamma,\tau)} \in \mathbb{Z}_2$ for all $(\gamma,\tau) \in \mathrm{Crit}(h)$ satisfying the finiteness condition (7.19). This finiteness condition must be imposed for the following reason. Assume that the critical point $(\gamma,\tau) \in \mathrm{Crit}(h)$ satisfies $\tau \neq 0$. Then the period-action equality in Remark 7.43 implies

$$\mathscr{A}_\varphi^H\left(\gamma^k, ((k-1)m+1)\tau\right) = ((k-1)m+1)\tau \qquad \forall k \in \mathbb{Z},$$

where $\gamma^k(t) := \gamma\left(((k-1)m+1)t\right)$ with $m := \mathrm{ord}(\varphi)$ denotes the k-th iterate of γ in $\mathscr{L}_\varphi \widehat{W}$. This iterate also belongs to $\mathrm{Crit}(\mathscr{A}_\varphi^H)$ for all $k \in \mathbb{Z}$ and possibly to $\mathrm{Crit}(h)$. Hence, the functional $\mathscr{A}_\varphi^H$ is unbounded from below on $\mathrm{Crit}(\mathscr{A}_\varphi^H)$ and so there might be infinitely many critical points that appear in the image $\partial(\gamma,\tau)$ of the boundary operator $\partial\colon \mathrm{RFC}^\varphi(\partial W, \widehat{W}) \to \mathrm{RFC}^\varphi(\partial W, \widehat{W})$. This phenomenon cannot happen in Hamiltonian Floer homology because $\gamma \in \mathrm{Crit}(\mathscr{A}_H)$ satisfies the Hamiltonian equation $\dot{\gamma}(t) = X_H(\gamma(t))$ for all $t \in \mathbb{T}$ and γ being a loop. Now the iteration γ^k for $m = 1$ satisfies the equation

$$\partial_t \gamma^k(t) = k X_H(\gamma^k(t)) \qquad \forall t \in \mathbb{T}, k \in \mathbb{Z},$$

which is not a critical point equation, implying $\gamma^k \notin \mathrm{Crit}(\mathscr{A}_H)$ if γ is not constant unless $k = 0, 1$.

Exercises and Problems of Chap. 8

8.2 In order to simplify the discussion, assume that $\varphi = \mathrm{id}_W$. Then a point $x \in \partial W$ is a leaf-wise intersection point if $\varphi_F(x) \in L_x = \{\phi_t^R(x) : t \in \mathbb{R}\}$ for some Hamiltonian symplectomorphism $F \in \mathrm{Ham}(\widehat{W}, \lambda)$. Assume that L_x is a loop, that is, the Reeb flow ϕ^R has at least one periodic orbit. This is the orbit of the satellite. The Hamiltonian symplectomorphism then introduces a time-dependent perturbation of the Hamiltonian system, and the energy is no longer conserved. At time one, this perturbation through F is over, and thus a leaf-wise intersection point $x \in \partial W$ is back on the periodic orbit L_x, which means that the satellite has returned to its initial orbit. By [14, Corollary 1.3], a generic perturbation in $\mathrm{Ham}_c(T^*\mathbb{S}^2)$ has infinitely many leaf-wise intersections for a generic perturbation of a regular energy hypersurface in the planar circular restricted three-body problem around Jupiter or Europa if the energy value is below the first critical value. This is again an example of a very abstract result. Of course, we could infer that there is a good chance that if a comet passes by our satellite, the satellite will return to an orbit nearby our initial one, unless the comet crashes into our satellite! The same conclusion is also valid for energy values slightly above the first critical value. In that case, the regular energy hypersurface becomes the connected sum $\mathbb{RP}^3 \# \mathbb{RP}^3$, and the satellite may now travel between Jupiter and Europa.

8.7 Let $\varphi_F \in \text{Ham}_c(\widehat{W}, \lambda)$ be a displacing Hamiltonian diffeomorphism for ∂W, that is, we have $\varphi_F(\partial W) \cap \partial W = \varnothing$. Assume that $\text{Crit}(\mathscr{A}_\varphi^{\mathfrak{M}}) \neq \varnothing$ and denote by (γ, τ) a critical point of $\mathscr{A}_\varphi^{\mathfrak{M}}$. By Proposition 8.6, the point $x := \gamma(\frac{1}{2}) \in \partial W$ is a twisted leaf-wise intersection point for φ_F, implying that $\varphi_F(x) \in \partial W$. But φ_F displaces the boundary from itself and we conclude $\text{Crit}(\mathscr{A}_\varphi^{\mathfrak{M}}) = \varnothing$.

8.10 The critical points of the standard Rabinowitz action functional satisfy

$$(\gamma, \tau) \in \text{Crit}(\mathscr{A}^H) \qquad \Leftrightarrow \qquad \begin{cases} \dot{\gamma}(t) = \tau R(\gamma(t)) \; \forall t \in \mathbb{T}, \\ \gamma(\mathbb{T}) \subseteq \partial W. \end{cases}$$

The Lagrange multiplier $\tau \in \mathbb{R}$ can be thought of as a generalised period of a Reeb orbit. Indeed, if $\tau > 0$, then γ is simply a τ-periodic Reeb orbit on ∂W. If $\tau < 0$, then γ is also a τ-periodic Reeb orbit, but it is traversed in the opposite direction. If $\tau = 0$, then γ is constant and thus just a point of ∂W. In the proof of Theorem 4.30, these generalised periods were not used as by assumption $\tau > 0$. However, in the definition of Rabinowitz–Floer homology, these generalised periods are crucial for proving the basic properties in Theorem 7.22. In particular, the inclusion $\partial W \subseteq \text{Crit}(\mathscr{A}^H)$ leads to a proof of the Weinstein conjecture for star-shaped hypersurfaces in $\mathbb{R}^{2n}$ in Corollary 8.9 using crucially Proposition 7.54!

8.12 For $m = 2$, the chain complex is given by

$$\dots \longrightarrow \mathbb{Z}_2^2 \xrightarrow{\mathbb{1}} \mathbb{Z}_2^2 \xrightarrow{A} \mathbb{Z}_2^2 \xrightarrow{\mathbb{1}} \mathbb{Z}_2^2 \xrightarrow{A} \mathbb{Z}_2^2 \xrightarrow{\mathbb{1}} \mathbb{Z}_2^2 \longrightarrow \dots$$

where $\mathbb{1} \in \mathbb{Z}_2^{2\times 2}$ has every entry equal to 1 and $A \in \mathbb{Z}_2^{2\times 2}$ is defined by

$$A := I_2 + e_{21} + e_{12} = \begin{pmatrix} 1 & 1 \\ 1 & 1 \end{pmatrix} = \mathbb{1}.$$

Passing to the quotient via the free $\mathbb{Z}_2$-action, we get the acyclic chain complex

$$\dots \longrightarrow \mathbb{Z}_2 \xrightarrow{0} \mathbb{Z}_2 \xrightarrow{0} \mathbb{Z}_2 \xrightarrow{0} \mathbb{Z}_2 \longrightarrow \dots$$

For $m = 3$, the chain complex is given by

$$\dots \longrightarrow \mathbb{Z}_2^3 \xrightarrow{\mathbb{1}} \mathbb{Z}_2^3 \xrightarrow{A} \mathbb{Z}_2^3 \xrightarrow{\mathbb{1}} \mathbb{Z}_2^3 \xrightarrow{A} \mathbb{Z}_2^3 \xrightarrow{\mathbb{1}} \mathbb{Z}_2^3 \longrightarrow \dots$$

where $\mathbb{1} \in \mathbb{Z}_2^{3\times 3}$ has every entry equal to 1 and $A \in \mathbb{Z}_2^{3\times 3}$ is defined by

$$A := I_3 + e_{21} + e_{32} + e_{13} = \begin{pmatrix} 1 & 0 & 1 \\ 1 & 1 & 0 \\ 0 & 1 & 1 \end{pmatrix}.$$

Passing to the quotient via the free $\mathbb{Z}_3$-action, we get the alternating chain complex

$$\dots \longrightarrow \mathbb{Z}_2 \xrightarrow{1} \mathbb{Z}_2 \xrightarrow{0} \mathbb{Z}_2 \xrightarrow{1} \mathbb{Z}_2 \xrightarrow{0} \mathbb{Z}_2 \xrightarrow{1} \mathbb{Z}_2 \longrightarrow \dots$$

8.14 Set $G := \mathbb{Z}_m$ and start with the case $A = \mathbb{Z}_2$. Then $A^G = \mathbb{Z}_2$ by the definition of a trivial $\mathbb{Z}_m$-module. Now, the groups $N\mathbb{Z}_2$ and $\{a \in \mathbb{Z}_2 : Na = 0\}$ depend on the parity of m. In fact, we see that

$$N\mathbb{Z}_2 = \begin{cases} 0 & m \text{ even}, \\ \mathbb{Z}_2 & m \text{ odd}, \end{cases} \qquad \text{and} \qquad \{a \in \mathbb{Z}_2 : Na = 0\} = \begin{cases} \mathbb{Z}_2 & m \text{ even}, \\ 0 & m \text{ odd}. \end{cases}$$

Consequently, we get by duality

$$\widehat{\mathrm{H}}_k(\mathbb{Z}_m; \mathbb{Z}_2) = \begin{cases} \mathbb{Z}_2 & m \text{ even}, \\ 0 & m \text{ odd}, \end{cases} \qquad \forall k \in \mathbb{Z}.$$

Next we consider the case $A = \mathbb{Z}$. Again, we have $A^G = \mathbb{Z}$ by the definition of a trivial $\mathbb{Z}_m$-module. Then we compute

$$\widehat{\mathrm{H}}^k(\mathbb{Z}_m; \mathbb{Z}) = \begin{cases} \mathbb{Z}/m\mathbb{Z} & k \text{ even}, \\ 0 & k \text{ odd}. \end{cases}$$

Consequently, we get by duality

$$\widehat{\mathrm{H}}_k(\mathbb{Z}_m; \mathbb{Z}) = \begin{cases} 0 & k \text{ even}, \\ \mathbb{Z}_m & k \text{ odd}. \end{cases}$$

8.19 We could use coefficients in $\mathbb{Z}$ instead of $\mathbb{Z}_2$. In fact, in that case, the Tate homology would not vanish even in the case where m is odd by Exercise 8.14. However, this would require defining Morse–Bott homology in the oriented case, already discussed in Remark 5.17. Moreover, one would also have to state and prove an analogue of Theorem 7.22 for coefficients in $\mathbb{Z}$, since the first and third property of twisted Rabinowitz–Floer homology is crucial in the proof of Theorem 8.11. In the special case of $\mathbb{R}^{2n}$, this should work but requires care!

8.22 Let $(\zeta_j) \subseteq \mathrm{RFC}^\varphi(\partial W, \widehat{W})$ be a sequence such that

$$[\zeta_j] = X \in \mathrm{RFH}^\varphi(\partial W, \widehat{W}) \setminus \{0\} \qquad \text{and} \qquad \lim_{j\to\infty} \mathscr{A}_\varphi^{\mathfrak{M}}(\zeta_j) = \sigma_{\mathfrak{M}}(X),$$

for all indices j. By definition of the spectral value $\sigma_{\mathfrak{M}}(X)$, there exists a sequence

$$(\gamma_j, \tau_j) \subseteq \mathrm{Crit}(\mathscr{A}_\varphi^{\mathfrak{M}}) \qquad \text{with} \qquad \mathscr{A}_\varphi^{\mathfrak{M}}(\gamma_j, \tau_j) = \mathscr{A}_\varphi^{\mathfrak{M}}(\zeta_j).$$

By the Fundamental Lemma 7.51, there exists a constant $C = C(\mathfrak{M}) > 0$ such that

$$|\tau_j| \leq C(|\mathscr{A}_\varphi^{\mathfrak{M}}(\zeta_j)| + 1) \to C(\sigma_{\mathfrak{M}}(X) + 1) \in \mathbb{R}.$$

as $j \to \infty$, which implies that the sequence $(\tau_j) \subseteq \mathbb{R}$ is uniformly bounded. By a similar argument as in the proof of Theorem 4.21, there exists $(\gamma, \tau) \in \mathrm{Crit}(\mathscr{A}_\varphi^{\mathfrak{M}})$ such that

$$(\gamma_j, \tau_j) \overset{C^\infty}{\to} (\gamma, \tau) \qquad \text{as} \qquad j \to \infty$$

modulo subsequences by Ascoli's Theorem and bootstrapping the critical point equation (8.1). By construction we have that

$$\mathscr{A}_\varphi^{\mathfrak{M}}(\gamma, \tau) = \lim_{j\to\infty} \mathscr{A}_\varphi^{\mathfrak{M}}(\gamma_j, \tau_j) = \lim_{j\to\infty} \mathscr{A}_\varphi^{\mathfrak{M}}(\zeta_j) = \sigma_{\mathfrak{M}}(X) \in \mathscr{A}_\varphi^{\mathfrak{M}}(\mathrm{Crit}(\mathscr{A}_\varphi^{\mathfrak{M}})).$$

8.23 Let $\varphi_F \in \mathrm{Ham}^\varphi(\mathbb{R}^{2n}, \lambda_0)$. Then $\varphi \circ \varphi_F = \varphi_F \circ \varphi$. In particular $0 \in \mathrm{Fix}(\varphi_F)$, and thus no element in $\mathrm{Ham}^\varphi(\mathbb{R}^{2n}, \lambda_0)$ can displace a star-shaped hypersurface as the origin in $\mathbb{R}^{2n}$ is fixed by φ_F.

8.25 Fix a twisted homotopy $(H_\sigma)_{\sigma\in[0,1]}$ from $\mathbb{S}^{2n-1}$ to Σ and consider the smooth family of twisted Rabinowitz action functionals

$$\mathscr{A}_\sigma \colon \mathscr{L}_\varphi\mathbb{R}^{2n} \times \mathbb{R} \to \mathbb{R}, \qquad \mathscr{A}_\sigma(\gamma, \tau) := \int_0^1 \gamma^*\lambda_0 - \tau \int_0^1 H_\sigma(\gamma(t))dt.$$

From the proof of Theorem 8.11 it follows that the set

$$\{\sigma(X) : X \in \overline{\mathrm{RFH}}_*^\varphi(\mathbb{S}^{2n-1}/\mathbb{Z}_m) \setminus \{0\}\}$$

is unbounded from above as m is even by assumption. Those are precisely the critical values of $\mathscr{A}_0$. Consequently, also $\mathscr{A}_1$ has critical points of arbitrarily large critical values. In particular, $\mathscr{A}_1$ admits infinitely many critical points, which are twisted periodic Reeb orbits on Σ, which project down to $\Sigma/\mathbb{Z}_m$ to give rise to noncontractible periodic Reeb orbits on the lens space $\Sigma/\mathbb{Z}_m$. However, all these orbits must not be geometrically distinct since they might arise from iterated generators.

8.31 We start with a twisted periodic loop $\gamma \in \mathscr{L}_\varphi M$. If $m := \mathrm{ord}(\varphi)$, then $\varphi^m(\gamma(0)) = \gamma(0) = x$, and so $v(t) := \gamma(mt)$, $t \in \mathbb{T}$, is a loop passing through the points $\{x, \varphi(x), \varphi^2(x), \varphi^3(x), \ldots, \varphi^{m-2}(x), \varphi^{m-1}(x)\}$. If the loop is contractible, then $v \in \Lambda_\varphi M$ belongs to the space of all contractible twisted periodic loops on M.

8.33 For $\omega = d\lambda$ we compute

$$\mathscr{A}_\varphi^H(v, \tau) = \int_0^1 \lambda(\dot{\gamma}(mt))dt - \tau \int_0^1 H(\gamma(mt))dt = \mathscr{A}^H(\gamma, \tau),$$

under the inclusion $\mathscr{L}_\varphi M \hookrightarrow \mathscr{L}M$.

8.36 Let Σ be a twisted stable hypersurface in a symplectically aspherical symplectic manifold. Then $\mathrm{Fix}(\varphi|_\Sigma) \subseteq \mathrm{Crit}\,\mathscr{A}^H_\varphi$ is a Morse–Bott component. Indeed, by the formula provided in Proposition 7.42, we have that

$$\ker \mathrm{Hess}\,\mathscr{A}^H_\varphi|_{(x,0)} \cong \ker(D\varphi_x - \mathrm{id}_{T_x\Sigma}) = T_x\,\mathrm{Fix}(\varphi|_\Sigma)$$

for all $x \in \mathrm{Fix}(\varphi|_\Sigma)$.

8.45 Assume that the twisted Rabinowitz–Floer homology $\mathrm{RFH}^\varphi(\Sigma, M)$ is well-defined. This is difficult to achieve for general stable hypersurfaces $\Sigma \subseteq M$ as discussed in [56], but it makes the argumentation more transparent. By Theorem 7.22, we compute

$$0 \cong \mathrm{RFH}^\varphi(\Sigma, M) \cong \mathrm{H}(\mathrm{Fix}(\varphi|_\Sigma); \mathbb{Z}_2) = \bigoplus_{j \geq 0} \mathrm{H}_j(\mathrm{Fix}(\varphi|_\Sigma); \mathbb{Z}_2) \neq 0,$$

if $\mathrm{Fix}(\varphi|_\Sigma) \neq \varnothing$. Consequently, there must exist a nonconstant twisted periodic Reeb orbit in the case where there exist constant twisted periodic Reeb orbits on the stable hypersurface Σ. However, if $\mathrm{Fix}(\varphi|_\Sigma) = \varnothing$, this argumentation no longer works, and one needs to assume the existence of a twisted periodic Reeb orbit to deduce the existence of another one under suitable assumptions, in general. In either case, the existence of a twisted periodic Reeb orbit on Σ "forces" the existence of another one. The author is not aware of an example of a stable hypersurface $\Sigma \subseteq M$ such that we have $\mathrm{Fix}(\varphi|_\Sigma) = \varnothing$ and there does not exist any twisted periodic Reeb orbit.

8.46 The statement immediately follows from the Forcing Theorem 8.43 by applying it to the Morse–Bott component $\mathrm{Fix}(\varphi|_\Sigma)$ from Exercise 8.36. The strict inequality follows from the proof of Theorem 8.43. Alternatively, see the proof of [114, Theorem D]. In general, also nonconstant contractible twisted periodic Reeb orbits might have vanishing symplectic area.

8.48

1. For every $(q, p) \in T^*Q$ and $v \in T_{(q,p)}T^*Q$ we compute

$$\begin{aligned}(\varphi^*_{-\theta}\lambda)_{(q,p)}(v) &= \lambda_{(q,p+\theta_q)}\left(D\varphi_{-\theta}(v)\right)\\ &= p\left(D\pi_{T^*Q}(v)\right) + \theta_q\left(D\pi_{T^*Q}(v)\right)\\ &= (\lambda + \pi^*_{T^*Q}\theta)_{(q,p)}(v),\end{aligned}$$

2. By the first part, the intersection of $\Sigma_k = H_\theta^{-1}(k)$ with $T_q^*\mathbb{T}^n$ is a sphere centred at θ_q for every $q \in \mathbb{T}^n$ and $k > 0$. For more details, see [5, Example 5.2]. Consequently, we have $e_0 = 0$. The computation of twisted periodic Reeb orbits is similar to [56, Proposition 6.3].

3. This follows from the Forcing Theorem 8.43, if we can show that $e(\Sigma_k) < +\infty$ for all $k < e_0(H)$. In particular, every energy hypersurface Σ_k is displaceable in the geometrically bounded and symplectically aspherical symplectic manifold $(T^*Q, dp \wedge dq)$ since T^*Q is an exact symplectic manifold with the canonical Liouville form pdq. As $k < e_0(H)$, we can displace Σ_k into the missing fibres. The explicit compactly supported Hamiltonian symplectomorphism achieving that is constructed in [61, Proposition 8.2], see also Problem 3.35.
4. The Mañé critical value c is infinite in this case because a nonzero ρ has no bounded primitives in $\mathbb{R}^n$. Moreover, if $k > c$ the hypersurface is not displaceable since the Rabinowitz–Floer homology does not vanish, see also Theorem 8.8.
5. Again, one does not assume that the stable energy hypersurface in the cotangent bundle is diplaceable. This assumption is restrictive in the nonmagnetic case, since the zero section is never displaceable in the cotangent bundle!

8.50 Here are some further steps in twisted Rabinowitz–Floer homology:

- Let (Q, m) be a compact connected Riemannian manifold. By [6] or [54, Theorem 1.10], we have

$$\mathrm{RFH}_k(ST^*Q, T^*Q) \cong \begin{cases} \mathrm{H}^{-k+1}(\mathscr{L}M) & k < 0, \\ \mathrm{H}_k(\mathscr{L}M) & k > 1. \end{cases}$$

  for the unit cotangent bundle $(ST^*Q, pdq|_{ST^*Q})$. In degrees $k = 0, 1$, the answer is known and depends on the Euler class [105, Section 3.2]. The proof uses a relation between Rabinowitz–Floer homology and symplectic homology as reviewed in Problem 7.61, respectively, symplectic cohomology. If $\varphi \in \mathrm{Aut}(DT^*Q, pdq|_{DT^*Q})$ is a Liouville automorphism for the unit disc cotangent bundle DT^*Q, then it is plausible to expect

$$\mathrm{RFH}_k^{\varphi}(ST^*Q, T^*Q) \cong \begin{cases} \mathrm{H}^{-k+1}(\mathscr{L}_\varphi M) & k < 0, \\ \mathrm{H}_k(\mathscr{L}_\varphi M) & k > 1. \end{cases}$$

  However, it would also be interesting to study the loop space homology $\mathrm{H}_*(\mathscr{L}_\varphi M)$ itself, because usually one computes the free loop space homology via Morse theory. For details, see [126, Chapter 2].
- Defining twisted Rabinowitz–Floer homology with coefficients in $\mathbb{Z}$. As seen many times, this would formally require defining Morse–Bott homology with coherent orientations. This can be achieved by imposing suitable orientability assumptions or using local coefficients. It should be interesting to come up with novel results, where this generalisation is used. Of course, the existence result in Theorem 8.17 is valid for all $m \geq 1$ independent of the parity of m by [10, Theorem 1.1]. So, only defining oriented twisted Rabinowitz–Floer homology to prove this generalisation would be an overkill, but there might be other interesting applications.

- Similarly to [10, Theorem 1.2], one could study the multiplicity results of twisted periodic Reeb orbits. A good starting point would be generalising the index theory developed in [118, Section 4]. A natural assumption would be to have some form of dynamical convexity, meaning that the Conley–Zehnder index of twisted periodic Reeb orbits is bounded from below by a suitable constant depending on the dimension of the hypersurface.
- Similarly to [138], one could consider paths $(\varphi_\sigma)_{\sigma\in[0,1]}$ connecting two Liouville automorphisms $\varphi_0, \varphi_1 \in \mathrm{Aut}(W,\lambda)$ for some Liouville domain (W,λ). Denote by $(\widehat{W},\lambda)$ the completion of (W,λ). Under some suitable assumptions on the path one should have the isotopy invariance of twisted Rabinowitz–Floer homology

$$\mathrm{RFH}^{\varphi_1}(\partial W, \widehat{W}) \cong \mathrm{RFH}^{\varphi_0}(\partial W, \widehat{W}).$$

  This was already partially used in the proof of Theorem 8.11. More abstractly, this could be related to the Seidel representation [136, Section 11.4] and could provide an analogous construction in the contact case.

8.53 We prove only the first part. Towards a contradiction, assume that Σ_c is a stable Hamiltonian manifold. By Problem 4.46 there exists a tubular neighbourhood, that is, a thickening of the stable hypersurface, such that the Hamiltonian line fields are conjugate to each other. Pick such a displaceable stable hypersurface Σ_k for $k < c$. By Schlenk's theorem [56, Theorem 4.9] there exists a contractible periodic orbit on Σ_k and thus by conjugacy also on Σ_c. This contradicts the assumption that Σ_c admits no contractible periodic orbits! Hence Σ_c is unstable, that is, Σ_c does not admit a stable Hamiltonian structure.

Exercises and Problems of Chap. 9

9.2 If $n \geq 2$ then $\mathrm{Vol}(Z^{2n}(r)) = +\infty$ for all $r > 0$. Thus, $c(A) := \mathrm{Vol}(A)^{1/n}$ never satisfies the normalisation property $c(Z^{2n}(r)) = \pi r^2$.

9.3 Let $x \in \mathbb{R}^{2n}$ and let $\varepsilon > 0$. The symplectomorphism

$$\varphi\colon \mathbb{R}^{2n} \to \mathbb{R}^{2n}, \qquad \varphi(z) := z - x,$$

satisfies $\varphi(x) \subseteq B^{2n}(\varepsilon)$ for all $\varepsilon > 0$. Therefore, we compute by the monotonicity and normalisation property

$$c(\{x\}) \leq c(B^{2n}(\varepsilon)) = \pi\varepsilon^2 \to 0$$

as $\varepsilon \to 0$.

9.5 The monotonicity and conformality properties for c_B and c_Z follow from the definitions of the ball and the cylindrical capacity, respectively. It is also clear by definition of the ball capacity c_B, that $c_B(Z^{2n}(R)) \geq \pi R^2$ for all $R > 0$. So, left to show, is that $c_B(Z^{2n}(R)) \leq \pi R^2$. However, this immediately follows from

Gromov's result since

$$B^{2n}(r) \hookrightarrow Z^{2n}(R) \qquad \Leftrightarrow \qquad r \leq R.$$

Showing the normalisation property for c_Z is similar. Assume that there are embeddings

$$B^{2n}(r) \hookrightarrow A \hookrightarrow Z^{2n}(R)$$

for some subset $A \subseteq \mathbb{R}^{2n}$. By the monotonicity and normalisation property of the symplectic capacity c we get

$$\pi r^2 = c(B^{2n}(r)) \leq c(A) \leq c(Z^{2n}(R)) = \pi R^2.$$

This readily implies $c_B(A) \leq c(A) \leq c_Z(A)$.

9.6 We consider the inequalities $e(B^{2n}(r)) \leq e(Z^{2n}(r)) \leq c_B(Z^{2n}(r)) = \pi r^2$ for all $r > 0$. The first inequality is immediate from the monotonicity property of the displacement energy by taking $\varphi = \mathrm{id}_{\mathbb{R}^{2n}}$. The last inequality is immediate from Exercise 9.5. So, the only inequality left to prove is

$$e(Z^{2n}(r)) \leq c_B(Z^{2n}(r)) \qquad \forall r > 0.$$

This is the content of [137, Exercise 12.3.7].

9.8 We compute

$$\begin{aligned}
\frac{d}{dt}\left(\phi_t^{F_t} \circ \phi_t^{G_t}\right) &= \frac{d}{dt}\left(\phi_t^{F_t}\right) \circ \phi_t^{G_t} + D\phi_t^{F_t} \circ \left(\frac{d}{dt}\phi_t^{G_t}\right) \\
&= X_{F_t} \circ \phi_t^{F_t} \circ \phi_t^{G_t} + D\phi_t^{F_t} \circ X_{G_t} \circ \phi_t^{G_t} \\
&= X_{F_t} \circ \phi_t^{F_t} \circ \phi_t^{G_t} + D\phi_t^{F_t} \circ X_{G_t} \circ \phi_t^{-F_t} \circ \phi_t^{F_t} \circ \phi_t^{G_t} \\
&= X_{F_t} \circ \phi_t^{F_t} \circ \phi_t^{G_t} + \left(\phi_t^{F_t}\right)_* X_{G_t} \circ \phi_t^{F_t} \circ \phi_t^{G_t} \\
&= X_{F_t} \circ \phi_t^{F_t} \circ \phi_t^{G_t} + X_{G_t \circ \phi_t^{-F_t}} \circ \phi_t^{F_t} \circ \phi_t^{G_t} \\
&= X_{F_t + G_t \circ \phi_t^{-F_t}} \circ \phi_t^{F_t} \circ \phi_t^{G_t} \\
&= X_{(F\#G)_t} \circ \phi_t^{F_t} \circ \phi_t^{G_t}.
\end{aligned}$$

Thus we conclude that $\phi_t^{(F\#G)_t} = \phi_t^{F_t} \circ \phi_t^{G_t}$, in particular $\varphi_{F\#G} = \varphi_F \circ \varphi_G$.

9.9 See [137, Proposition 12.3.1] and [137, Theorem 12.3.3].

9.12

1. Let (Σ, ω, α) be a compact contact manifold. Then we define the systole by

$$\operatorname{sys}(\Sigma, \omega, \alpha) := \inf\{\tau > 0 : \exists\, (\gamma, \tau) \in \mathscr{P}(\Sigma, \omega, \alpha)\},$$

   where $\mathscr{P}(\Sigma, \omega, \alpha)$ denotes the set of all prametrised periodic Reeb orbits on (Σ, ω, α). Be careful not to use a minimum instead of an infimum, since a priori it is not clear that $\operatorname{sys}(\Sigma, \omega, \alpha)$ is non-zero or not equal to infinity.
2. For star-shaped hypersurfaces $\Sigma \subseteq \mathbb{R}^{2n}$ we already know that $\mathscr{P}(\Sigma) \neq \varnothing$ by the Rabinowitz result (Corollary 8.9). For general compact contact manifolds, $\mathscr{P}(\Sigma, \omega, \alpha) \neq \varnothing$ is only conjectured by Weinstein. Certainly, if Σ is three-dimensional, one can say more as explained in Remark 4.16. So we can formulate the following statement: *If* $\mathscr{P}(\Sigma, \omega, \alpha) \neq \varnothing$, then there exists $(\gamma, \tau) \in \mathscr{P}(\Sigma, \omega, \alpha)$ such that $\tau = \operatorname{sys}(\Sigma, \omega, \alpha)$ holds. In particular, we have that $\operatorname{sys}(\Sigma) > 0$ and

$$\operatorname{sys}(\Sigma, \omega, \alpha) = \min\{\tau > 0 : \exists\, (\gamma, \tau) \in \mathscr{P}(\Sigma, \omega, \alpha)\},$$

3. We claim that $(DT^*Q, \lambda|_{DT^*Q})$ is Liouville domain for the unit disc bundle

$$DT^*Q := \{(q, p) \in T^*Q : m_*(p, p) \leq 1\}.$$

   The Liouville vector field $X = X_\lambda \in \mathfrak{X}(T^*Q)$ is locally given by the fibrewise radial vector field $X = p\partial_p$. Now note that the boundary $\partial DT^*Q = ST^*Q$ admits the defining Hamiltonian function

$$H\colon T^*Q \to \mathbb{R}, \qquad H(q, p) := \frac{1}{2}\|p\|^2_{m^*}.$$

   We compute $dH(X)|_{ST^*Q} = 2H|_{ST^*Q} = 1$, and thus the Liouville vector field X is outward-pointing along ST^*Q. This computation also implies

$$\lambda(X_H)|_{ST^*Q} = d\lambda(X, X_H) = dH(X) = 1,$$

   implying $R = X_H|_{ST^*Q}$. By Problem 3.34, we conclude that a curve $\gamma \in C^\infty(\mathbb{R}, Q)$ is a geodesic for m of unit speed, if and only if $\Phi_m \circ \dot{\gamma} \in C^\infty(\mathbb{R}, T^*Q)$ is a Reeb orbit for $(ST^*Q, \lambda|_{ST^*Q})$, where $\Phi_m\colon TQ \to T^*Q$ denotes the canonical tangent-cotangent bundle isomorphism $\Phi_m(\dot{q}) := m_q(\dot{q}, \cdot)$. Moreover, γ is periodic if and only if $\Phi_m \circ \dot{\gamma}$ is periodic, and the m-length of γ is equal to the period of $\Phi_m \circ \dot{\gamma}$ since γ is of unit speed. Unfortunately, we cannot make use of Schlenk's theorem for finding periodic geodesics on Q since the unit sphere bundle ST^*Q contains the zero-section and is therefore not displaceable in the cotangent bundle T^*Q.

9.15 Fix $\varepsilon > 0$ and choose a Hamiltonian function $F \in C_c^\infty(\mathbb{R}^2 \times [0,1])$ for the boundary $\Sigma = \partial A$ satisfying

$$F_t = 0, \quad \forall t \in [0, \tfrac{1}{2}], \qquad \|F\| < e(\Sigma) + \varepsilon \qquad \text{and} \qquad \varphi_F(\Sigma) \cap \Sigma = \varnothing.$$

Choose a smooth family $(\beta_r)_{r \in [0,+\infty)}$ of cut-off functions $\beta_r \in C^\infty(\mathbb{R}, [0,1])$ with

$$\begin{cases} \beta_r(s) = 0 & |s| \geq r, \\ \beta_r(s) = 1 & |s| \leq r - 1, \\ s\beta_r'(s) \leq 0 & \forall s \in \mathbb{R}, \end{cases}$$

for all $r \in [0, +\infty)$. Define a family of functionals $\mathscr{A}_r \colon \mathscr{L} \times \mathbb{R} \times \mathbb{R} \to \mathbb{R}$ for the free loop space $\mathscr{L} := C^\infty(\mathbb{T}, \mathbb{R}^2)$ by

$$\mathscr{A}_r(\gamma, \tau, s) := \int_0^1 \gamma^*\lambda - \tau \int_0^1 H_\Sigma(\gamma(t))dt - \beta_r(s) \int_0^1 F_t(\gamma(t))dt$$

for all $r \in [0, +\infty)$. The functionals $\mathscr{A}_r$ can be interpreted as follows: For $r \in [0,1)$, we have that $\mathscr{A}_r = \mathscr{A}^H$ is the standard Rabinowitz action functional. If $r \geq 1$ increases, the perturbation by F is gradually switched on, resulting in the perturbed Rabinowitz action functional $\mathscr{A}^{(H,F)}$ in the limit $r \to +\infty$. The critical manifold $\mathrm{Crit}(\mathscr{A}_0)$ consists of Σ together with all $\mathbb{T}$-reparametrisations of iterations $(\Gamma^k, k\,\mathrm{sys}(\Sigma))$ for $k \in \mathbb{Z}$, since one can choose any starting point on Σ. Moreover, this implies that $\mathrm{Crit}(\mathscr{A}_0) \cong \mathbb{T} \times \mathbb{Z}$ is a stack of circles. Fix $x \in \Sigma$ and define the moduli space $\mathscr{M}$ consisting of all $(u, \tau, r) \in C^\infty(\mathbb{R}, \mathscr{L} \times \mathbb{R}) \times [0, +\infty)$ satisfying

$$\begin{cases} \partial_s(u, \tau) = \nabla\mathscr{A}_r(u(s), \tau(s), s) \quad \forall s \in \mathbb{R}, \\ \lim\limits_{s \to -\infty} (u(s), \tau(s)) = (x, 0), \\ \lim\limits_{s \to +\infty} (u(s), \tau(s)) \in \Sigma \times \{0\}. \end{cases}$$

The first equation is a gradient flow equation, whereas the two asymptotic conditions can be thought of as boundary conditions. By the definition of $\mathscr{M}$, we have $(x, 0, 0)$ belongs to $\mathscr{M}$. We claim that if $\tau = \mathscr{A}_0(\gamma, \tau) > \|F\|$ for all $\forall(\gamma, \tau) \in \mathrm{Crit}(\mathscr{A}_0) \setminus \Sigma$, then $\mathscr{M}$ is compact. Before proving this result, we explain how the equality $e(A) = \mathrm{Area}(A)$ follows from this. The moduli space $\mathscr{M} = \sigma^{-1}(0)$ is the zero-level set of the Fredholm section $\sigma(u, \tau, r) := \partial_s(u, \tau) - \nabla\mathscr{A}_r(u, \tau, s)$ of a suitable Banach bundle. Fredholm operators are essentially as well behaved as finite-dimensional operators, thus allowing an infinite-dimensional implicit function theorem to hold as described in [136, Section A.3]. One can check that the derivative $D\sigma(x, 0, 0)$ is surjective and, by compactness guaranteed by Lemma 8.49, we can therefore perturb σ away from the boundary $\partial\mathscr{M}$ so that $D\sigma(u, \tau, r)$ is surjective for all $(u, \tau, r) \in \sigma^{-1}(0)$. Hence, the moduli space $\mathscr{M}$ is a compact smooth manifold with a boundary that consists precisely of one point. This is absurd, and we conclude that there must exists a critical point $(\gamma, \tau) \in \mathrm{Crit}(\mathscr{A}_0) \setminus \Sigma$ with

$$0 < |\mathscr{A}_0(\gamma, \tau)| \leq \|F\| < e(\Sigma) + \varepsilon \leq \mathrm{Area}(A) + \varepsilon.$$

But by Lemma 9.13, the only critical point that satisfies this inequality is the one representing the systole $\mathrm{sys}(\partial A)$, coinciding with $\mathrm{Area}(A)$.

9.24 If A is symmetric, then necessarily also the boundary ∂A. But the boundary ∂A is parametrised by Γ, which is therefore invariant under φ. The systole is just the period of going once around Γ, so it also coincides with the symmetric systole, and the symmetric ratio is equal to one.

9.29 Let $x \in \Sigma$ be a translated point for Σ. Define $\gamma \in C^\infty(\mathbb{R}, \Sigma)$ by $\gamma(t) := \phi^R_{\tau t}(x)$ for the period $\tau > 0$ of x. Then we compute

$$\gamma(t+1) = \phi^R_{\tau(t+1)}(x) = \phi^R_{\tau t}\left(\phi^R_\tau(x)\right) = \phi^R_{\tau t}\left(\varphi(x)\right) = \varphi\left(\phi^R_{\tau t}(x)\right) = \varphi(\gamma(t))$$

for all $t \in \mathbb{R}$, where the second last inequality follows from Exercise 7.39, since φ is a strict contactomorphism. Thus γ is a twisted Reeb orbit satisfying

$$\dot{\gamma}(t) = \frac{d}{dt}\phi^R_{\tau t}(x) = \tau R\left(\phi^R_{\tau t}(x)\right) = \tau R(\gamma(t))$$

for all $t \in \mathbb{R}$.

9.30

1. Define $\widehat{\varphi}(x, r) := (\varphi(x), r - f(x))$, where $\varphi^*\alpha = e^f\alpha$. Assume that $x \in \Sigma$ is a translated point for φ. Then x is a leaf-wise intersection point for $\widehat{\varphi}$. In fact, by the definition of a translated point, there exists $\tau > 0$ with $\varphi(x) = \phi^R_\tau(x) \in L_x$, where L_x denotes the leaf through x. By definition of $\widehat{\varphi}$, we compute

$$\widehat{\varphi}(x) = (\varphi(x), -f(x)) = (\varphi(x), 0) \in \Sigma,$$

since by definition of a translated point $\varphi^*\alpha|_x = \alpha_x = e^0\alpha_x = e^{f(x)}\alpha_x$. Note that we can cut-off the Hamiltonian symplectomorphism to make it compactly supported without changing the leaf-wise intersection points.
2. The result in Problem 8.54 can be rephrased as follows. Let φ be a Hamiltonian symplectomorphism of $\mathbb{R}^{2n}$ with Hofer norm strictly smaller than the displacement energy of the sphere $\mathbb{S}^{2n-1}$. Then there exists a leaf-wise intersection point for φ. If the Hamiltonian symplectomorphism is chosen generically, then there exist at least two leaf-wise intersection points by [12, Theorem B]. Thus, Sandon's result gives a different class of Hamiltonian symplectomorphisms of $\mathbb{R}^{2n}$ for which the same result is true, namely those Hamiltonian symplectomorphisms which are the lift to $\mathbb{R}^{2n}$ of contactomorphisms of $\mathbb{S}^{2n-1}$. Theorem 8.24 implies

the existence of infinitely many translated points or there exists a translated point on a closed leaf for every contactomorphism of any lens space $L_m(k_1, \dots, k_n)$, for $m \geq 1$ even. In particular, this holds for the real projective space $\mathbb{RP}^{2n-1}$, which is generalised by [165, Theorem 1.1 (ii)].

9.31 See [137, Lemma 2.2.1].

9.33

1. Let (γ, τ) be a parametrised periodic Reeb orbit in (Σ, α). For simplicity, assume that γ is contractible. Then one can consider the path
$$\Psi\colon [0,1] \to \mathrm{Sp}(2n-2), \qquad \Psi_t := \Phi^{-1}_{e^{2\pi i t}} \circ D^{\xi}\phi^R_{\tau t} \circ \Phi_0,$$
where $\Phi_t\colon \mathbb{D} \times \mathbb{R}^{2n-2} \to \bar{\gamma}^*\xi$ is a symplectic trivialisation of $\bar{\gamma}^*\xi$, $\xi := \ker\alpha$, for a filling disc $\bar{\gamma} \in C^\infty(\mathbb{D}, \Sigma)$ of γ. To any such path, one can associate the Conley–Zehnder index $\mu_{\mathrm{CZ}}(\Psi) \in \mathbb{Z}$, which a priori depends on the explicit filling disc $\bar{\gamma}$. However, it is definitively possible to ask for a lower bound on this index for any contact manifold. So, the notion of dynamical convexity makes sense on arbitrary contact manifolds.
2. By observing Eliashberg's dichotomy in Problem 9.36 for the sphere $\mathbb{S}^3$, we see that the standard contact structure on $\mathbb{S}^3$ is tight. But since star-shaped deformations of $\mathbb{S}^3$ induce the standard contact structure by Exercise 4.20, we conclude that an appropriate generalisation of Conjecture 9.20 is indeed Conjecture 9.32. Of course, one could consider Conjecture 9.32 also for overtwisted contact structures on more general contact manifolds.

9.34

1. The restriction $\lambda|_{\partial U} = i_X\omega|_{\partial U} \in \Omega^1(\partial U)$ is nowhere vanishing in the line bundle $\ker\omega|_{\partial U}$ since it is a volume form. See, for example, [129, Proposition 15.21].
2. Let $\Gamma \in C^\infty(\mathbb{T}, \partial U)$ be the solution of $\dot{\Gamma}(t) = \tau R(\Gamma(t))$ for all $t \in \mathbb{T}$ with a minimal period $\tau > 0$. Using Stoke's Theorem we compute
$$\tau = \mathrm{sys}(\partial U) = \tau\int_0^1 \lambda(R(\Gamma(t)))dt = \int_0^1 \Gamma^*\lambda = \int_{\partial U}\lambda = \int_U \omega_0 = \mathrm{Area}(U).$$
So, the systole of ∂U coincides with the area of U for every simply connected open subset $U \subset \mathbb{R}^2$.
3. By homotopy invariance of the Maslov index [137, Theorem 2.2.12] and formula $\mu_{\mathrm{CZ}} = 2\mu_{\mathrm{M}}$ for a loop, we get $\mu_{\mathrm{CZ}}(\gamma) \geq 2$ for all periodic orbits of period $\tau > 0$ on the boundary ∂U by the normalisation property of the Maslov index. This means that ∂U is dynamically convex as defined in Problem 8.51 since $n = 1$. Hence, by part two of this exercise we already know that $\mathrm{sys}(\partial U) = \mathrm{Area}(U)$

and by [174, Theorem 4.2] or [166, Corollary C.10 (ii)] we also have $e(U) = \text{Area}(U)$, implying Conjecture 9.32. For $n \geq 2$, this conjecture is wide open to the authors' knowledge, since already Conjecture 9.20 is wide open unless $n = 1, 2$. It should be quite interesting to perform numerical experiments to find conditions when the property of being dynamically convex of a star-shaped hypersurface stops along a homotopy of star-shaped hypersurfaces in $\mathbb{R}^{2n}$. Even showing that any convex domain is dynamically convex is relatively involved.

Exercises and Problems of Appendix A

A.4

1. For any subvectorspace $U \subseteq V$ we define a map

$$\Phi\colon V \to U^*, \qquad \Phi(v)(u) := \omega(v, u).$$

We claim that Φ is surjective. Let $\varphi \in U^*$ be a linear functional and consider an extension $\widetilde{\varphi} \in V^*$. As ω is nondegenerate, there exists $v \in V$ such that $\widetilde{\varphi} = \omega(v, \cdot)$. It follows that $\Phi(v) = \widetilde{\varphi}$ and thus Φ is indeed surjective. Moreover, note that

$$\ker \Phi = U^{\perp}.$$

Thus by the standard rank-nullity formula for finite-dimensional vector spaces, it follows that

$$\dim V = \dim \ker \Phi + \dim \operatorname{im} \Phi = \dim U^{\perp} + \dim U^* = \dim U^{\perp} + \dim U.$$

2. By definition of the symplectic orthogonal complement, we see that $U \subseteq (U^{\perp})^{\perp}$. But by the first part. of this exercise we have that

$$\dim(U^{\perp})^{\perp} = \dim V - \dim U^{\perp} = \dim V - (\dim V - \dim U) = \dim U,$$

and thus $U = (U^{\perp})^{\perp}$.

A.5

1. If $\dim V = 0$ there is nothing to show. So we can assume $\dim V \geq 1$. Thus there exists a vector $x \neq 0$ in V. Set $x_1 := \frac{x}{\sqrt{\langle x, x\rangle}}$ and consider the subvectorspace defined by $U := \text{span}(x_1)$. We claim that

$$V = U \oplus U^{\perp},$$

where similar to the symplectic orthogonal complement we define

$$U^{\perp} := \{v \in V : \langle v, u \rangle = 0 \, \forall u \in U\}.$$

For showing the direct sum decomposition, we have to show two different properties.

- $\underline{U \cap U^{\perp} = \{0\}}$. Assume that $v \in U \cap U^{\perp}$. Hence there exists $a \in \mathbb{R}$ such that

  $$v = ax_1 \in U.$$

  As also $v \in U^{\perp}$ by assumption, we conclude

  $$0 = \langle v, x_1 \rangle = a \langle x_1, x_1 \rangle.$$

  With $a \neq 0$ and by using the positive-definiteness of the inner product, it follows that $x_1 = 0$. A contradiction to the assumption that $x_1 \neq 0$. So $a = 0$ must hold.
- $\underline{U + U^{\perp} = V}$. Let $v \in V$. Then we can write

  $$v = \underbrace{\langle v, x_1 \rangle x_1}_{=:u} + \underbrace{v - \langle v, x_1 \rangle x_1}_{=:u^{\perp}}.$$

  By definition, we have that $u \in U$ and we claim $u^{\perp} \in U^{\perp}$. Indeed, we compute

  $$\langle u^{\perp}, x_1 \rangle = 0.$$

Next, we inductively decompose $U^{\perp}$ as described above, and after finitely many steps we obtain the decomposition

$$V = \bigoplus_{i=1}^{n} U_i, \qquad U_i = \operatorname{span}(x_i).$$

By construction we have that

$$\omega(x_i, x_j) = \delta_{ij}.$$

2. Let V be a finite-dimensional real vector space with $\dim V = n$. Then there exists an isomorphism

$$\varphi \colon V \to \mathbb{R}^n$$

given by the choice of a basis for V. Since $\mathbb{R}^n$ is a Euclidean vector space with for example the standard inner product $\langle\cdot,\cdot\rangle$, we can equip V with the inner product

$$\langle x, y\rangle_\varphi := \langle\varphi(x), \varphi(y)\rangle \qquad \forall x, y \in V.$$

Assume that (V, ω) symplectic vector space. Then we infer from Lemma A.3, that $n = \dim V$ must be even. In particular, if n is odd, there cannot exist any symplectic structure on V.

A.6 For all $A \in \mathrm{Sp}(2n)$ we compute

$$\det(J_0) = \det(A^T J_0 A) = \det(A^T)\det(J_0)\det(A) = \det(J_0)\det(A)^2.$$

We conclude $\det(A)^2 = 1$, since $\det(J_0) \neq 0$. Using that $\det(A) \in \mathbb{R}$ we get $\det(A) = \pm 1$.

A.7 By Exercise A.6, it is enough to show that $\mathrm{Sp}(2n)$ is a subgroup of the general linear group $\mathrm{GL}(2n)$ of invertible $2n \times 2n$-matrices with real entries.

- $\underline{\mathrm{Sp}(2n) \neq \varnothing}$. We have $I_{2n} \in \mathrm{Sp}(2n)$ for the identity matrix I_{2n}.
- $\underline{AB \in \mathrm{Sp}(2n) \text{ for all } A, B \in \mathrm{Sp}(2n)}$. We compute

$$(AB)^T J_0 AB = B^T A^T J_0 AB = B^T J_0 B = J_0.$$

- $\underline{A^{-1} \in \mathrm{Sp}(2n) \text{ for all } A \in \mathrm{Sp}(2n)}$. It holds that

$$A^T J_0 A = J_0 \qquad \Leftrightarrow \qquad J_0 = (A^{-1})^T J_0 A^{-1}.$$

We conclude that $\mathrm{Sp}(2n)$ is a group.

A.9

1. From the definition of the set $\mathfrak{sp}(2n)$, it follows that $0_{2n} \in \mathfrak{sp}(2n)$ for the zero matrix 0_{2n} and that $aA + bB \in \mathfrak{sp}(2n)$ for all $a, b \in \mathbb{R}$ and $A, B \in \mathfrak{sp}(2n)$. Thus, $\mathfrak{sp}(2n)$ is a subvectorspace of the real vector space $\mathfrak{gl}(2n)$ of all $2n \times 2n$-matrices with real entries. In order to compute the dimension of $\mathfrak{sp}(2n)$, we write

$$\left(\begin{array}{c|c} A & B \\ \hline C & D \end{array}\right) \in \mathfrak{sp}(2n)$$

with $A, B, C, D \in \mathbb{R}^{n\times n}$. From the requirement that this block matrix is in $\mathfrak{sp}(2n)$, we immediately deduce the equations

$$A = -D^T, \qquad B = B^T \qquad \text{and} \qquad C = C^T.$$

From this it follow that

$$\dim \mathfrak{sp}(2n) = 2 \cdot \frac{n(n+1)}{2} + n^2 = 2n^2 + n.$$

2. By [129, Problem 20-4] we have the formula

$$\det(e^A) = e^{\operatorname{tr} A}.$$

Thus it is enough to show that $\operatorname{tr}(A) = 0$ for all $A \in \mathfrak{sp}(2n)$. We observe

$$\begin{aligned} A^T J_0 + J_0 A = 0 \quad &\Leftrightarrow \quad J_0^T A^T J_0 + J_0^T J_0 A = 0 \\ &\Leftrightarrow \quad J_0^T A^T J_0 + A = 0 \\ &\Leftrightarrow \quad A = -J_0^T A^T J_0 \\ &\Leftrightarrow \quad A = J_0 A^T J_0. \end{aligned}$$

Thus we compute

$$\operatorname{tr}(A) = \operatorname{tr}(J_0 A^T J_0) = \operatorname{tr}(J_0^2 A^T) = -\operatorname{tr}(A^T) = -\operatorname{tr}(A),$$

by using that the trace of a product is independent of the order of the product. Thus, we conclude that $\operatorname{tr}(A) = 0$ since the field $\mathbb{R}$ is not of characteristic two.
3. For all $A \in \mathfrak{sp}(2n)$ we compute

$$(e^A)^T J_0 e^A = e^{A^T} J_0 e^A = e^{J_0 A J_0} J_0 e^A = -J_0 e^{-A} J_0^2 e^A = J_0 e^{-A} e^A = J_0.$$

Thus it follows that $e^A \in \mathrm{Sp}(2n)$.

A.10 See [200, Proposition V.5].

A.12 By the regular level set theorem [129, Corollary 5.14] it is enough to show that the differential

$$D(\det)_A \colon \mathbb{R}^{2n \times 2n} \to \mathbb{R}$$

is surjective for all matrices $A \in \mathrm{SL}(2n)$. Let $B \in \mathbb{R}^{2n \times 2n}$ be an arbitrary matrix. Then we compute

$$\begin{aligned} D(\det)_A(B) &= \frac{d}{dt}\bigg|_{t=0} \det(A + tB) \\ &= \frac{d}{dt}\bigg|_{t=0} \det(A^{-1}) \det(A + tB) \end{aligned}$$

$$= \frac{d}{dt}\bigg|_{t=0} \det(I_{2n} + tA^{-1}B)$$
$$= \operatorname{tr}(A^{-1}B).$$

Thus we conclude for all $\lambda \in \mathbb{R}$, that

$$D(\det)_A\left(\frac{\lambda}{2n}A\right) = \frac{\lambda}{2n}\operatorname{tr}(I_{2n}) = \lambda.$$

We deduce that $\mathrm{SL}(2n) \subseteq \mathbb{R}^{2n\times 2n}$ is a properly embedded submanifold of dimension

$$\dim \mathrm{SL}(2n) = \dim \det^{-1}(1) = 4n^2 - 1,$$

with the corresponding Lie algebra

$$T_{I_{2n}}\,\mathrm{SL}(2n) = \ker D(\det)_{I_{2n}} = \{B \in \mathfrak{gl}(2n) : \operatorname{tr}(B) = 0\} = \mathfrak{sl}(2n).$$

A.13 Similarly to Exercise A.12, our aim is to write the symplectic group $\mathrm{Sp}(2n)$ as a preimage of a regular value of a suitable map. One such choice is the map

$$f\colon \mathbb{R}^{2n\times 2n} \to \mathbb{R}^{2n\times 2n}, \qquad f(A) := A^T J_0 A,$$

because then $\mathrm{Sp}(2n) = f^{-1}(J_0)$. We observe that

$$f(A)^T = A^T J_0^T A = -A^T J_0 A = -f(A).$$

So, the image of the map f is contained in the Lie algebra $\mathfrak{o}(2n)$ of the orthogonal group $\mathrm{O}(2n)$ defined by $\mathfrak{o}(2n) := \{A \in \mathfrak{gl}(2n) : A^T + A = 0\}$, and we get a well-defined map $f\colon \mathbb{R}^{2n\times 2n} \to \mathfrak{o}(2n)$. The differential $Df_A\colon \mathbb{R}^{2n\times 2n} \to \mathfrak{o}(2n)$ for all matrices $A \in \mathbb{R}^{2n\times 2n}$ is given by

$$Df_A(B) = \frac{d}{dt}\bigg|_{t=0}(A+tB)^T J_0(A+tB) = B^T J_0 A + A^T J_0 B.$$

We claim that for every matrix $A \in \mathrm{Sp}(2n)$ the differential Df_A is surjective. For every $B \in \mathfrak{o}(2n)$ we compute

$$Df_A\left(-\frac{1}{2}AJ_0B\right) = \frac{1}{2}B^T J_0 A^T J_0 A - \frac{1}{2}A^T J_0 A J_0 B$$
$$= \frac{1}{2}B^T J_0^2 - \frac{1}{2}J_0^2 B$$

$$
\begin{aligned}
&= -\frac{1}{2}B^T + \frac{1}{2}B \\
&= \frac{1}{2}B + \frac{1}{2}B \\
&= B.
\end{aligned}
$$

Therefore, by the regular level set theorem [129, Corollary 5.14], $\mathrm{Sp}(2n) \subseteq \mathbb{R}^{2n\times 2n}$ is a properly embedded smooth submanifold of dimension

$$
\dim \mathrm{Sp}(2n) = 4n^2 - \dim \mathfrak{o}(2n) = 4n^2 - \frac{2n(2n-1)}{2} = 2n^2 + n,
$$

with the corresponding Lie algebra

$$
T_{I_{2n}} \mathrm{Sp}(2n) = \ker Df_{I_{2n}} = \{B \in \mathfrak{o}(2n) : B^T J_0 + J_0 B = 0\} = \mathfrak{sp}(2n).
$$

A.14 We already know by Lemma A.8, that $\det(A) = 1$ for all $A \in \mathrm{Sp}(2n)$. Thus, in particular, we have $\mathrm{Sp}(2n) \subseteq \mathrm{SL}(2n)$. For $n = 1$ we conclude from the dimension formulas provided in Exercises A.12 and A.13, that $\dim \mathrm{SL}(2) = \dim \mathrm{Sp}(2)$, and thus the two Lie groups coincide by connectedness. In general, we have that

$$
\dim \mathrm{SL}(2n) = 4n^2 - 1 \qquad \text{and} \qquad \dim \mathrm{Sp}(2n) = 2n^2 + n
$$

by Exercises A.12 and A.13 implying

$$
\dim \mathrm{Sp}(2n) < \dim \mathrm{SL}(2n)
$$

for all $n \geq 2$.

References

1. Abbas, C., Hofer, H.: Holomorphic Curves and Global Questions in Contact Geometry. Birkhäuser Advanced Texts Basler Lehrbücher. Springer International Publishing, New York (2019)
2. Abbondandolo, A.: Lectures on the free period lagrangian action functional. J. Fixed Point Theory Appl. **13**(2), 397–430 (2013)
3. Abbondandolo, A., Kang, J.: Symplectic homology of convex domains and Clarke's duality. Duke Math. J. **171**(3), 739–830 (2022)
4. Abbondandolo, A., Merry, W.J.: Floer homology on the time-energy extended phase space. J. Symplectic Geom. **16**(2), 279–355 (2018)
5. Abbondandolo, A., Schlenk, F.: Floer homologies, with applications. Jahresber. Dtsch. Math. Ver. **121**(3), 155–238 (2019)
6. Abbondandolo, A., Schwarz, M.: Estimates and Computations in Rabinowitz–Floer Homology. J. Topol. Anal. **01**(04), 307–405 (2009)
7. Abbondandolo, A., Benedetti, G., Edtmair, O.: Symplectic capacities of domains close to the ball and Banach–Mazur geodesics in the space of contact forms (2023). https://arxiv.org/abs/2312.07363
8. Abbondandolo, A., Edtmair, O., Kang, J.: On closed characteristics of minimal action on a convex three-sphere (2024). https://arxiv.org/abs/2412.01777
9. Abraham, R., Marsden, J.E.: Foundations of Mechanics, 2nd edn. AMS Chelsea Publishing, Providence (1978)
10. Abreu, M., Liu, H., Macarini, L.: Symmetric periodic Reeb orbits on the sphere. Trans. Am. Math. Soc. **377**(9), 6751–6770 (2024)
11. Akopyan, A., Karasev, R.: Estimating symplectic capacities from lengths of closed curves on the unit spheres (2017). https://arxiv.org/abs/1801.00242
12. Albers, P., Frauenfelder, U.: Leaf-wise intersections and Rabinowitz Floer homology. J. Topol. Anal. **02**, 77–98 (2010)
13. Albers, P., Frauenfelder, U.: Spectral invariants in Rabinowitz-Floer homology and global Hamiltonian perturbations. J. Mod. Dyn. **4**(2), 329–357 (2010)
14. Albers, P., Frauenfelder, U.: Infinitely many leaf-wise intersections on cotangent bundles. Expo. Math. **30**(2), 168–181 (2012)
15. Albers, P., Frauenfelder, U.: On a theorem by Ekeland–Hofer. Isr. J. Math. **187**(1), 485–491 (2012)
16. Albers, P., Frauenfelder, U.: Rabinowitz Floer homology: a survey. In: Bär, C., Lohkamp, J., Schwarz, M. (eds.) Global Differential Geometry, pp. 437–461. Springer, Berlin (2012)
17. Albers, P., Wysocki, K.: M-Polyfolds in Morse Theory (2013). Unpublished manuscript
18. Albers, P., Frauenfelder, U., Koert, O.V., Paternain, G.P.: Contact geometry of the restricted three-body problem. Commun. Pure Appl. Math. **65**(2), 229–263 (2012)
19. Albers, P., Fuchs, U., Merry, W.: Orderability and the Weinstein conjecture. Compo. Math. **151**(12), 2251–2272 (2015)

Y. Bähni, *Lectures on Twisted Rabinowitz-Floer Homology*, Pathways in Mathematics, https://doi.org/10.1007/978-3-032-10673-5

20. Allgower, E.L., Georg, K.: Introduction to Numerical Continuation Methods. Classics in Applied Mathematics, No. 45. Society for Industrial and Applied Mathematics, Philadelphia (2003)
21. Arnol'd, V.I.: What is mathematical physics? Physics-Uspekhi **47**(12), 1281 (2004). https://dx.doi.org/10.1070/PU2004v047n12ABEH002089
22. Arnold, V.I., Kozlov, V.V., Neishtadt, A.I.: Mathematical Aspects of Classical and Celestial Mechanics. Encyclopedia of Mathematical Sciences, vol. 3, 3rd edn. Springer, Berlin (2006)
23. Aubert, C.: Categories for me, and you? (2019). https://arxiv.org/abs/1910.05172
24. Audin, M., Damian, M.: Morse Theory and Floer Homology. Universitext. Springer, Berlin (2014)
25. Aydin, C.: From Babylonian lunar observations to Floquet multipliers and Conley–Zehnder indices. J. Math. Phys. **64**(8), 082902 (2023)
26. Aydin, C.: The Conley–Zehnder indices of the spatial Hill three-body problem. Celestial Mech. Dyn. Astron. **135**(3), 32 (2023)
27. Bähni, Y.: First steps in twisted Rabinowitz–Floer homology. J. Symplectic Geom. **21**(1), 111–158 (2023)
28. Bähni, Y.: On a theorem by Schlenk. Calc. Var. Partial Differ. Equ. **63**(124) (2024)
29. Banyaga, A., Hurtubise, D.: Lectures on Morse Homology. Kluwer Texts in the Mathematical Sciences. Springer, Dordrecht (2004)
30. Banyaga, A., Hurtubise, D.E.: Cascades and perturbed Morse–Bott functions. Algebr. Geom. Topol. **13**(1), 237–275 (2013)
31. Banyaga, A., Hurtubise, D., Spaeth, P.: Twisted Morse Complexes - Morse Homology and Cohomology with Local Coefficients. Lecture Notes in Mathematics, vol. 2361. Springer, Berlin (2024)
32. Baumgartner, V., Daguati, S., Trninic, D., Akveld, M., Caspar, A., Hungerbühler, N., Kapur, M.: Problem-solving before instruction for learning linear algebra in university mathematics. Instruct. Sci. **53**, 1573–1602 (2025)
33. Beckschulte, F., Datta, I., Seifert, I., Vocke, A.M., Wehrheim, K.: A Polyfold Proof of Gromov's Non-squeezing Theorem, pp. 1–45. Springer International Publishing, Cham (2021)
34. Belbruno, E., Frauenfelder, U., van Koert, O.: A family of periodic orbits in the three-dimensional lunar problem. Celestial Mech. Dyn. Astron. **131**(7) (2019)
35. Belbruno, E., Frauenfelder, U., van Koert, O.: The omega limit set of a family of chords. J. Topol. Anal. **13**(01), 49–73 (2021)
36. Benedetti, G.: First steps into the world of systolic inequalities: from Riemannian to symplectic geometry (2021). https://arxiv.org/abs/2103.09356
37. Berger, M.: A Panoramic View of Riemannian Geometry. Springer, Berlin (2007)
38. Betancourt, M.J.: Adiabatic Monte Carlo (2015). https://arxiv.org/abs/1405.3489
39. Betancourt, M.: A Conceptual Introduction to Hamiltonian Monte Carlo (2018). https://arxiv.org/abs/1701.02434
40. Betancourt, M., Byrne, S., Livingstone, S., Girolami, M.: The geometric foundations of Hamiltonian Monte Carlo. Bernoulli **23**(4A), 2257–2298 (2017)
41. Bimmermann, J., Maier, L.: Magnetic billiards and the Hofer-Zehnder capacity of disk tangent bundles of lens spaces (2024). https://arxiv.org/abs/2403.06761
42. Bott, R.: Lectures on Morse Theory, Old and New. Bull. (New Series) Am. Math. Soc. **7**(2), 331–358 (1982)
43. Brezis, H.: Functional Analysis, Sobolev Spaces and Partial Differential Equations. Universitext. Springer, New York (2011)
44. Brown, K.: Cohomology of Groups. Graduate Texts in Mathematics. Springer, New York (1982)
45. Bürkner, P.C.: brms: an R package for Bayesian multilevel models using Stan. J. Stat. Softw. **80**(1), 1–28 (2017)
46. Bürkner, P.C.: The brms Book: applied Bayesian regression modelling using R and Stan (2024). https://paulbuerkner.com/software/brms-book/brms-book.pdf

47. Chekanov, Y.V.: Lagrangian intersections, symplectic energy, and areas of holomorphic curves. Duke Math. J. **95**(1) (1998)
48. Chevalley, C.C., Eilenberg, S.: Cohomology theory of Lie groups and Lie algebras. Trans. Am. Math. Soc. **63**, 85–124 (1948)
49. Cieliebak, K.: Nonlinear functional analysis (2018). https://www.uni-augsburg.de/en/fakultaet/mntf/math/prof/geom/books-scripts-videos/
50. Cieliebak, K.: Lagrange Multiplier Functionals and Their Applications in Symplectic Geometry and String Topology, vol. 4, pp. 2504–2528. EMS Press, Berlin (2022)
51. Cieliebak, K., Frauenfelder, U.: A Floer homology for exact contact embeddings. Pac. J. Math. **239**(2), 251–316 (2009)
52. Cieliebak, K., Frauenfelder, U.: A variational characterization of Einstein–Brillouin–Keller quantization (2024). https://arxiv.org/abs/2401.14223
53. Cieliebak, K., Ginzburg, V.L., Kerman, E.: Symplectic homology and periodic orbits near symplectic submanifolds. Commentarii Math. Helv. **79**, 554–581 (2004)
54. Cieliebak, K., Frauenfelder, U., Oancea, A.: Rabinowitz Floer homology and symplectic homology. Ann. Sci, Ecole Normale Supérieure Ser. 4, **43**(6), 957–1015 (2010)
55. Cieliebak, K., Frauenfelder, U., Paternain, G.P.: Stability is not open. Ann. Inst. Fourier **60**(7), 2449–2459 (2010)
56. Cieliebak, K., Frauenfelder, U., Paternain, G.P.: Symplectic topology of Mañé's critical values. Geom. Topol. **14**(3), 1765–1870 (2010)
57. Cieliebak, K., Volkov, E.: First steps in stable Hamiltonian topology. J. Eur. Math. Soc. **017**(2), 321–404 (2015)
58. Cieliebak, K., Frauenfelder, U., van Koert, O.: Periodic orbits in the restricted three-body problem and Arnold's J^+-invariant. Regul. Chaotic Dyn. **22**(4), 408–434 (2017)
59. Cieliebak, K., Frauenfelder, U., Zhao, L.: J^+ -invariants for planar two-center Stark–Zeeman systems. Ergodic Theory Dyn. Sys. **43**(7), 2258–2292 (2023)
60. Cineli, E., Ginzburg, V.L., Gurel, B.Z., Mazzucchelli, M.: On the Barcode Entropy of Reeb Flows (2024). https://arxiv.org/abs/2401.01421
61. Contreras, G.: The Palais–Smale condition on contact type energy levels for convex Lagrangian systems. Calc. Var. Partial Differ. Equ. **27**(3), 321–395 (2006)
62. Cristofaro-Gardiner, D., Humilière, V., Seyfaddini, S.: Proof of the Simplicity Conjecture, vol. 199. Department of Mathematics, Princeton University, Princeton (2024)
63. Cushman, R.M., Bates, L.M.: Global Aspects of Classical Integrable Systems, 2nd edn. Birkhäuser, Basel (2015)
64. da Silva, A.C.: Lectures on Symplectic Geometry, corrected 2nd edn. Lecture Notes in Mathematics, No. 1764. Springer, Berlin (2008)
65. Diez, T.: Slice theorem for Fréchet group actions and covariant symplectic field theory (2014). https://arxiv.org/abs/1405.2249
66. Diogo, L., Lisi, S.T.: Morse–Bott split symplectic homology. J. Fixed Point Theory Appl. **21**(3), 77 (2019)
67. Dostoglou, S., Salamon, D.: Instanton Homology and Symplectic Fixed Points. London Mathematical Society Lecture Note Series, pp. 57–94. Cambridge University Press, Cambridge (1994)
68. Einsiedler, M., Ward, T.: Ergodic Theory: With a View Towards Number Theory. Graduate Texts in Mathematics. Springer, London (2010)
69. Fabert, O., Fish, J.W., Golovko, R., Wehrheim, K.: Polyfolds: a first and second look. EMS Surv. Math. Sci. **3**(2), 131–208 (2016)
70. Fauck, A.: Rabinowitz–Floer homology on Brieskorn spheres. Int. Math. Res. Notices **2015**(14), 5874–5906 (2014)
71. Filippenko, B., Wehrheim, K.: A polyfold proof of the Arnold conjecture. Sel. Math. **28**(11) (2022)
72. Floer, A., Hofer, H.: Coherent orientations for periodic orbit problems in symplectic geometry. Math. Z. **212**(1), 13–38 (1993)

73. Frauenfelder, U.: The Arnold-Givental conjecture and moment Floer homology. Int. Math. Res. Notices **2004**(42), 2179–2269 (2004)
74. Frauenfelder, U.: The Stark problem as a concave toric domain. Geom. Dedic. **217**(10) (2023)
75. Frauenfelder, U., Nicholls, R.: The moduli space of gradient flow lines and Morse homology (2020). https://arxiv.org/abs/2005.10799
76. Frauenfelder, U., Schlenk, F.: S^1-equivariant Rabinowitz–Floer homology. Hokkaido Math. J. **45**(3), 293–323 (2016)
77. Frauenfelder, U., van Koert, O.: The Restricted Three-Body Problem and Holomorphic Curves. Pathways in Mathematics. Birkhäuser, Basel (2018)
78. Frauenfelder, U., Weber, J.: The shift map on Floer trajectory spaces. J. Symplectic Geom. **19**(2), 351–397 (2021)
79. Frauenfelder, U., Weber, J.: Local gluing. Bull. Braz. Math. Soc. New Ser. **56**(3), 44 (2025)
80. Frauenfelder, U., Zhao, L.: Existence of either a periodic collisional orbit or infinitely many consecutive collision orbits in the planar circular restricted three-body problem. Math. Z. **291**(1), 215–225 (2019). https://doi.org/10.1007/s00209-018-2080-7
81. Fukaya, K., Oh, Y.G., Ohta, H., Ono, K.: Kuranishi Structures and Virtual Fundamental Chains. Springer Monographs in Mathematics. Springer, Singapore (2020)
82. Fukaya, K., Oh, Y.G., Ohta, H., Ono, K.: Construction of a linear K-system in Hamiltonian Floer theory. J. Fixed Point Theory Appl. **24**(39) (2022)
83. Geiges, H.: An Introduction to Contact Topology. Cambridge Studies in Advanced Mathematics, No. 109. Cambridge University Press, Cambridge (2008)
84. Geiges, H., Zehmisch, K.: Course on Holomorphic Discs. Birkhäuser Advanced Texts Basler Lehrbücher. Springer Nature Switzerland (2023)
85. Ghiggini, P., Niederkrüger-Eid, K.: On the symplectic fillings of standard real projective spaces. J. Fixed Point Theory Appl. **24**(2), 37 (2022)
86. Gil, A.F., Litteri, W., Rodriguez-Fernandez, V., Camacho, D., Vasile, M.: Generative Design of Periodic Orbits in the Restricted Three-Body Problem (2024). https://arxiv.org/abs/2408.03691
87. Gilbarg, D., Trudinger, N.: Elliptic Partial Differential Equations of Second Order. Grundlehren der mathematischen Wissenschaften. Springer, Berlin (2001)
88. Ginzburg, V.L.: On the existence and non-existence of closed trajectories for some Hamiltonian flows. Math. Z. **223**(1), 397–409 (1996)
89. Ginzburg, V.L.: The Hamiltonian Seifert conjecture: examples and open problems. In: Casacuberta, C., Miró-Roig, R.M., Verdera, J., Xambó-Descamps, S. (eds.) European Congress of Mathematics, pp. 547–555. Birkhäuser Basel, Basel (2001)
90. Ginzburg, V.L., Gürel, B.Z.: A C^2-smooth counterexample to the Hamiltonian Seifert conjecture in $\mathbb{R}^4$. Ann. Math. **158**(3), 953–976 (2003)
91. Gonchenko, S., Kazakov, A., Turaev, D., Shilnikov, A.L.: Leonid Shilnikov and mathematical theory of dynamical chaos. Chaos: Interdiscip. J. Nonlinear Sci. **32**(1), 010402 (2022)
92. Gotay, M.J., Isenberg, J.A.: The symplectization of science - symplectic geometry lies at the very foundations of physics and mathematics. Gazette Math. **54**, 57–79 (1992)
93. Greutmann, P., Saalbach, H., Stern, E.: Professionelles Handlungswissen für Lehrerinnen und Lehrer: Lernen - Lehren - Können. Kohlhammer Verlag, Stuttgart (2021)
94. Groman, Y.: Floer theory and reduced cohomology on open manifolds. J. Geom. Topol. **27**(4), 1273–1390 (2023)
95. Gromov, M.: Filling Riemannian manifolds. J. Differ. Geom. **18**(1), 1–147 (1983)
96. Gromov, M.: Pseudo holomorphic curves in symplectic manifolds. Invent. Math. **82**(2), 307–347 (1985)
97. Gromov, M.: The Abel Prize 2008–2012, pp. 127–246. Springer, Berlin (2014)
98. Guillemin, V., Sternberg, S.: Symplectic Techniques in Physics, 1st edn. Cambridge University Press, Cambridge (1984)
99. Gutt, S.: Remarks on Symplectic Connections. Lett. Math. Phys. **78**(3), 307–328 (2006)
100. Gutt, J., Hutchings, M., Ramos, V.G.B.: Examples around the strong Viterbo conjecture. J. Fixed Point Theory Appl. **24**(2), 41 (2022)

101. Haim-Kislev, P., Ostrover, Y.: A Counterexample to Viterbo's Conjecture (2024). https://arxiv.org/abs/2405.16513
102. Hairer, E., Lubich, C., Wanner, G.: Geometric Numerical Integration: Structure-Preserving Algorithms for Ordinary Differential Equations, 2nd edn. Springer Series in Computational Mathematics, No. 31. Springer, Berlin (2013)
103. Hall, B.C.: Lie Groups, Lie Algebras, and Representations: An Elementary Introduction, 2nd edn. Graduate Texts in Mathematics, No. 222. Springer International Publishing Switzerland (2015)
104. Hatcher, A.: Algebraic Topology. Cambridge University Press, Cambridge (2001)
105. Hatcher, A.: Vector Bundles and K-Theory (2017). https://pi.math.cornell.edu/~hatcher/VBKT/VB.pdf
106. Hirvonen, V.: Lagrangian Mechanics for the Non-Physicist. Profound Physis. Profound Physics (2023)
107. Hofer, H., Zehnder, E.: Symplectic Invariants and Hamiltonian Dynamics. Birkhäuser, Basel (1994)
108. Hofer, H., Taubes, C.H., Weinstein, A., Zehnder, E. (eds.): The Floer Memorial Volume. Progress in Systems and Control Theory. Birkhäuser Verlag, Basel (1995)
109. Hofer, H., Wysocki, K., Zehnder, E.: Polyfold and Fredholm Theory. Ergebnisse der Mathematik und ihrer Grenzgebiete, vol. 72. Springer, Berlin (2021)
110. Honda, K.: On the classification of tight contact structures I. Geom. Topol. **4**(1), 309–368 (2000)
111. Hurtubise, D.E.: Three approaches to Morse-Bott homology. Afr. Diaspora J. Math. New Ser. **14**(2), 145–177 (2012)
112. Hutchings, M., Taubes, C.H.: The Weinstein conjecture for stable Hamiltonian structures. Geom. Topol. **13**(2), 901–941 (2009)
113. Joung, C., Koert, O.v.: Computational symplectic topology and symmetric orbits in the restricted three-body problem. Nonlinearity **38**(2) (2025)
114. Kang, J.: Generalized Rabinowitz Floer homology and coisotropic intersections. Int. Math. Res. Notices **2013**(10), 2271–2322 (2012)
115. Katz, M.G.: Systolic Geometry and Topology. Mathematical Surveys and Monographs. American Mathematical Society, Providence (2007)
116. Katz, M.G., Nowik, T.: A systolic inequality with remainder in the real projective plane. Open Math. **18**(1), 902–906 (2020)
117. Kim, J., Kim, S., Kwon, M.: Bifurcations of symmetric periodic orbits via Floer homology. Calc. Var. Partial Differ. Equ. **59**(3) (2020)
118. Kim, J., Kim, S., Kwon, M.: Equivariant wrapped Floer homology and symmetric periodic Reeb orbits. Ergodic Theory Dyn. Syst. **42**(5), 1708–1763 (2022)
119. Kim, J., Kim, S., Kwon, M.: Remarks on the systoles of symmetric convex hypersurfaces and symplectic capacities. J. Fixed Point Theory Appl. **24**(2), 28 (2022)
120. Knauf, A.: Mathematical Physics: Classical Mechanics. Unitext, vol. 109. Springer, Berlin (2018)
121. Knörrer, H.: Geometrie: Ein Lehrbuch für Mathematik- und Physikstudierende. Vieweg studium; Aufbaukurs Mathematik. Vieweg+Teubner Verlag (2007)
122. Koon, W.S., Lo, M.W., Marsden, J.E., Ross, S.D.: Dynamical Systems, the Three-Body Problem and Space Mission Design, pp. 1167–1181. World Scientific, Singapore (2000)
123. Kwon, M., van Koert, O.: Brieskorn manifolds in contact topology. Bull. Lond. Math. Soc. **48**(2), 173–241 (2016)
124. Lalonde, F., McDuff, D.: The geometry of symplectic energy. Ann. Math. **141**(2), 349–371 (1995)
125. Lang, S.: Fundamentals of Differential Geometry. Graduate Texts in Mathematics. Springer, New York (1999)
126. Latschev, J., Oancea, A. (eds.): Free Loop Spaces in Geometry and Topology: Including the Monograph Symplectic Cohomology and Viterbo's Theorem by Mohammed Abouzaid. IRMA Lectures in Mathematics and Theoretical Physics. European Mathematical Society, Berlin (2015)

127. Laudenbach, F., Sikorav, J.C.: Hamiltonian disjunction and limits of Lagrangian submanifolds. Int. Math. Res. Notices **1994**(4), 161–168 (1994)
128. Lee, J.M.: Introduction to Topological Manifolds, 2nd edn. Springer Science+Business Media, Berlin (2011)
129. Lee, J.M.: Introduction to Smooth Manifolds, 2nd edn. Graduate Texts in Mathematics. Springer, Berlin (2012)
130. Lee, J.M.: Introduction to Riemannian Manifolds, 2nd edn. Graduate Texts in Mathematics. Springer International Publishing, Cham (2018)
131. Liu, C., Liu, K.: The Three-Body Problem. Netflix Tie-In: Now a Major Netflix Series. The Three-Body Problem Series. Head of Zeus (2024)
132. Liu, H., Zhang, H.: Multiple p-cyclic symmetric closed characteristics on compact convex p-cyclic symmetric hypersurfaces in $\mathbb{R}^{2n}$. Front. Math. China **15**(6), 1155–1173 (2020)
133. Lockwood, O., Weiss, P., Aronshtein, F., Verdon, G.: Quantum dynamical Hamiltonian Monte Carlo. Phys. Rev. Res. **6**, 033142 (2024)
134. Macarini, L., Paternain, G.P.: On the stability of Mañé critical hypersurfaces. Calc. Var. Partial Differ. Equ. **39**(3), 579–591 (2010)
135. Mazzucchelli, M.: Critical Point Theory for Lagrangian Systems. Progress in Mathematics, No. 293. Birkhäuser, Basel (2012)
136. McDuff, D., Salamon, D.: J-holomorphic Curves and Symplectic Topology. Colloquium Publications, vol. 52, 2nd edn. American Mathematical Society, Providence (2012)
137. McDuff, D., Salamon, D.: Introduction to Symplectic Topology, 3rd edn. Oxford Graduate Texts in Mathematics, No. 27. Oxford University Press, Oxford (2017)
138. Meiwes, M., Naef, K.: Translated points on hypertight contact manifolds. J. Topol. Anal. **10**(02), 289–322 (2018)
139. Merry, W.J.: Lagrangian Rabinowitz–Floer homology and twisted cotangent bundles. Geom. Dedic. **171**(1), 345–386 (2014)
140. Milnor, J.: Morse Theory. Princeton University Press, Princeton (1963)
141. Moreno, A.: Contact geometry in the restricted three-body problem: a survey. J. Fixed Point Theory Appl. **24**(2), 29 (2022)
142. Moreno, A.: The Symplectic Geometry of the Three-Body Problem (2024). https://arxiv.org/abs/2101.04438
143. Moreno, A., Aydin, C., van Koert, O., Frauenfelder, U., Koh, D.: Bifurcation graphs for the cr3bp via symplectic methods. J. Astronaut. Sci. **71**(6), 51 (2024)
144. Moser, J.: Periodic orbits near an equilibrium and a theorem by Alan Weinstein. Commun. Pure Appl. Math. **29**(6), 727–747 (1976)
145. Munkres, J.R.: Topology, 2nd edn. Pearson Education, Inc., London (2000)
146. Naef, K.: Translated points on dynamically convex contact manifolds. Ph.D. thesis, ETH Zurich, Zurich (2018)
147. NASA Jet Propulsion Laboratory: Three-Body Periodic Orbits. https://ssd.jpl.nasa.gov/tools/periodic_orbits.html
148. Nelson, J.: From Dynamics to Contact and Symplectic Topology and Back (2016). https://arxiv.org/abs/1611.02676
149. Nicolaescu, L.: An Invitation to Morse Theory, 2nd edn. Universitext. Springer, Berlin (2011)
150. Niculescu, C., Persson, L.: Convex Functions and Their Applications: A Contemporary Approach. CMS Books in Mathematics. Springer International Publishing, New York (2018)
151. Noether, E.: Invariant variation problems. Transp. Theory Stat. Phys. **1**(3), 186–207 (1971)
152. Palis, J., Manning, A., de Melo, W.: Geometric Theory of Dynamical Systems: An Introduction. Springer, New York (2012)
153. Pardon, J.: An algebraic approach to virtual fundamental cycles on moduli spaces of pseudo-holomorphic curves. Geom. Topol. **20**, 779–1034 (2016)
154. Polterovich, L.: An Obstacle to Non-Lagrangian Intersections, pp. 575–586. Birkhäuser Basel, Basel (1995)
155. Polterovich, L.: The Geometry of the Group of Symplectic Diffeomorphism. Springer Basel AG, Basel (2001)

156. Polterovich, L., Rosen, D., Samvelyan, K., Zhang, J.: Topological Persistence in Geometry and Analysis. University Lecture Series. American Mathematical Society, Providence (2020)
157. Rabinowitz, P.H.: Periodic solutions of Hamiltonian systems. Commun. Pure Appl. Math. **31**(2), 157–184 (1978)
158. Remmert, R., Schumacher, G.: Funktionentheorie 1, 5. edn. Grundwissen Mathematik. Springer, Berlin (2002)
159. Ritoré, M., Miquel, V., Sinestrari, C.: Mean Curvature Flow and Isoperimetric Inequalities. Advanced Courses in Mathematics - CRM Barcelona. Birkhäuser, Basel (2009)
160. Rittberg, C.J., Tanswell, F.S., Van Bendegem, J.P.: Epistemic injustice in mathematics. Synthese **197**(9), 3875–3904 (2020)
161. Rot, T.O.: The Morse–Bott inequalities, orientations, and the Thom isomorphism in Morse homology. C. R. Acad. Sci. Paris **354**, 1026–1028 (2016)
162. Ruck, K.: Tate homology and powered flybys. J. Symplectic Geom. **22**(2), 197–221 (2024)
163. Russo, A., Lax, G.: Using artificial intelligence for space challenges: a survey. Appl. Sci. **12**(10), 5106 (2022)
164. Salamon, D.: Lectures on Floer homology. Symplectic Geom. Topol. **7**, 143–229 (1999)
165. Sandon, S.: A Morse estimate for translated points of contactomorphisms of spheres and projective spaces. Geom. Dedic. **165**(1), 95–110 (2013)
166. Schlenk, F.: Embedding Problems in Symplectic Geometry. De Gruyter Expositions in Mathematics, No. 40. Walter de Gruyter, Berlin (2005)
167. Schlenk, F.: Applications of Hofer's geometry to Hamiltonian dynamics. Commentarii Math. Helv. **81**(1), 105–121 (2006)
168. Schneider, A.: Global Surfaces of Section for Dynamically Convex Reeb Flows on Lens Spaces . Trans. Am. Math. Soc. **373**(4), 2775–2803 (2020)
169. Schumacher, R., Stern, E.: Intelligentes Wissen – und wie man es fördert. Springer Spektrum Berlin, Heidelberg (2022)
170. Schwarz, M.: Morse Homology. Progress in Nonlinear Differential Equations and Their Applications. Springer, Berlin (1993)
171. Schwarzbach, B., Kosmann-Schwarzbach, Y.: The Noether Theorems: Invariance and Conservation Laws in the Twentieth Century. Sources and Studies in the History of Mathematics and Physical Sciences. Springer, New York (2010)
172. Shilnikov, A.: Complete dynamical analysis of a neuron model. Nonlinear Dyn. **68**(3), 305–328 (2012)
173. Shilnikov, L.P., Shilnikov, A.L., Turaev, D.V.: Showcase of Blue Sky catastrophes. Int. J. Bifurcat. Chaos **24**(08) (2014)
174. Siburg, K.F.: Symplectic capacities in two dimensions. Manuscripta Math. **78**(2), 149–163 (1993)
175. Szebehely, V.: Theory of Orbit - The Restricted Problem of Three Bodies. Academic Press, Cambridge (1967)
176. Takhtajan, L.A.: Quantum Mechanics for Mathematicians. Graduate Studies in Mathematics, vol. 95. American Mathematical Society, Providence (2008)
177. Taubes, C.H.: The Seiberg–Witten equations and the Weinstein conjecture. Geom. Topol. **11**(4), 2117–2202 (2007)
178. Teschl, G.: Ordinary Differential Equations and Dynamical Systems. Graduate Studies in Mathematics, vol. 140. American Mathematical Society, Providence (2012)
179. Teschl, G.: Mathematical Methods in Quantum Mechanics. Graduate Studies in Mathematics. American Mathematical Society, Providence (2014)
180. The Simons Foundation: The Scientific Live of Mikhail Gromov (2014). https://www.simonsfoundation.org/2014/12/22/mikhail-gromov/
181. Tipler, P.A., Mosca, G.: Physik, 9. edn. Springer Spektrum (2024)
182. Trninic, D., Wagner, R., Kapur, M.: Rethinking failure in mathematics education: a historical appeal. Think. Skills Creat. **30**, 76–89 (2018). The Role of Failure in Promoting Thinking Skills and Creativity

183. Tu, L.W.: Introductory Lectures on Equivariant Cohomology. Annals of Mathematics Studies, No. 204. Princeton University Press, Princeton (2020)
184. Uebele, P.: Periodic Reeb flows and products in symplectic homology. J. Symplectic Geom. **17**(4), 1201–1250 (2019)
185. Uljarevic, I.: Floer homology of automorphisms of Liouville domains. J. Symplectic Geom. **15**(3), 861–903 (2017)
186. Uriot, T., Izzo, D., Simões, L.F., Abay, R., Einecke, N., Rebhan, S., Martinez-Heras, J., Letizia, F., Siminski, J., Merz, K.: Spacecraft collision avoidance challenge: design and results of a machine learning competition. Astrodynamics **6**(2), 121–140 (2022)
187. van de Schoot, R., Depaoli, S., King, R., Kramer, B., Märtens, K., Tadesse, M.G., Vannucci, M., Gelman, A., Veen, D., Willemsen, J., Yau, C.: Bayesian statistics and modelling. Nat. Rev. Methods Primers **1**(1), 1 (2021)
188. Viavattene, G., Devereux, E., Snelling, D., Payne, N., Wokes, S., Ceriotti, M.: Design of multiple space debris removal missions using machine learning. Acta Astronaut. **193**, 277–286 (2022)
189. Weber, J.: The Morse–Witten complex via dynamical systems. Expo. Math. **24**(2), 127–159 (2006)
190. Weber, J.: Stable foliations and semi-flow Morse homology. Ann. Sc. Norm. Super. Pisa Cl. Sci. **XVII**(5), 853–909 (2017)
191. Weber, J.: Topological Methods in the Quest for Periodic Orbits (2018). https://arxiv.org/abs/1802.06435
192. Weibel, C.A.: An Introduction to Homological Algebra. Cambridge Studies in Advanced Mathematics, vol. 38. Cambridge University Press, Cambridge (1994)
193. Weinstein, A.: Periodic orbits for convex Hamiltonian systems. Ann. Math. **108**(3), 507–518 (1978)
194. Weinstein, A.: On the hypotheses of Rabinowitz' periodic orbit theorems. J. Differ. Equ. **33**(3), 353–358 (1979)
195. Wendl, C.: Lectures on symplectic field theory (2016)
196. Weyl, H.: The Classical Groups: Their Invariants and Representations. No. 1, Teil 1 in Princeton Landmarks in Mathematics and Physics. Princeton University Press, Princeton (1946)
197. Wittmann, J.: The Banach manifold $C^k(M, N)$. Differ. Geom. Appl. **63**, 166–185 (2019)
198. Xing, J., Yang, X., Li, Y.: Global Weinstein Type Theorem on Multiple Rotating Periodic Solutions for Hamiltonian Systems (2023). https://arxiv.org/abs/2212.12736
199. Zalamea, F.: The twofold role of observables in classical and quantum kinematics. Found. Phys. **48**(9), 1061–1091 (2018)
200. Zehnder, E.: Lectures on Dynamical Systems. European Mathematical Society, Helsinki (2010)
201. Zeilinger, A.: Einsteins Spuk, 14 edn. Goldmann, Munich (2007)

Index

Y. Bähni, *Lectures on Twisted Rabinowitz-Floer Homology*, Pathways in Mathematics, https://doi.org/10.1007/978-3-032-10673-5

MIX
Papier aus verantwortungsvollen Quellen
Paper from responsible sources
FSC® C105338

If you have any concerns about our products, you can contact us on
ProductSafety@springernature.com

In case Publisher is established outside the EU, the EU authorized representative is:
Springer Nature Customer Service Center GmbH
Europaplatz 3, 69115 Heidelberg, Germany

Printed by Libri Plureos GmbH
in Hamburg, Germany